Robustness in Statistical Forecasting

Yuriy Kharin

Robustness in Statistical Forecasting

Springer

Yuriy Kharin
Department of Mathematical Modeling
and Data Analysis
Belarusian State University
Minsk
Belarus

ISBN 978-3-319-34568-0 ISBN 978-3-319-00840-0 (eBook)
DOI 10.1007/978-3-319-00840-0
Springer Cham Heidelberg New York Dordrecht London

Mathematics Subject Classification (2010): 62-02, 62M20, 62M10, 62G35, 62-07, 62F35, 62C20, 62P20

Printed on acid-free paper

Springer is part of Springer Science+Business Media (www.springer.com)

Preface

Statistical forecasting procedures are used to solve many applied problems in engineering, economics, finance, medicine, environmental studies, etc. For the majority of the developed statistical forecasting procedures, optimality (or asymptotic optimality as observation time increases) is proved w.r.t. the mean square forecast risk under the assumptions of an underlying hypothetical model. In practice, however, the observed data usually deviates from hypothetical models: random observation errors may be non-Gaussian, correlated, or inhomogeneous; the data may be contaminated by outliers, level shifts, or missing values; trend, regression, and autoregression functions do not necessarily belong to the declared parametric families, etc. Unfortunately, the forecasting procedures which have been proved to be optimal under the hypothetical model often become unstable under even small model distortions, resulting in forecast risks or mean square errors which are significantly higher than the theoretical values obtained in the absence of distortion. This necessitates the development of *robust* statistical algorithms, which are designed to retain most of their properties under small deviations from model assumptions.

The available textbooks on the subjects of statistical forecasting and robust statistical methods can be split into two distinct clusters. *The first cluster* includes books on theoretical and applied aspects of statistical forecasting where little or no attention is paid to robustness. The focus of these books is on various hypothetical models, methods, and computer algorithms used in forecasting, as well as their performance in the absence of model distortions.

The second cluster includes books on robust statistics which are dedicated to such diverse subjects as robust statistical parameter estimation, robust hypothesis testing in parametric (e.g., shift-scale) families of probability distributions, regression analysis, discriminant analysis, cluster analysis, time series analysis, etc. However, the topic of robustness in statistical forecasting remains barely touched upon, and little or no information is provided on such important aspects of forecasting as analysis of risk increments due to different types and levels of distortion, estimation of critical distortion levels for the traditional forecasting procedures, and development of robust forecasting procedures tailored to the distortion types that are commonly encountered in applications.

This monograph is an attempt to fill the described gap in the literature by going beyond the fundamental subjects of robust statistical estimation and robust statistical hypothesis testing and presenting a systematic collection of known and new results related to the following topical problems:

- Construction of mathematical models and descriptions of typical distortions in applied forecasting problems;
- Quantitative evaluation of the robustness of traditional forecasting procedures;
- Evaluation of critical distortion levels;
- Construction of new robust forecasting procedures satisfying the minimax-risk criterion.

Solving these problems answers the following questions, which are highly relevant to both theoretical and applied aspects of statistical forecasting:

- Which distortion types can be accommodated by forecasting procedures?
- What are the maximal distortion levels allowing for "safe" use of the traditional forecasting algorithms?
- How can we estimate the effect of distortions on the mean square risk of traditional forecasting algorithms?
- Which robust forecasting statistics are the most suitable under different types of distortions?

The monograph is organized into ten chapters. Chapter 1 serves as a general introduction to the subject of statistical forecasting, presenting its history and some of the possible applications. Chapter 2 describes the decision-theoretic approach to forecasting, which is different from the general statistical approach used in Chaps. 3–10. Chapter 3 presents mathematical models of the time series commonly used in statistical forecasting. Chapter 4 classifies types of model distortions and defines metrics for optimality and robustness in statistical forecasting. Chapter 5 presents methods for optimal parametric and nonparametric time series regression forecasting. In Chap. 6, robustness of these methods is evaluated, and robust forecasting statistics are constructed. A similar treatment of the ARIMA(p, d, q) autoregressive integrated moving average time series model is presented in Chap. 7. Chapter 8 presents an analysis of optimality and robustness of forecasting based on autoregressive time series models under missing values. Robustness of multivariate time series forecasting based on systems of simultaneous equations is investigated in Chaps. 9, and 10 discusses forecasting of discrete time series.

The interdependence of the chapters is illustrated in Fig. 1 below; solid lines represent prerequisites, and dashed lines indicate weaker relations.

Presentation of the material within the chapters follows the pattern "model $\rightarrow$ method $\rightarrow$ algorithm $\rightarrow$ computation results based on simulated or real-world data." The theoretical results are illustrated by computer experiments using real-world statistical data (eight instances) and simulated data (ten instances).

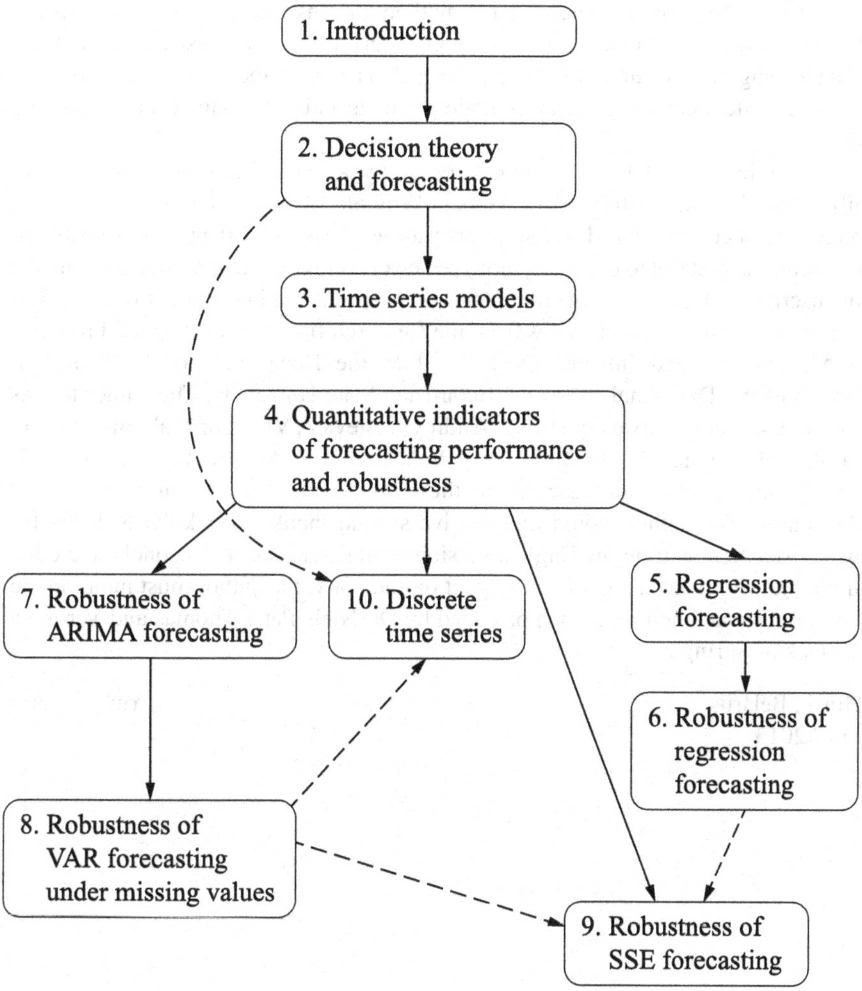

Fig. 1 The interdependence of the chapters

The robust forecasting algorithms described in the monograph have been implemented as computer programs; however, the developed software package cannot be made available to the reader (in particular, only the Russian-language user interface has been designed). The author has intentionally deferred from connecting the material presented in this book to any specific software package or framework. Instead, the methods and algorithms are presented in the most general form, allowing the reader to implement them within a statistical package of their choice (e.g., S-PLUS or R). This also ensures that software developers can easily incorporate the developed methods into their products.

The book is primarily intended for mathematicians, statisticians, and software developers in applied mathematics, computer science, data analysis, econometrics, financial engineering, and biometrics. It can also be recommended as a textbook for a one-semester course for advanced undergraduate and postgraduate students of the mentioned disciplines.

The author would like to thank Serguei Ayvazyan, Christophe Croux, Peter Filzmoser, Ursula Gather, Peter Huber, William Mallios, Helmut Rieder, and Peter Rousseeuw for the discussions that aroused his interest in robust statistics. The author is grateful to the anonymous reviewers for their attention in analyzing the manuscript, and the resulting comments and proposals on improving the book. The author also thanks his colleagues from the Research Institute for Applied Problems of Mathematics and Informatics, as well as the Department of Mathematical Modeling and Data Analysis of the Belarusian State University. The author thanks his Ph.D. students Alexander Huryn, Andrey Kostevich, Vladimir Malugin, Maksim Pashkevich, Andrey Piatlistkiy, Valery Voloshko, and Andrey Yarmola for their contributions to the joint research in the field of robust statistical methods and algorithms. The author would like to give special thanks to Nikolai Kalosha for his assistance in editing the English version of the manuscript. To conclude the list of people who made a significant impact on the book, the author must mention the wise counsel and helpful advice provided by Dr. Niels Peter Thomas and Veronika Rosteck of Springer.

Minsk, Belarus
April 2013

Yuriy Kharin

Contents

Symbols and Abbreviations

$\emptyset$	Empty set
$\mathbb{C}$	Set of complex numbers
$\mathbb{N}$	Set of positive integers
$\mathbb{R}$	Set of real numbers
$\mathbb{R}^N$	N-dimensional Euclidean space
$\mathbb{R}^{m\times n}$	Space of $(m \times n)$ real matrices
$\mathbb{Z}$	Set of integers
$[x]$	Integer part of x
$\|X\|$	Modulus of X if it represents a number cardinality of X if it represents a set determinant of X if it represents a matrix
$\delta_{x,y}$	Kronecker symbol: $\delta_{x,y} = \{1,\ x = y;\quad 0,\ x \neq y\}$
$\max A$	Maximal element of the set A
$\min A$	Minimal element of the set A
$\mathbf{1}(z)$	Unit step function
$\mathbf{0}_n, \mathbf{0}_{n\times m}$	Respectively, n-vector and $(n \times m)$ matrix of zeros
$\mathbf{1}_n, \mathbf{1}_{n\times m}$	Respectively, n-vector and $(n \times m)$ matrix of ones
$\mathbf{I}_n$	$(n \times n)$ identity matrix
$\left[\begin{array}{c\|c} A & B \\ \hline C & D \end{array}\right]$	Block matrix composed of submatrices A, B, C, D
$\mathbf{Det}(M) = \|M\|$	Determinant of a matrix M
$\mathrm{tr}\,(M)$	Trace of the matrix M
$\lambda_{\max}(M)$	Maximal absolute eigenvalue of the matrix M
M'	Transpose of the matrix M
M^{-1}	(Multiplicative) inverse of the matrix M
$(\Omega, \mathcal{F}, \mathbb{P})$	Base probability space
$\mathbb{P}\{A\}$	Probability of a random event $A \in \mathcal{F}$
$\mathbf{I}_A$	Indicator function of the event A
$\mathbb{D}\{\xi\}$	Variance of a random variable ξ
$\mathbf{Corr}\{\xi, \eta\}$	Correlation coefficient of random variables ξ and η

$\mathcal{L}\{X\}$	Probability distribution law of the random vector X
$\mathbb{E}\{X\}$	Expectation of a random vector X
$\mathbf{Cov}\{X, Y\}$	Covariance matrix of random vectors X and Y
$\varphi(x) = \frac{1}{\sqrt{2\pi}} e^{-x^2/2}$	Standard normal probability density function
$\Phi(x) = \int_{-\infty}^{x} \varphi(z)dz$	Standard normal cumulative distribution function
$\mathcal{N}(\mu, \sigma^2)$	Univariate normal (Gaussian) distribution with the mean value μ and the variance σ^2
$\mathcal{N}_n(\mu, \Sigma)$	Multivariate normal (Gaussian) distribution with the mean n-vector μ and the covariance $(n \times n)$ matrix Σ
χ_n^2	Chi-squared distribution with n degrees of freedom
t_n	Student's t-distribution with n degrees of freedom
$F_{m,n}$	Fischer–Snedecor distribution with m, n degrees of freedom
$\text{Bi}(n, p)$	Binomial distribution
$\xrightarrow{\text{a.s.}}$	Almost sure convergence
$\xrightarrow{\text{m.s.}}$	Mean square convergence
$\xrightarrow{\text{D}}$	Convergence in distribution
$\xrightarrow{\text{P}}$	Convergence in probability
$O(f(T))$	Big-O (Landau) notation; denotes a sequence limited from above by the sequence $f(T)$, $T \in \mathbb{N}$, as $T \to +\infty$
$o(f(T))$	Little-o (Landau) notation; denotes a sequence that is infinitely small compared to the sequence $f(T)$, $T \in \mathbb{N}$, as $T \to +\infty$
BDR	Bayesian decision rule
LSE	Least squares estimator
ML	Maximum likelihood
MLE	Maximum likelihood estimator
AR(p)	Autoregressive model of order p
ARMA(p, q)	Autoregressive moving average model of orders p, q
ARIMA(p, d, q)	Autoregressive integrated moving average model of orders p, d, q
VAR(p)	Vector autoregressive model of order p
□	End of proof

Chapter 1
Introduction

Abstract The introduction defines statistical forecasting and discusses its history and applications.

Webster encyclopedic dictionary [17] defines *forecasting* as "an activity aimed at computing or predicting some future events or conditions based on rational analysis of the relevant data." Let us illustrate this definition by using macro-level forecasting of economic indicators in a national economy S. Let X be multivariate macroeconomic data (e.g., the quarterly GDP values, investments, state expenditures, and unemployment levels) observed in the system S up to and including a given time point T, and Y be a collection of variables representing macroeconomic indicators for a future time point $t = T + \tau$, $\tau > 0$. The aim of mathematical forecasting is to construct a functional mapping

$$\hat{Y} = f(X),$$

where $\hat{Y}$ is the computed prediction (estimator) for Y called the *forecast* or the *forecast value* and $f(\cdot)$ is the *forecasting function (statistic)*.

Note that the term "forecasting" is sometimes generalized by ordering the data X collected in the system S not over time t, but over some other variable (e.g., a coordinate, an environment variable).

Statistical forecasting is widely used in applications, including decision making and planning at top executive level. Let us give some practical examples of forecasting problems [1, 2, 6–9, 14, 19, 20]:

1. *Economics, finance, and business:*

- Planning and state regulation of national economies;
- Forecasting the GDP, commodity prices, and various other economical indicators;
- Forecasting future interest rates;
- Credit scoring in banking;

Y. Kharin, *Robustness in Statistical Forecasting*, DOI 10.1007/978-3-319-00840-0_1,

- Forecasting stock market rates and currency exchange rates;
- Forecasting future claims on insurance policies.

2. *Marketing:*

- Forecasting sales and projecting raw material expenditure;
- Forecasting total product demand, as well as its distribution across different regions and consumer groups.

3. *Engineering:*

- Forecasting reliability of complex technological systems;
- Forecasting product quality based on various production factors.

4. *Geology:* forecasting deposits of natural resources.
5. *Environmental studies:* forecasting various environmental variables.
6. *Meteorology:* weather forecasting.
7. *Medicine:* forecasting efficiency of medical procedures.
8. *Psychology:* forecasting the respondent's professional qualities based on the results of psychological tests.
9. *Sports:* forecasting the future performance of an adult sportsman based on results as a junior.

Based on the nature of the investigated process and the chosen theoretical model, mathematical forecasting can be based on either *deterministic* or *stochastic* methods. This book follows the stochastic approach, which is more relevant to applications due to a high degree of uncertainty present in most practical problems.

As a rule, *methods of statistical forecasting* are based on the plug-in principle and consist of two steps:

Step 1. Construct an adequate model for the investigated process based on the collected statistical data X and prior knowledge;
Step 2. Based on the model constructed in Step 1, compute the forecast $\hat{Y}$ which is optimal w.r.t. a certain criterion.

A general diagram of the forecasting process is shown in Fig. 1.1.

Forecast optimality is usually understood in the sense of the minimum forecast error which is measured by averaging over some suitable metric $\|\cdot\|$:

$$r = r(f) = \mathbb{E}\left\{\|\hat{Y} - Y\|\right\} \to \min_{f(\cdot)},$$

where $\mathbb{E}\{\cdot\}$ is the mathematical expectation. The solution $f^*(\cdot)$ of this optimization problem defines the optimal forecasting algorithm:

$$\hat{Y} = f^*(X).$$

Many applied forecasting problems in economics, finance, business, and other fields are reduced to one of the fundamental problems in mathematical statistics—forecasting discrete time random processes, which are also called random sequences

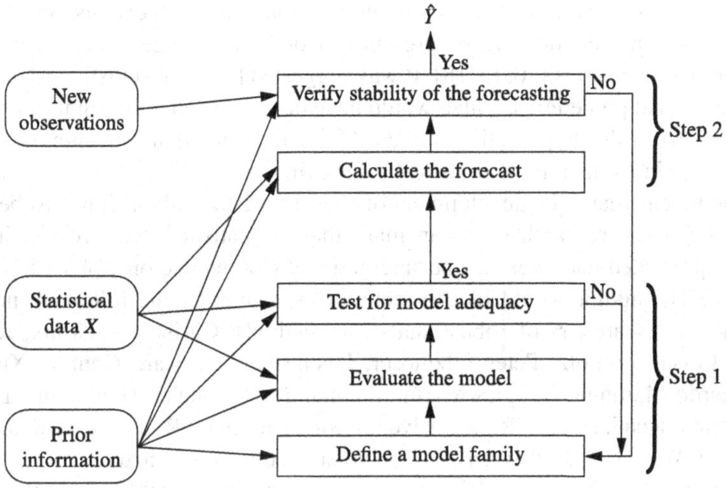

Fig. 1.1 Diagram of the forecasting process

or time series. From a mathematical point of view, forecasting is equivalent to solving the following problem: for some $\tau \in \mathbb{N}$, estimate the τ-step-ahead future value $x_{T+\tau} \in \mathbb{R}^d$ (assuming the general case of a d-variate random process) using a realization of this process $\{x_1, \ldots, x_T\} \subset \mathbb{R}^d$ observed over $T \in \mathbb{N}$ preceding time units as sample data.

Let us briefly discuss the history of statistical forecasting. Rigorous mathematical studies of stochastic forecasting were started in the 1930s by the founder of modern probability theory Andrey Kolmogorov [15, 16]. Two stages can be identified in the development of statistical forecasting techniques. *The first stage* took place before 1974 and was characterized by construction of forecasting statistics (algorithms or procedures) that minimized the forecast risk (the mean square error) for a number of simple time series models, such as stationary models with fixed spectral densities, stationary models with trends belonging to a given function family, autoregressive models, and moving average models. Results of this type have been surveyed in [3–5].

In the 1970s it was found that applying many of the developed "optimal" forecasting procedures to real-world data often resulted in much higher risks compared to the expected theoretical values. The reason for this phenomenon was pointed out by Peter Huber (Swiss Federal Institute of Technology) in his talk [11] at the 1974 International Congress of Mathematicians (Vancouver, Canada): "Statistical inferences are based only in part upon the observations. An equally important base is formed by prior assumptions about the underlying situation". These assumptions form the hypothetical model M_0 of the process being investigated. In applications, the behavior of the investigated processes often deviates from the model M_0, leading to instability of forecasting statistics. The following main types of deviations from hypothetical models have been identified: non-normal observation errors, dependence (or correlation) of observation errors, inhomogeneous observation errors, model specification errors, presence of outliers, change points, or missing values in the time series [10, 13, 18]. It was suggested [12] that statisticians develop *robust* statistical procedures, which would have been affected only slightly by small deviations from the hypothetical model M_0. This marked the beginning of the *second stage* in the history of statistical forecasting.

In the recent years, the development of robust statistical algorithms has become one of the major research topics in mathematical statistics. New results in this field are presented each year at the International Conference on Robust Statistics (ICORS). The author would like to mention the names of the following notable modern-day researchers of robust statistical methods: Christophe Croux, Rudolf Dutter, Luisa Fernholz, Peter Filzmoser, Ursula Gather, Marc Genton, Xuming He, Ricardo Maronna, Douglas Martin, Stephan Morgenthaler, Hannu Oja, Daniel Peña, Marco Riani, Helmut Rieder, Elvezio Ronchetti, Peter Rousseeuw, Stefan Van Aelst, Roy Welsch, Gert Willems, Victor Yohai, and Ruben Zamar.

The main motivation for writing this monograph was a notable gap in the available books on robust statistics, which are mainly devoted to parameter estimation and hypothesis testing under various types of distortions. However, little attention is paid to robustness of statistical forecasting. This work attempts to fill this gap, focusing on the following problems:

(a) mathematical description of the distortion types common to applied forecasting problems;
(b) sensitivity analysis and quantitative robustness estimation of traditional forecasting algorithms under distortions;
(c) construction and mathematical evaluation of new robust forecasting statistics.

References

1. Abraham, B., Ledolter, J.: Statistical Methods for Forecasting. Wiley, New York (1989)
2. Aitchison, J., Dunsmore, J.: Statistical Prediction Analysis. CUP, Cambridge (1975)
3. Anderson, T.: The Statistical Analysis of Time Series. Wiley, New York (1994)
4. Bowerman, B., O'Connel, R.: Forecasting and Time Series. Duxbury Press, Belmont (1993)
5. Box, G., Jenkins, G., Reinsel, G.: Time Series Analysis: Forecasting and Control. Wiley-Blackwell, New York (2008)
6. Brockwell, P., Davis, R.: Introduction to Time Series and Forecasting. Springer, New York (2002)
7. Christensen, R.: Bayesian Ideas and Data Analysis: An Introduction for Scientists and Statisticians. CRC Press, Boca Raton (2011)
8. Davison, A.: Statistical Models. Cambridge University Press, Cambridge (2009)
9. Greene, W.: Econometric Analysis. Macmillan, New York (2000)
10. Hampel, F., Ronchetti, E., Rousseeuw, P., Stahel, W.: Robust Statistics: The Approach Based on Influence Functions. Wiley, New York (1986)
11. Huber, P.: Some mathematical problems arising in robust statistics. In: Proceedings of the International Congress of Mathematicians, pp. 821–824. University of Vancouver, Vancouver (1974)

12. Huber, P.: Robust Statistics. Wiley, New York (1981)
13. Kharin, Yu.: Robustness in Statistical Pattern Recognition. Kluwer Academic, Dordrecht (1996)
14. Kharin, Yu., Charemza, W., Malugin, V.: On modeling of Russian and Belarusian economies by the LAM-3 econometric model (in Russian). Appl. Econom. **1**(2), 124–139 (2006)
15. Kolmogorov, A.: On the use of statistically estimated forecasting formulae (in Russian). Geophys. J. **3**, 78–82 (1933)
16. Kolmogorov, A.: Interpolation and extrapolation of stationary stochastic series (in Russian). Izv. Akad. Nauk SSSR Ser. Mat **5**(1), 3–14 (1941)
17. MacLaren, A.: Consolidated-Webster Encyclopedic Dictionary. CBP, New York (1946)
18. Maronna, R., Zamar, R.: Robust multivariate estimates for high-dimensional datasets. Technometrics **44**, 307–317 (2002)
19. Newbold, P.: Introductory Business Forecasting. South-Western Publishing, Dallas (1990)
20. Taylor, S.: Modelling Financial Time Series. Wiley, New York (1986)

Chapter 2
A Decision-Theoretic Approach to Forecasting

Abstract Statistical forecasting is prediction of future states of a certain process based on the available stochastic observations as well as the available prior model assumptions made about this process. This chapter describes a general (universal) approach to statistical forecasting based on mathematical decision theory, including a brief discussion of discriminant analysis. The following fundamental notions are introduced: optimal and suboptimal forecasts, loss function, risk functional, minimax, admissible, and Bayesian decision rules (BDRs), Bayesian forecast density, decision rule randomization, plug-in principle.

2.1 The Mathematical Model of Decision Making

A generalized mathematical model of decision making has been formulated by Abraham Wald [14] as a generalization of the models used for hypothesis testing and parameter estimation to obtain an adequate description of settings that include stochastic processes. The high degree of uncertainty present in most applied forecasting problems makes the decision-making approach extremely relevant to statistical forecasting.

A general decision-making model contains two abstract objects: the environment (**E**) and the decision maker (**DM**), as well as the following six mathematical objects:

$$(\Theta, \mathcal{Y}, \mathcal{X}, w(\cdot), F(\cdot), D).$$

Here $\Theta \subseteq \mathbb{R}^m$ is the parameter space containing all possible states $\theta \in \Theta$ of the environment **E**, which includes a certain "actual state of **E**" denoted as $\theta^0 \in \Theta$ (this actual state is assumed to be unknown to the **DM** at the moment when the decision is made); $\mathcal{Y} \subseteq \mathbb{R}^m$ is the decision space (each element $Y \in \mathcal{Y}$ is a possible decision of the **DM**); $w(\cdot)$ is the loss function

$$w = w(\theta, Y), \quad \theta \in \Theta, \quad Y \in \mathcal{Y}, \quad w \in \mathbb{R}^1,$$

Y. Kharin, *Robustness in Statistical Forecasting*, DOI 10.1007/978-3-319-00840-0_2,

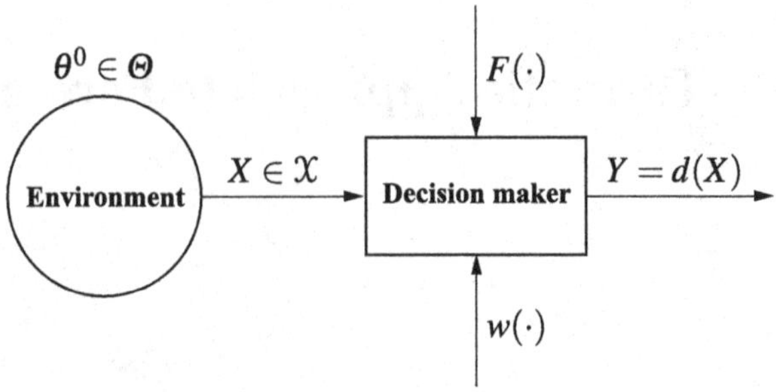

Fig. 2.1 The process of decision making

where w is the loss taken by the **DM** for $\theta^0 = \theta$ and the decision Y; the function $u = u(\theta, Y) = -w(\theta, Y)$ is usually called the utility function; $\mathcal{X} = \mathcal{B}(\mathbb{R}^N)$ is the sample space (a Borel σ-algebra defined over an N-dimensional Euclidean space) where statistical data is observed; the random N-vector of observations $X \in \mathcal{X}$ is defined over the probability space $(\Omega, \mathcal{F}, \mathbb{P})$, and $F(X; \theta^0) : \mathcal{X} \times \Theta \to [0, 1]$ is the N-dimensional distribution function of X which depends on the parameter θ^0; D is the decision rule space consisting of all Borel maps $d(\cdot) : \mathcal{X} \to \mathcal{Y}$:

$$D = \{Y = d(X) : X \in \mathcal{X},\ \ Y \in \mathcal{Y}\}.$$

Decision making within this model is illustrated in Fig. 2.1. At the moment when the decision $Y \in \mathcal{Y}$ is being made, the **DM** doesn't know the "actual state of **E**" $\theta^0 \in \Theta$, and therefore the actual loss $w(\theta^0, Y)$ is also unknown. However, the **DM** knows the possible loss $w = w(\theta, Y)$ for *every* possible situation $(\theta, Y) \in \Theta \times \mathcal{Y}$. In order to reduce the uncertainty of θ^0, the **DM** collects statistical data in the form of an observation $X \in \mathcal{X}$, which has the probability distribution defined by θ^0. Based on the knowledge of the loss function $w(\cdot)$, the distribution function $F(\cdot)$, and the collected statistical data X, the **DM** uses a certain performance criterion to choose the optimal decision rule $d_{opt}(\cdot) \in D$ and to make the best possible decision $\hat{Y} = d_{opt}(X)$ by following this rule.

2.2 Minimax, Admissible, and Bayesian Families of Decision Rules

Under a generalized decision-making model presented in Sect. 2.1, consider the problem of constructing the optimal decision rule $d_{opt}(\cdot) \in D$. Let us define a criterion of decision rule optimality [10].

Definition 2.1. The conditional risk of a decision rule $d(\cdot) \in D$ for $\theta^0 = \theta$ is defined as the conditional expectation of the loss function:

$$r = r(d(\cdot);\theta) = \mathbb{E}_\theta\{w(\theta, d(X))\} = \int\limits_{R^N} w(\theta, d(X))dF(X;\theta), \quad \theta \in \Theta, \quad r \in \mathbb{R}^1. \tag{2.1}$$

Smaller values of the functional (2.1) correspond to more effective decision rules. It follows from the definition that the uncertainty of the value θ^0 complicates the minimization of the risk functional.

Definition 2.2. A minimax decision rule $Y = d^*(X)$ is defined as a decision rule minimizing the supremum of the risk functional (2.1):

$$r_+(d^*(\cdot)) = \inf_{d(\cdot)\in D} r_+(d(\cdot)), \qquad r_+(d(\cdot)) = \sup_{\theta\in\Theta} r(d(\cdot);\theta) \tag{2.2}$$

where $r_+(d(\cdot))$ is the guaranteed (upper) risk, i.e., the maximum possible value of the risk functional for the decision rule $d(\cdot)$.

The guaranteed risk corresponds to the least favorable state of the environment **E**, and thus the minimax decision rule (2.2) is often called "pessimistic."

Another popular approach to decision making is the Bayesian approach [10] which is based on the assumption that there exists an a priori known m-dimensional probability distribution function $G(\theta)$ of the random vector $\theta^0 \in \Theta$ defining the state of the environment **E**.

Definition 2.3. Under the assumptions of the decision-making model defined earlier, let $\theta^0 \in \Theta \subseteq R^m$ be a random m-vector characterized by a prior distribution function $G(\theta)$. Then the Bayesian (unconditional) decision risk is defined as the following functional:

$$r = r(d(\cdot)) = \mathbb{E}\{r(d(\cdot);\theta^0)\} = \int\limits_{\mathbb{R}^m} r(d(\cdot);\theta)dG(\theta), \quad d(\cdot) \in D, \quad r \in \mathbb{R}^1, \tag{2.3}$$

or equivalently

$$r = r(d(\cdot)) = \mathbb{E}\{w(\theta^0, d(X))\} = \int\limits_{\mathbb{R}^m}\int\limits_{\mathbb{R}^N} w(\theta, d(X))dF(X;\theta)dG(\theta),$$

which follows from (2.1) and the total expectation formula.

Definition 2.4. A Bayesian decision rule (BDR) is a decision rule $Y = d_0(X)$ that minimizes the Bayesian risk (2.3):

$$r(d_0(\cdot)) = \inf_{d(\cdot)\in D} r(d(\cdot)). \tag{2.4}$$

Let us introduce the last type of decision rules that will be discussed in this chapter—the admissible decision rules.

Definition 2.5. It is said that a decision rule $d'(\cdot)$ dominates a decision rule $d''(\cdot)$, where $d'(\cdot), d''(\cdot) \in D$, if

$$r(d'(\cdot);\theta) \leq r(d''(\cdot);\theta) \quad \forall \theta \in \Theta, \tag{2.5}$$

and there exists a $\theta \in \Theta$ such that the inequality in (2.5) is strict. A decision rule $\tilde{d}(\cdot)$ is said to be admissible if no other decision rule $d(\cdot) \in D$ dominates $\tilde{d}(\cdot)$.

Definition 2.6. Decision rules $d_1(\cdot), d_2(\cdot) \in D$ are said to be equivalent w.r.t. the Bayesian decision risk if they have the same Bayesian risk values:

$$r(d_1(\cdot)) = r(d_2(\cdot)).$$

Equivalence w.r.t. the guaranteed risk is defined similarly:

$$r_+(d_1(\cdot)) = r_+(d_2(\cdot)).$$

Let us establish some properties of the above decision rules (see [1] for a more systematic treatment).

Properties of Bayesian, Minimax, and Admissible Decision Rules

Property 2.1. A BDR $d_0(\cdot)$ minimizes the posterior mean loss $w(Y \mid X)$:

$$\hat{Y} = d_0(X) = \arg\min_{Y \in \mathcal{Y}} w(Y \mid X), \tag{2.6}$$

where

$$w(Y \mid X) = \mathbb{E}\{w(\theta^0, Y) \mid X\} = \int_{\mathbb{R}^m} w(\theta, Y) dG(\theta \mid X), \tag{2.7}$$

and $G(\theta \mid X)$ is the posterior probability distribution function of the random parameter θ^0 given the observation X.

Proof. Using (2.3) and the total expectation formula, let us rewrite the Bayesian risk as follows:

$$r(d(\cdot)) = \mathbb{E}\{w(\theta^0, d(X))\} = \mathbb{E}\{\mathbb{E}\{w(\theta^0, d(X)) \mid X\}\},$$

where the outer expectation is computed w.r.t. the unconditional distribution of the random vector X with a distribution function

$$F(X) = \int_{\mathbb{R}^m} F(X;\theta)dG(\theta);$$

the inner conditional expectation defines the posterior loss (2.7). Thus,

$$r(d(\cdot)) = \mathbb{E}\{w(d(X) \mid X)\} = \int_{\mathbb{R}^N} w(d(X) \mid X)dF(X) \geq r_0 ::= \int_{\mathbb{R}^N} \min_{Y \in \mathcal{Y}} w(Y \mid X)dF(X),$$

and it is obvious that the lower bound r_0 is attained for the decision rule defined by (2.6), (2.7). From Definition 2.4, this decision rule is a BDR. □

Property 2.2. If a BDR $Y = d_0(X)$ is unique, it is also admissible.

Proof. The statement will be proved by contradiction. Suppose that the BDR $d_0(\cdot)$ is not admissible. Then, by Definition 2.5, there exists a decision rule $d'(\cdot) \in D$, $d'(\cdot) \neq d(\cdot)$, such that

$$r(d'(\cdot);\theta) \leq r(d_0(\cdot);\theta) \quad \forall\theta \in \Theta, \quad \exists\theta' \in \Theta: \quad r(d'(\cdot);\theta') < r(d_0(\cdot);\theta).$$

Integrating the first inequality over the probability distribution $G(\theta)$ of θ and applying (2.3) yields the inequality

$$r(d'(\cdot)) \leq r(d_0(\cdot)).$$

Strictness of this inequality would contradict the definition of a BDR given in (2.4). However, an equality is also impossible, since in that case $d'(\cdot)$ would be a different BDR, contradicting the uniqueness of the BDR. This contradiction concludes the proof. □

Property 2.3. Given that the parameter space is finite, $\Theta = \{\theta^{(1)}, \ldots, \theta^{(K)}\}$, $K < \infty$, and that the prior probability distribution of $\theta^0 \in \Theta$ is nonsingular,

$$p_k = \mathbb{P}\{\theta^0 = \theta^{(k)}\} > 0, \quad k = 1, \ldots, K,$$

the BDR $Y = d_0(X)$ is admissible.

Proof. Assume the opposite: $d_0(\cdot)$ is not an admissible decision rule. Then (2.5) implies that there exist a decision rule $d(\cdot) \in D$ and an index $i^* \in \{1, \ldots, K\}$ such that

$$r(d(\cdot);\theta_k) \leq r(d_0(\cdot);\theta_k), \quad k \neq i^*; \quad r(d(\cdot);\theta_{i*}) < r(d_0(\cdot);\theta_{i*}).$$

Multiplying both sides of these inequalities by $p_k > 0$ and $p_{i^*} > 0$, respectively, taking a sum, and applying the equality (2.3) yield

$$r(d(\cdot)) = \sum_{k=1}^{K} p_k r(d(\cdot); \theta^{(k)}) < \sum_{k=1}^{K} p_k r(d_0(\cdot); \theta^{(k)}) = r(d_0(\cdot)).$$

This inequality contradicts the definition of the BDR (2.4). □

Property 2.4. If the parameter space is finite, $\Theta = \{\theta^{(1)}, \dots, \theta^{(K)}\}$, $K < \infty$, and $d(\cdot)$ is an admissible decision rule, then there exists a prior distribution

$$p_k = P\{\theta^{(0)} = \theta^{(k)}\}, \quad k = 1, \dots, K,$$

such that the decision rule $d(\cdot)$ is the BDR w.r.t. the prior distribution $\{p_k\}$. In other words, in this case the set of admissible decision rules is included in the set of BDRs.

Proof. The proof can be obtained by repeating the argument of the previous proof. □

Property 2.5. If a minimax decision rule $d^*(\cdot)$ is unique, then it is also admissible.

Proof. Let us assume the opposite: there exists a different decision rule $d(\cdot) \in D$, $d(\cdot) \neq d^*(\cdot)$, such that

$$r(d(\cdot); \theta) \leq r(d^*(\cdot); \theta) \quad \forall \theta \in \Theta, \quad \exists \theta' \in \Theta : \quad r(d(\cdot); \theta') < r(d^*(\cdot); \theta').$$

From (2.2), this also yields the inequality

$$r_+(d(\cdot)) \leq r_+(d^*(\cdot)).$$

This contradicts the condition that $d^*(\cdot)$ is a unique minimax decision rule. □

Property 2.6. Given that $d(\cdot)$ is an admissible decision rule and that the corresponding risk function (2.1) doesn't depend on $\theta \in \Theta$, i.e., $r(d(\cdot); \theta) = \text{const.}$, the decision rule $d(\cdot)$ is also a minimax decision rule.

Proof. Assume that the minimax condition isn't satisfied for the decision rule $d(\cdot)$ and that there exists a different minimax decision rule $d'(\cdot) \neq d(\cdot)$:

$$r_+(d'(\cdot)) < r_+(d(\cdot)).$$

However, we have also assumed that $r_+(d(\cdot)) \equiv r_+(d(\cdot); \theta)$, and thus

$$r(d'(\cdot); \theta) < r(d(\cdot); \theta) \quad \forall \theta \in \Theta,$$

which contradicts the admissibility of $d(\cdot)$. □

2.3 The Bayesian Forecast Density

Randomization of the decision rule is a commonly used decision-theoretic technique of reducing the decision risk [1].

Definition 2.7. A *randomized decision rule* is a family of random variables

$$Y = d(X, \omega) : \mathfrak{X} \times \Omega \to \mathfrak{Y},$$

lying in the basic probability space $(\Omega, \mathfrak{F}, \mathbb{P})$ and defined by a *critical function* $\pi(Y; X)$. For a discrete decision space $\mathfrak{Y}$, the function $\pi(Y; X)$ is defined as

$$\pi = \pi(Y; X) ::= \mathbb{P}\{d(X, \omega) = Y \mid X\}, \quad Y \in \mathfrak{Y},$$

and we have

$$0 \leq \pi(Y; X) \leq 1, \quad \sum_{Y \in \mathfrak{Y}} \pi(Y; X) \equiv 1.$$

In the continuous case, where $\mathfrak{Y} \subset \mathbb{R}^M$, and the Lebesgue measure $\text{mes}_M(\mathfrak{Y})$ is positive, $\pi = \pi(Y; X)$ is defined as the M-dimensional probability density of the random variable $d(X, \omega)$, and we have

$$\pi(Y; X) \geq 0, \quad \int_{\mathbb{R}^M} \pi(Y; X) dY = 1.$$

Let us consider applications of randomized decision rules in statistical forecasting. Assume that a forecast is constructed for a random M-vector $Y \in \mathfrak{Y}$ that describes an unknown future state of the process or the phenomenon that is being investigated. Its probability density $g(Y \mid \theta)$ depends on a parameter $\theta \in \Theta \subseteq \mathbb{R}^m$ with an unknown true value $\theta^0 \in \Theta$. Following the Bayesian paradigm, it is assumed that θ^0 is a random m-vector with a given prior probability density $q(\theta)$. Let $X \in \mathfrak{X} \subseteq \mathbb{R}^N$ be statistical data describing past and current states of the process with a conditional probability density $p(X \mid \theta)$ given $\theta^0 = \theta$. Thus, the random parameter vector θ^0 is stochastically dependent not only on the past and current states X but also on the future states Y. This allows forecasting of Y based on the collected statistical data X under prior probabilistic uncertainty of θ^0.

The problem of constructing a forecast for Y based on X using the randomized decision rule $\hat{Y} = d(X, \omega)$ lies in finding the critical function $\pi(Y; X)$. Following the Bayesian approach outlined above, one of the methods of constructing the critical function is the use of the posterior probability density of Y given the observation X:

$$\pi(Y; X) = p(Y \mid X), \quad Y \in \mathfrak{Y}. \tag{2.8}$$

Fig. 2.2 Stochastic dependence between X, θ^0, and Y

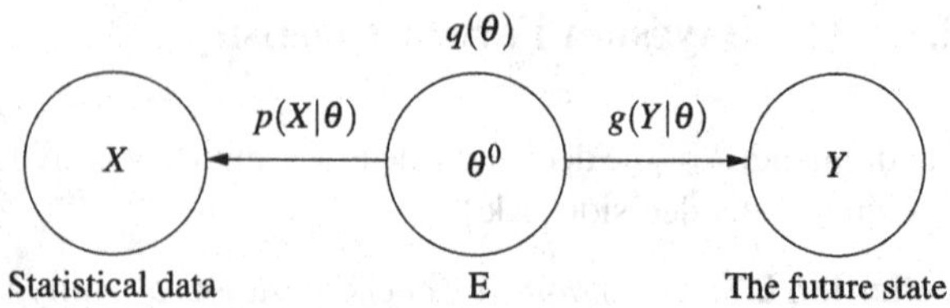

The conditional probability density (2.8) used in forecasting is called the *Bayesian forecast density*.

Following the accepted stochastic model (Fig. 2.2) of the dependence between X, θ^0, and Y, Bayes formulae, together with certain well-known properties of multivariate probability densities, imply that

$$\pi(Y; X) = \int_{\mathbb{R}^m} g(Y \mid \theta) p(\theta \mid X) d\theta, \tag{2.9}$$

where

$$p(\theta \mid X) = p(X \mid \theta) q(\theta) \left(\int_{\mathbb{R}^m} p(X \mid \theta') q(\theta') d\theta' \right)^{-1} \tag{2.10}$$

is the posterior probability density of the random vector θ^0 given a fixed value of the random vector X.

The Bayesian forecast density (2.9), (2.10) allows us not only to compute the randomized forecast $\hat{Y} \in \mathcal{Y}$ as a result of simulating a random M-vector with the probability density $\pi(Y; X)$, $Y \in \mathcal{Y}$ but also to compute the traditional (nonrandomized) point and interval forecasts. Numerical characteristics of the Bayesian forecast density $\pi(Y; X)$ can be used as point forecasts of Y:

- Posterior expected forecast

$$\hat{Y}_0 = \int_{\mathbb{R}^m} Y \pi(Y; X) dY; \tag{2.11}$$

- Posterior mode forecast

$$\hat{Y}_1 = \arg\max_{Y \in \mathcal{Y}} \pi(Y; X); \tag{2.12}$$

- Posterior median forecast (for $M = 1$): $\hat{Y}_2$ is defined as a root of the equation

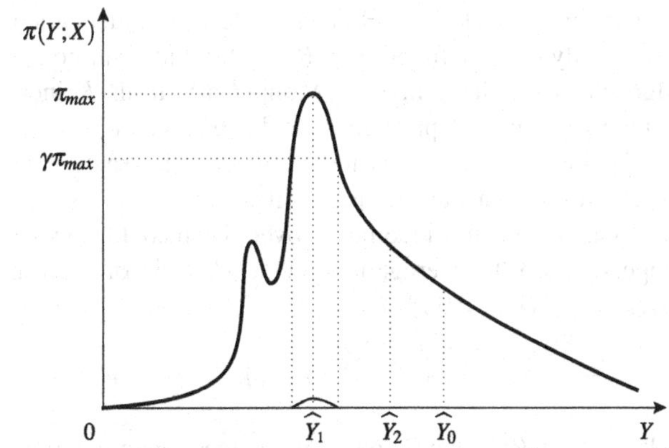

Fig. 2.3 Construction of point and interval forecasts from the Bayesian forecast density

$$\int_{-\infty}^{\hat{Y}_2} \pi(Y;X)dY = \int_{\hat{Y}_2}^{+\infty} \pi(Y;X)dY. \tag{2.13}$$

Figure 2.3 above presents an example of using the Bayesian forecast density to construct point and interval forecasts in the univariate case ($M = 1$).

The following two techniques can be proposed for *set (interval) forecasting*. Let the *domain of γ-maximal Bayesian forecast density* be a subset of the possible forecasts defined as

$$\mathcal{Y}_\gamma = \{Y \in \mathcal{Y} : \pi(Y;X) \geq \gamma \pi_{max}\}, \tag{2.14}$$

where $\pi_{max} = \max_{Y \in \mathcal{Y}} \pi(Y;X)$ and the parameter $\gamma \in (0.5, 1)$ defines the size of the domain $\mathcal{Y}_\gamma$, i.e., its M-dimensional volume $\text{mes}_M(\mathcal{Y}_\gamma)$. Following the theory of statistical interval estimation, let us define the posterior γ-confidence region $\mathcal{Y}_\gamma$ as the solution of the following conditional minimization problem:

$$\text{mes}_M(\mathcal{Y}_\gamma) \to \min, \quad \int_{\mathcal{Y}_\gamma} \pi(Y;X)dY = \gamma. \tag{2.15}$$

In order to simplify the computations, it is often advisable to consider a parametric family of possible confidence regions.

It should be noted that computation of Bayesian density forecasts (2.9), (2.10) is often complicated by the necessity of multiple integration over $\theta \in \Theta \subseteq \mathbb{R}^m$. If analytic computation of the integrals in (2.9) and (2.10) appears to be unfeasible, Monte Carlo numeric integration may be used:

$$\pi(Y;X) \approx \frac{1}{K} \sum_{i=1}^{K} g(Y \mid \theta^{(i)}).$$

Here $\theta^{(1)}, \ldots, \theta^{(K)} \in \Theta$ is a sample of K independent random vectors with the probability density function $p(\theta \mid X)$, which can be simulated by using standard algorithms. As the number of Monte Carlo trials K increases to infinity, the mean square error of this approximation decreases as K^{-1}.

To illustrate the notions and methods of this section, let us consider a problem of forecasting a financial company's income.

Assume that the income Y over the next business day is a random variable depending on the average increment $\theta^0 \in \mathbb{R}^1$ of a certain currency exchange rate over the previous day:

$$Y = \mu_0 + k\,\theta^0 + \varepsilon,$$

where $\mu_0 \in R^1$ is the (known) guaranteed mean income that doesn't depend on the currency exchange market, $k\,\theta^0$ is the income depending on θ^0 (here $k > 0$ is a known proportionality coefficient); ε is a random variation of the income modeled by a normally distributed random variable,

$$\mathcal{L}\{\varepsilon\} = N(0, \Delta^2),$$

with a known variance Δ^2. The parameter θ^0 is unknown, but statistical data $x_1, \ldots, x_N$ representing the 1-day exchange rate increments over the previous day (offered by N commercial banks) has been collected, where $X = (x_i) \in \mathbb{R}^N$ is assumed to be a random sample of size N taken from a normal probability distribution, $\mathcal{L}\{x_i\} = N(0, \sigma^2)$, with a known variance σ^2.

We would like to make point and interval forecasts of the income Y based on statistical data X, the above model assumptions, and a prior assumption that θ^0 is uniformly distributed over a given interval $[a, b]$ (for example, we can assume that the minimum and maximum exchange rates a and b have been set by a central bank).

Model assumptions yield the following expressions:

$$\begin{aligned} g(Y \mid \theta) &= \frac{1}{\sqrt{2\pi}\,\Delta} \exp\left(-\frac{1}{2\Delta^2}(Y - \mu_0 - k\theta)^2\right), \\ p(X \mid \theta) &= (2\pi\sigma^2)^{-N/2} \exp\left(-\frac{1}{2\sigma^2}\sum_{i=1}^{N}(x_i - \theta)^2\right), \\ q(\theta) &= \frac{1}{b-a}\mathbf{1}_{[a,b]}(\theta), \end{aligned}$$

where $\mathbf{1}_A(\theta) = \{1, \theta \in A; 0, \theta \notin A\}$ is the indicator function of the set A. Applying (2.10) results in the Bayesian forecast density

$$p(\theta \mid X) = \frac{\exp\left(-N(2\sigma^2)^{-1}(\theta - \overline{x})^2\right)}{\sqrt{2\pi}\frac{\sigma}{\sqrt{N}}\left(\Phi\left(\sqrt{N}\frac{b-\overline{x}}{\sigma}\right) - \Phi\left(\sqrt{N}\frac{a-\overline{x}}{\sigma}\right)\right)}\mathbf{1}_{[a,b]}(\theta),$$

which is the normal probability density function $N(\overline{x}, \sigma^2/N)$ constrained to $[a, b]$, where $\overline{x} = N^{-1} \sum_{i=1}^{N} x_i$ is the sample mean. This, together with (2.9), leads to the equation

$$\pi(Y; X) = \left(2\pi \frac{\sigma\Delta}{\sqrt{N}} \left(\Phi\left(\sqrt{N}\frac{b-\overline{x}}{\sigma}\right) - \Phi\left(\sqrt{N}\frac{a-\overline{x}}{\sigma}\right)\right)\right)^{-1} \times$$

$$\times \int_a^b \exp\left(-\frac{1}{2}\left(\frac{N}{\sigma^2}(\theta-\overline{x})^2 + \frac{1}{\Delta^2}(k\theta+\mu_0-Y)^2\right)\right) d\theta,$$

where the right-hand side integral can be rewritten using the standard normal distribution function $\Phi(\cdot)$ by performing a substitution of the variables. Applying this equation to (2.11)–(2.15) yields the desired forecasts.

2.4 Forecasting Discrete States by Discriminant Analysis

2.4.1 The Mathematical Model

In applications, the underlying process can often be described by a discrete stochastic model [2, 5–8, 11]:

$$\nu \in S = \{1, 2, \ldots, L\},$$

where ν is the future unknown state of the system and $2 \leq L < +\infty$ is the number of possible values of ν (i.e., the number of possible forecasts). Let us consider some examples.

Example 2.1. A bank scores a prospective client (a certain company) applying for a loan. The financial circumstances of the client are characterized by N business indicators $X = (x_i) \in \mathbb{R}^N$ (for instance, x_1 is the total annual income, x_2 is the demand for the products made by the company, and x_3 characterizes the dynamics of the company's bank accounts). Based on statistical data X, the bank makes a forecast $\hat{\nu} = d(X)$, where $\hat{\nu} = 1$ stands for a "reliable client" bringing a profit to the bank and $\hat{\nu} = 2$—an "unreliable client" failing to repay the loan and causing a loss (in this example, $L = 2$).

Example 2.2. Let $\hat{\nu} = d(X)$ be a success forecast for a certain clinical treatment based on a patient's medical test results $X \in \mathbb{R}^N$; $\hat{\nu} = 0$ means that the patient's health will remain unchanged, $\hat{\nu} = 1$ corresponds to a health improvement, and $\hat{\nu} = 2$—a health deterioration (in this example, $L = 3$).

The statistical classification model or, to be precise, the discriminant analysis model [3, 9, 11] can be used to solve this type of applied problems. Discriminant analysis is a branch of statistical data analysis devoted to models and methods of identifying the observed data as belonging to one of the given populations (classes, patterns, etc.), i.e., classification of statistical observations.

Let us interpret a classification problem as a forecasting problem defined earlier. Assume that a random observation $x = (x_k) \in \mathbb{R}^N$ belongs to one of the $L \geq 2$ classes $\Omega_1, \dots, \Omega_L$, and let the possible forecasts be the indices of these classes: a forecast $\nu = i$ corresponds to the class Ω_i and vice versa. Let an observation belonging to the class Ω_i be a random N-vector $X_i \in \mathbb{R}^N$ with a conditional probability density $p_i^0(x)$, $i \in S$. We are going to assume the knowledge of prior class probabilities $\pi_1, \dots, \pi_L$:

$$\pi_i = \mathbb{P}\{\nu = i\} > 0, \quad \sum_{i \in S} \pi_i = 1.$$

We also assume prior knowledge of the $(L \times L)$ forecasting (classification) loss matrix $W = (w_{il})$, where $w_{il} \geq 0$ is the loss taken if an observation belonging to the class Ω_i is classified as belonging to the class Ω_l, i.e., if $\nu = i$, but $\hat{\nu} = l$. For example, a *(0–1) loss matrix* W is defined as follows:

$$w_{il} = 1 - \delta_{il}, \quad i, l \in S, \tag{2.16}$$

where δ_{il} is the Kronecker delta.

Under this model, optimal forecasting, as defined in Sects. 2.1 and 2.3, is equivalent to constructing a BDR

$$\hat{\nu} = d_0(x) : \mathbb{R}^N \to S, \tag{2.17}$$

that minimizes the mean loss resulting from the forecast. This problem is solved differently depending on the available prior knowledge of probabilistic characteristics of the classes $\{\pi_i, p_i^0(\cdot)\}$.

2.4.2 Complete Prior Knowledge of $\{\pi_i, p_i^0(\cdot)\}$

Let us introduce the following notation:

$$f_j(x; \{p_i^0(\cdot)\}) = \sum_{i \in S} \pi_i p_i^0(x) w_{ij}; \quad p(x) = \sum_{i \in S} \pi_i p_i^0(x). \tag{2.18}$$

Here $p(x)$ is the unconditional probability density function of the random observation $X \in \mathbb{R}^N$ determined by the stochastic model of the investigated process. From the Bayes formula,

$$\mathbb{P}\{\nu = j \mid X = x\} = \frac{\pi_j p_j^0(x)}{p(x)}, \tag{2.19}$$

and thus (2.18) can be rewritten as

$$\frac{f_j(x; \{p_i^0(\cdot)\})}{p(x)} = \sum_{i \in S} \mathbb{P}\{\nu = i \mid X = x\} w_{ij} = \mathbb{E}\{w_{\nu j} \mid X = x\}. \tag{2.20}$$

The relation (2.20) means that, to a multiplier $p(x)$ not depending on j, the function f_j in (2.18) defines the posterior mean forecast loss $\hat{\nu} = j$ given the observation vector $X = x$.

Theorem 2.1. *Under prior knowledge of the probability distributions $\{\pi_i, p_i^0(\cdot)\}$, assume that for all $i, k, l \in S$, $k \neq l$, the condition*

$$\mathbb{P}_{\theta_i^0}\{f_k(X; \{p_j^0(\cdot)\}) - f_l(X; \{p_j^0(\cdot)\}) = 0\} = 0 \tag{2.21}$$

is satisfied. Then the BDR (2.17) *is unique and, up to a set of Lebesgue measure zero, has the form*

$$\hat{\nu} = d = d_0(x) = \arg\min_{j \in S} f_j(x; \{p_i^0(\cdot)\}), \quad x \in \mathbb{R}^N, \quad d \in S. \tag{2.22}$$

This BDR minimizes the mean loss of forecasting (the Bayesian risk):

$$r_0 = \int_{\mathbb{R}^N} \min_{j \in S} f_j(x; \{p_i^0(\cdot)\}) dx. \tag{2.23}$$

Proof. Taking into account (2.3) and (2.18), the Bayesian risk functional can be rewritten as

$$r = r(d(\cdot); \{p_i^0(\cdot)\}) = \sum_{i \in S} \pi_i \int_{\mathbb{R}^N} p_i^0(x) w_{id(x)} = \int_{\mathbb{R}^N} f_{d(x)}(x; \{p_i^0(\cdot)\}) dx, \quad d(\cdot) \in D. \tag{2.24}$$

Looking at the form of the functional (2.24), it is easy to find the lower bound of the Bayesian risk over all possible decision rules $d(\cdot) \in D$:

$$r(d(\cdot); \{p_i^0(\cdot)\}) \geq \min_{d(\cdot) \in D} \int_{\mathbb{R}^N} f_{d(x)}(x; \{p_i^0(\cdot)\}) dx \geq \int_{\mathbb{R}^N} \min_{j \in S} f_j(x; \{p_i^0(\cdot)\}) dx = r_0.$$

Therefore, the above inequality becomes an equality after substituting the decision rule (2.22). □

Observe that, as in Sect. 2.2 (see Property 2.1), the obtained BDR (2.22) minimizes the posterior mean loss.

Corollary 2.1. *For a (0–1) loss matrix* (2.16)*, the Bayesian risk can be interpreted as the probability of making an incorrect forecast*

$$r_0 = \inf_{d(\cdot)\in D} \mathbb{P}\{\hat{\nu} \neq \nu\} = 1 - \int_{\mathbb{R}^N} \max_{i\in S}(\pi_i p_i^0(x))dx,$$

and the expression for the BDR can be written in a simplified form:

$$\hat{\nu} = d = d_0(x) = \arg\max_{i\in S}(\pi_i p_i^0(x)), \quad x \in \mathbb{R}^N, \quad d \in S. \tag{2.25}$$

Proof. Let us substitute (2.16) into (2.18), (2.22), and (2.24). Taking normalization into account, we obtain

$$f_j(x; \{p_i^0(\cdot)\}) = p(x) - \pi_j p_j^0(x),$$
$$r(d(\cdot); \{p_i^0(\cdot)\}) = 1 - \int_{\mathbb{R}^N} \pi_{d(x)} p_{d(x)}^0(x)dx,$$

and (2.25) follows immediately. □

Corollary 2.2. *Under the assumptions of Corollary 2.1, let the observations be described by an N-dimensional normal (Gaussian) model:*

$$p_i^0(x) = n_N(x \mid \mu_{i,\Sigma_i}) = (2\pi)^{-\frac{N}{2}} |\Sigma_i|^{-\frac{1}{2}} \exp\left(-\frac{1}{2}(x-\mu_i)'\Sigma_i^{-1}(x-\mu_i)\right), \tag{2.26}$$

where $\mu_i = (\mu_{ij}) \in \mathbb{R}^N$ is the mean vector and $\Sigma_i = (\sigma_{ijk}) \in \mathbb{R}^{N\times N}$ is the nonsingular covariance matrix of the random vector $X_i \in \mathbb{R}^N$. Then the BDR is quadratic:

$$\hat{\nu} = d = d_0(x) = \arg\min_{i\in S}\left((x-\mu_i)'\Sigma_i^{-1}(x-\mu_i) + \ln(|\Sigma_i|/\pi_i^2)\right). \tag{2.27}$$

Proof. Substitute (2.26) into (2.25) and perform the obvious transformations. □

Corollary 2.3. *Under the assumptions of Corollary 2.1, take Fisher's model [3]:*

$$\Sigma_1 = \Sigma_2 = \cdots = \Sigma_L = \Sigma \tag{2.28}$$

with two classes ($L = 2$). In that case the BDR is linear:

$$
\begin{aligned}
\hat{\nu} &= d = d_0(x) = \mathbf{1}(l(x)) + 1, \quad x \in R^N, \\
l(x) &= b'x + \beta, \\
b &= \Sigma^{-1}(\mu_2 - \mu_1), \\
\beta &= (\mu_1' \Sigma^{-1} \mu_1 - \mu_2' \Sigma^{-1} \mu_2)/2 + \ln(\pi_2/\pi_1), \\
r_0 &= 1 - \left(\pi_1 \Phi\left(\frac{\Delta}{2} + \frac{1}{\Delta} \ln \frac{\pi_1}{\pi_2} \right) + \pi_2 \Phi\left(\frac{\Delta}{2} - \frac{1}{\Delta} \ln \frac{\pi_1}{\pi_2} \right) \right),
\end{aligned}
\tag{2.29}
$$

where $\Phi(\cdot)$ is the standard normal $N(0, 1)$ distribution function and

$$\Delta = \sqrt{(\mu_2 - \mu_1)' \Sigma^{-1} (\mu_2 - \mu_1)} \geq 0$$

is the so-called Mahalanobis distance between classes [3].

Proof. Rewriting the BDR as (2.27) for $L = 2$ and taking into account the notation (2.28), (2.29) yields

$$\hat{\nu} = d = d_0(x) = \arg\min_{i \in S} \left(-\mu_i' \Sigma^{-1} x + \frac{1}{2} \mu_i' \Sigma^{-1} \mu_i - \ln \pi_i \right) \equiv \mathbf{1}(l(x)) + 1, \quad x \in \mathbb{R}^N,$$

which is the first expression of (2.29).

Now let us compute the Bayesian risk (i.e., the unconditional probability of a forecast error) for the BDR (2.29):

$$r_0 = \pi_1 P_1 + \pi_2 P_2, \tag{2.30}$$

where

$$P_i = \mathbb{P}\{\hat{\nu} \neq i \mid \nu = i\}, \quad i \in S,$$

is the conditional probability of a forecast error given that the true number of the class equals $\nu = i$. Due to (2.29), we have

$$P_1 = \mathbb{P}\{\hat{\nu} = 2 \mid \nu = 1\} = \mathbb{P}\{l(X_1) \geq 0\} = 1 - F_{l_1}(0), \tag{2.31}$$

where $l_i = l(X_i) = b'X_i + \beta$ is a random variable and $F_{l_i}(z)$, $z \in \mathbb{R}^1$, is the distribution function of the random variable l_i, $i \in S$. From the condition (2.26), the probability distribution of X_i can be written as

$$\mathcal{L}\{X_i\} = \mathcal{N}_N(\mu_i, \Sigma_i),$$

and the linear transformation theorem for normal random vectors [3] yields

$$\mathcal{L}\{l_i\} = \mathcal{N}_1(m_i, \zeta_i),$$
$$m_i = b'\mu_i + \beta = (-1)^i \Delta^2/2 + \ln \frac{\pi_2}{\pi_1},$$
$$\zeta_i = b'\Sigma b = (\mu_2 - \mu_1)'\Sigma^{-1}(\mu_2 - \mu_1) = \Delta^2,$$

where the variance $\zeta_i = \Delta^2$ doesn't depend on $i \in S$. Therefore,

$$F_{l_i}(z) = \Phi\left(\frac{z - m_i}{\Delta}\right), \quad i \in S.$$

Substituting this equality into (2.31) results in the expression

$$P_1 = 1 - \Phi\left(\frac{\Delta}{2} + \frac{1}{\Delta}\ln\frac{\pi_1}{\pi_2}\right).$$

Similarly, we have

$$P_2 = 1 - \Phi\left(\frac{\Delta}{2} - \frac{1}{\Delta}\ln\frac{\pi_1}{\pi_2}\right).$$

Substituting P_1, P_2 into (2.30) yields (2.29). □

Definition 2.8. Fisher's linear discriminant function is defined as $l(x) = b'x + \beta$ (as implied by (2.29), its sign determines the forecast). The set

$$\Gamma_0 = \{x : b'x + \beta = 0\} \subset \mathbb{R}^N$$

is called Fisher's discriminant hyperplane.

Figure 2.4 illustrates Fisher's linear decision rule for $N = 2$.

To conclude this subsection, let us observe that, as in Sect. 2.3, it is possible to construct a randomized decision rule $\hat{\upsilon} = \tilde{d}(x, \omega)$ which is going to be described by a Bayesian forecast distribution (2.9), (2.10) defined on the set of possible forecasts $A = S = \{1, 2, \ldots, L\}$:

$$\pi(i; x) = \mathbb{P}\{\hat{\upsilon} = i \mid X = x\} = \frac{\pi_i p_i^0(x)}{p(x)}, \quad i \in S.$$

The nonrandomized forecast (2.25) is, in fact, equal to the posterior mode (2.12).

Some applications require interval forecasts $H_\gamma \subseteq S$ defined by (2.15). In the discrete case, this definition can be rewritten as

$$\sum_{i \in H_\gamma} \pi(i, x) \geq \gamma, \quad |H_\gamma| \to \min. \tag{2.32}$$

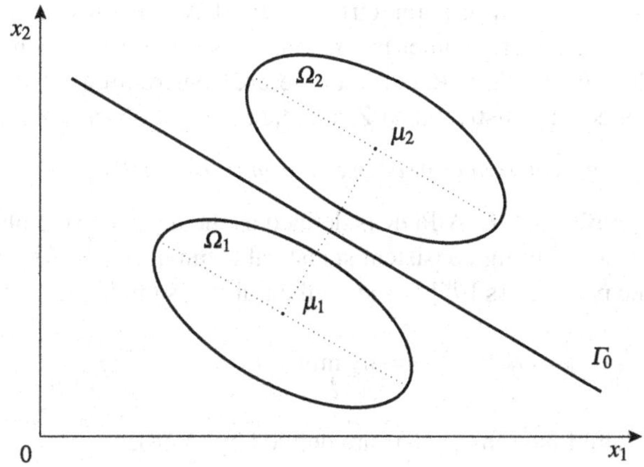

Fig. 2.4 Fisher's linear decision rule

Interval forecasts defined by (2.32) become very useful in the rather common case, where the number of classes is large ($L >> 1$).

2.4.3 *Prior Uncertainty*

Consider a setting with a priori unknown conditional probability densities of the observations $\{p_i^0(\cdot)\}$. To overcome this prior uncertainty, we can use a so-called classified training sample $Z \subset \mathbb{R}^N$ of total size $n = n_1 + \cdots + n_L$, which consists of L independent subsamples:

$$Z = \underset{i \in S}{\cup} Z_i, \quad Z_i \cap Z_j = \emptyset, \quad j \neq i.$$

Here

$$Z_i = \{z_{ij} \in \mathbb{R}^N : j = 1, \ldots, n_i\}$$

is a random subsample of size n_i taken from the class Ω_i (i.e., a subset of statistical data corresponding to the forecast value $\nu = i$).

Let us start by considering the case of *parametric prior uncertainty*, where the densities $\{p_i^0(\cdot)\}$, $i \in S$, belong to a given family of probability distributions, but the distribution parameters remain unknown:

$$p_i^0(x) = q(x; \theta_i^0), \quad x \in \mathbb{R}^N, \quad \theta_i^0 \in \Theta,$$

where

$$Q = \{q(x; \theta), \ x \in \mathbb{R}^N : \theta \in \Theta \subseteq \mathbb{R}^m\}$$

is some given m-parametric family of N-dimensional probability densities. Forecasting under parametric uncertainty is usually based on one of the two approaches described below. Recall that we are constructing a forecast $\hat{\nu} \in S$ based on the collected statistical data Z and the recorded observation $x \in \mathbb{R}^N$.

A. Construction of plug-in decision rules (PDRs).

Definition 2.9. A PDR is defined as the decision rule obtained from a BDR (2.22) by substituting consistent statistical estimators $\{\hat{\theta}_i\}$ for the unknown true values of the parameters $\{\theta_i^0\}$ based on the training sample Z:

$$\tilde{\nu} = d_1(x; Z) = \arg\min_{j \in S} f_j(x; \{q(x; \hat{\theta}_i)\}), \quad x \in \mathbb{R}^N, \quad \tilde{\nu} \in S, \tag{2.33}$$

where functions $\{f_j(\cdot)\}$ are defined by (2.18).

The estimators $\{\hat{\theta}_i\}$ are usually the maximum likelihood estimators (MLEs):

$$\hat{\theta}_i = \arg\max_{\theta \in \Theta} \frac{1}{n_i} \sum_{j=1}^{n_i} \ln q(z_{ij}; \theta), \quad i \in S. \tag{2.34}$$

Theorem 2.2. *If the parametric family Q of probability densities satisfies the classical regularity conditions [4], then the forecast $\tilde{\nu}$ defined by the PDR* (2.33), (2.34), *converges in probability to the forecast $\hat{\nu}$ defined by the BDR* (2.22)*:*

$$d_1(x; Z) \xrightarrow{\mathbf{P}} d_0(x), \quad x \in \mathbb{R}^N, \tag{2.35}$$

for

$$n_0 = \min_{i \in S} n_i \to \infty.$$

Proof. Regularity conditions together with certain well-known asymptotic properties of MLEs [4] imply that

$$\hat{\theta}_i \xrightarrow{\mathbf{P}} \theta_i^0, \quad i \in S.$$

Notation (2.18) and well-known results on functional transformations of convergent sequences [4, 13] yield the relations

$$q(x; \hat{\theta}_i) \xrightarrow{\mathbf{P}} q(x; \theta_i^0),$$
$$f_j(x; \{q(x; \hat{\theta}_i)\}) \xrightarrow{\mathbf{P}} f_j(x; \{q(x; \theta_i^0)\}), \quad i, j \in S, \quad x \in \mathbb{R}^N.$$

Since S is a finite set, the convergence of the objective functions

$$f_j(\cdot), \quad j \in S,$$

implies that the minimum points also converge:

$$\arg\min_{j\in S} f_j(x;\{q(x;\hat{\theta}_i)\}) \xrightarrow{\mathbf{P}} \arg\min_{j\in S} f_j(x;\{q(x;\theta_i^0)\}), \quad x\in\mathbb{R}^N.$$

This, together with (2.33) and (2.22), proves the convergence in (2.35). □

Let us define the unconditional Bayesian risk of a PDR $d_1(\cdot)$ similarly to (2.24):

$$r(d_1(\cdot)) = \mathbb{E}\left\{\int_{\mathbb{R}^N} f_{d_1(x;Z)}(x;\{p_i^0(x)\})dx\right\}, \tag{2.36}$$

where the expectation $\mathbb{E}\{\cdot\}$ is computed w.r.t. the probability distribution of the random sample Z. Known asymptotic expansions of the deviations $\{\hat{\theta}_i - \theta_i^0\}$ [9, 13] lead to the following asymptotic expansion for the unconditional risk (2.36):

$$r(d_1(\cdot)) = r_0 + \sum_{i=1}^{L} \frac{\varrho_i}{n_i} + O\left(n_0^{-3/2}\right), \tag{2.37}$$

where the coefficients $\{\varrho_i\}$ above satisfy the condition

$$\varrho_i = \varrho_i\left(N, \{\pi_i\}, \{q(\cdot;\theta_i^0)\}, \{w_{ij}\}\right) \geq 0.$$

It is easy to see from (2.37) that for $n_0 \to \infty$ the PDR risk (2.33) converges to the minimal Bayesian risk (2.23):

$$r(d_1) \to r_0, \tag{2.38}$$

and therefore in practice (2.33) is often called the *suboptimal decision rule*.

In practical applications, it is important to choose sufficiently large training sample sizes $n_1, n_2, \ldots, n_L$ that guarantee a minor relative increase in the forecast risk due to the uncertainty of $\{\theta_i^0\}$. The relation (2.37) can be used to evaluate this increment:

$$\frac{r(d_1) - r_0}{r_0} \approx \sum_{i=1}^{L} \frac{\varrho_i}{n_i r_0} \leq \delta. \tag{2.39}$$

B. Using the Bayesian forecast distribution.

Define an Lm-dimensional composite column vector of parameters for the probability distributions

$$p_i^0 = q(x;\theta_i^0), \quad x\in\mathbb{R}^N, \quad \theta_i^0 \in \Theta \subseteq \mathbb{R}^m,$$

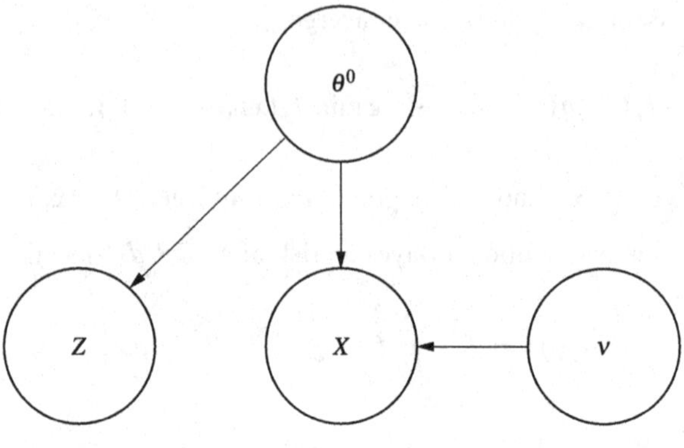

Fig. 2.5 Stochastic dependence between model components

where $i \in S$, as

$$\theta^0 = \left(\theta_1^{0\prime} \vdots \theta_2^{0\prime} \vdots \ldots \vdots \theta_L^{0\prime}\right)' \in \mathbb{R}^{Lm},$$

and assume that θ^0 is a random vector with an a priori given probability density function $q(\theta)$, $\theta \in \mathbb{R}^{Lm}$.

To define a Bayesian forecast distribution and construct a randomized decision rule

$$\tilde{\nu} = d_2(x; Z, \omega),$$

we are going to use the diagram of the stochastic dependence between ν, X, θ^0, and Z presented in Fig. 2.5.

The Bayesian forecast distribution is defined on the decision space S in the following way:

$$\pi(i; x, Z) = \mathbb{P}\{\nu = i \mid X = x, Z\} = \int\limits_{\mathbb{R}^{Lm}} \frac{\pi_i q(x; \theta_i)}{\sum\limits_{j \in S} \pi_j q(x; \theta_j)} p(\theta \mid Z) d\theta,$$

$$p(\theta \mid Z) = \frac{q(\theta) \prod\limits_{i=1}^{L} \prod\limits_{j=1}^{n_i} q(z_{ij}; \theta_i)}{\int\limits_{\mathbb{R}^{Lm}} q(\theta') \prod\limits_{i=1}^{L} \prod\limits_{j=1}^{n_i} q(z_{ij}; \theta_i') d\theta'}. \tag{2.40}$$

As in Sect. 2.3, (2.40) can be used to construct point and interval forecasts of ν.

To conclude the section, let us briefly discuss the case of models with *nonparametric prior uncertainty*, where the N-dimensional probability densities lie in a distribution family $\mathcal{P}$ which doesn't allow for a finite parameterization:

$$p_1^0(\cdot), p_2^0(\cdot), \ldots, p_L^0(\cdot) \in \mathcal{P}.$$

In this setting, the approach A is still valid, requiring only a modified construction of admissible estimators $\{\hat{p}_i(\cdot)\}$ for $\{p_i^0(\cdot)\}$.

Two types of nonparametric estimators $\{\hat{p}_i(\cdot)\}$ are the most relevant to applications: the Rosenblatt–Parzen estimators and the k-Nearest-Neighbor estimators.

A nonparametric (kernel) Rosenblatt–Parzen estimator [12] of the density $p_i^0(\cdot)$ based on the sample Z_i is defined as the statistic

$$\hat{p}_i(x) = \frac{1}{n_i |H_i|} \sum_{j=1}^{n_i} K(H_i^{-1}(x - z_{ij})), \quad x = (x_l) \in \mathbb{R}^N. \tag{2.41}$$

In this definition,

$$K(x) = \prod_{l=1}^{N} K_l(x_l)$$

is an *N-dimensional kernel*, and each $K_l(y)$, $y \in \mathbb{R}^1$, is a *one-dimensional kernel*—a nonnegative bounded differentiable even function such that $K_l(|y|)$ is nonincreasing in $|y|$, the conditions

$$\int_0^{+\infty} y^m K_l(y)dy < +\infty \quad (m > 0), \qquad \int_{-\infty}^{+\infty} y^2 K_l(y)dy = 1;$$

are satisfied, and $H_i = \text{diag}\{h_{il}\}$ is a diagonal $(N \times N)$-matrix. Diagonal elements of H_i are known as smoothing coefficients; they are strictly positive, $h_{il} > 0$. Given the convergence

$$h_{il} = h_{il}(n_i) \to 0, \quad n_i |H_i| \to \infty$$

as $n_i \to \infty$, the estimator (2.41) is consistent [12].

It has been proved [9] that if $\mathcal{P}$ is a family of thrice differentiable densities, the sizes of training samples $\{n_i\}$ are comparable:

$$n_0 = \min_{i \in S} n_i \to \infty, \quad n_i = c_i n_0, \quad 1 \le c_i < +\infty,$$

where $\{c_i\}$ are certain constants, and the smoothing coefficients can be written asymptotically as

$$h_{il} = b_{il} n_i^{-\frac{1}{N+4}}, \quad l = 1, \ldots, N,$$

then the unconditional Bayesian risk (2.36) satisfies an asymptotic expansion similar to (2.37):

$$r(d_1(\cdot)) = r_0 + \frac{q}{n_0^{4/(N+4)}} + o\left(n_0^{-4/(N+4)}\right), \tag{2.42}$$

where $q = q(N, \{\pi_i\}, \{p_i^0(\cdot)\}, \{w_{ij}\})$ is a known coefficient of the asymptotic formula. The asymptotic expansion (2.42) implies that a PDR $d_1(\cdot)$ constructed from Rosenblatt–Parzen estimators satisfies (2.38) and is, therefore, suboptimal. Similarly to the parametric case (2.39), the asymptotic expansion (2.42) yields the following explicit relation between n_0 and δ:

$$n_0 \geq \left(\frac{q}{r_0 \delta}\right)^{\frac{N}{4}+1}. \tag{2.43}$$

A generalized nonparametric k-Nearest-Neighbor (k-NN) estimator of the density $p_i^0(\cdot)$ based on the sample Z_i is defined as the following statistic [9]:

$$\hat{p}_i(x) = \frac{1}{n_i \varrho_i^N} \sum_{j=1}^{n_i} L_i\left(\frac{x - z_{ij}}{\varrho_i}\right), \quad x \in \mathbb{R}^N, \quad i \in S, \tag{2.44}$$

where $\varrho_i = \varrho_i(x; Z_i) > 0$ is the Euclidean distance between a point $x \in \mathbb{R}^N$ and the k_ith nearest neighbor of the point x in Z_i; each number of neighbors k_i, $2 \leq k_i \leq n_i$, is a positive integer parameter of the estimator; the function $L_i(u)$, $u = (u_k) \in \mathbb{R}^N$, is a bounded integrable weight function such that

$$\int\limits_{\mathbb{R}^N} L_i(u)du = 1, \quad \int\limits_{\mathbb{R}^N} |u|^3 L_i(u)du < \infty, \quad \int\limits_{\mathbb{R}^N} u_k L_i(u)du = 0, \quad k = 1, \ldots, N.$$

In applications, a uniform weight function is used most frequently:

$$L_i(u) = (2\pi^{N/2})^{-1} N \Gamma(N/2) \mathbf{1}_{[0,1]}(|u|),$$

where $\Gamma(\cdot)$ is the gamma function. Assuming the convergences $k_i = k_i(n_i) \to \infty$, $k_i(n_i)/n_i \to 0$ as $n_i \to \infty$, the estimator (2.44) is consistent.

It is known [9] that if $\mathcal{P}$ is a family of thrice differentiable densities, then a PDR based on k-NN estimators (2.44) with the coefficients k_i defined as

$$k_i = [b_i n_i^{4/(N+4)}], \quad i = 1, \ldots, N,$$

where $\{b_i\}$ are some constants, is suboptimal. The unconditional Bayesian risk of this PDR satisfies the asymptotic formula (2.42) with a different value of the coefficient q.

The above analysis shows that parametric prior uncertainty leads to a risk increment that can be estimated as $O(n_0^{-1})$, and nonparametric uncertainty results in a much larger increment—$O\left(n_0^{-4/(N+4)}\right)$. The difference between risks of nonparametric and parametric forecasting becomes higher as N (the number of dimensions of the observation space, or, equivalently, the number of quantities characterizing the investigated process) increases.

To conclude the chapter, let us note that an even higher level of prior uncertainty may be considered, where the training sample Z is assumed to be unclassified. In that case, a forecasting algorithm can be constructed by applying methods of cluster analysis [9].

References

1. Aitchison, J., Dunsmore, J.: Statistical Prediction Analysis. CUP, Cambridge (1975)
2. Amagor, H.: A Markov analysis of DNA sequences. J. Theor. Biol. **104**, 633–642 (1983)
3. Anderson, T.: An Introduction to Multivariate Statistical Analysis. Wiley, Hoboken (2003)
4. Borovkov, A.: Mathematical Statistics. Gordon & Breach, Amsterdam (1998)
5. Collet, D.: Modeling Binary Data. Chapman & Hall, London (2002)
6. Fokianos, K., Fried, R.: Interventions in ingarch models. J. Time Ser. Anal. **31**, 210–225 (2010)
7. Fokianos, K., Kedem, B.: Prediction and classification of nonstationary categorical time series. J. Multivar. Anal. **67**, 277–296 (1998)
8. Fokianos, K., Kedem, B.: Regression theory for categorical time series. Stat. Sci. **18**, 357–376 (2003)
9. Kharin, Yu.: Robustness in Statistical Pattern Recognition. Kluwer Academic, Dordrecht (1996)
10. Lloyd, E.: Handbook of Applicable Mathematics, vol. 6. Wiley, Chichester (1994)
11. McLachlan, G.: Discriminant Analysis and Statistical Pattern Recognition. Wiley, New York (1992)
12. Parzen, E.: On the estimation of a probability density function and the mode. Ann. Math. Stat. **40**, 1063–1076 (1962)
13. Serfling, R.: Approximation Theorems of Mathematical Statistics. Wiley, New York (1980)
14. Wald, A.: Sequential Analysis. Wiley, New York (1947)

Chapter 3
Time Series Models of Statistical Forecasting

Abstract This chapter introduces time series models which are most commonly used in statistical forecasting: regression models, including trend models, stationary time series models, the ARIMA(p, d, q) model, nonlinear models, multivariate time series models (including VARMA(p, q) and simultaneous equations models), as well as models of discrete time series with a specific focus on high-order Markov chains.

3.1 Regression Models of Time Series

Construction of a sufficiently adequate mathematical model of the process that is being forecast is a necessary stage in the development of forecasting techniques. A universal model that is most widely used in applications is the *random process* model

$$x_t = x_t(\omega) : \Omega \times \mathcal{T} \to \mathcal{X}, \tag{3.1}$$

which is defined on a certain probability space $(\Omega, \mathcal{F}, \mathbb{P})$. Here $\Omega = \{\omega\}$ is the set of all elementary events, $\mathcal{F}$ is a σ-algebra of subsets (random events) of Ω, $P = \mathbb{P}(A)$, $A \in \mathcal{F}$, is a certain probability measure defined on $\mathcal{F}$; $t \in \mathcal{T}$ is a parameter of the parametric family (3.1) of random vectors $\{x_t\}$ which is interpreted as time; $\mathcal{T} \subseteq \mathbb{R}^1$ is the set of observation times; $\mathcal{X} \subseteq \mathbb{R}^N$ is the state space of the investigated process. If $\mathcal{T}$ is a discrete set (without loss of generality, it can be assumed that $\mathcal{T} = \mathbb{Z} = \{\ldots, -1, 0, +1, \ldots\}$), then x_t is called a discrete time random process or a *random sequence*. If $\mathcal{X}$ is a discrete set, we have a random process with a discrete (finite or countable) state space, or a *discrete random process*. For $N > 1$, the process x_t is called a *vector* or a multivariate random process, and for $N = 1$—a scalar random process, or simply a random process. A *time series* is defined as a sequence of observations ordered by observation time t and recorded continuously or intermittently (as a rule, with the same time interval

Y. Kharin, *Robustness in Statistical Forecasting*, DOI 10.1007/978-3-319-00840-0_3,

between consecutive observations). With this definition of a time series, we have generalized the notion of a statistical sample by paying particular attention to the dynamics of the underlying process.

For simplicity, further it is assumed that discrete time observations are separated by identical time intervals. The sequence of observations is modeled by a discrete time random process (3.1):

$$x_t = x_t(\omega), \quad t \in \mathbb{Z}.$$

This model is usually referred to as a *random time series* (similarly to a random sample).

Let us assume that

$$z_t \in \mathbb{R}^M, \quad t \in \mathbb{Z},$$

is a given nonrandom sequence of vectors;

$$u_t = u_t(\omega) \in \mathbb{R}^N, \quad t \in \mathbb{Z}$$

is a sequence of independent random vectors defined on $(\Omega, \mathfrak{F}, \mathbb{P})$ with zero vector expectations and a covariance matrix $\Sigma = (\sigma_{jk}) \in \mathbb{R}^{N \times N}$:

$$\mathbb{E}\{u_t\} = \mathbf{0}_N, \quad \mathbf{Cov}\{u_t, u_{t'}\} = \mathbb{E}\{u_t u'_{t'}\} = \delta_{tt'} \Sigma; \tag{3.2}$$

also let

$$y = F(z; \theta^0) : \mathbb{R}^M \times \Theta \to \mathbb{R}^N \tag{3.3}$$

be a fixed function depending on a vector parameter $\theta^0 \in \Theta \subseteq \mathbb{R}^m$.

Definition 3.1. A regression model of a time series defined on the probability space $(\Omega, \mathfrak{F}, \mathbb{P})$ is given by the following stochastic equation:

$$x_t = F(z_t; \theta^0) + u_t, \quad t \in \mathbb{Z}. \tag{3.4}$$

Here $F(\cdot)$ is the *regression function*, z_t is the *regressor vector* (or the factor vector), and u_t is the random observation error.

By (3.2) and (3.4) we have

$$\mathbb{E}\{x_t\} = F(z_t; \theta^0),$$

and thus the regression function (3.3) determines the mean trend, i.e., the dependence between the expectation of a regression process and the time t.

The stochastic regression model (3.4) is defined by the function $F(\cdot)$, the vector parameter θ^0, and the covariance matrix Σ.

If $F(\cdot)$ is linear w.r.t. θ^0, we have a *linear regression time series*:

$$x_t = \theta^0 z_t + u_t, \quad t \in \mathbb{Z},$$

where $\theta^0 \in \mathbb{R}^{N\times M}$ is the matrix of regression coefficients ($m = M^2$); otherwise we have a *nonlinear regression* model.

Definition 3.2. A *trend model* is defined as a special case of the model (3.4), where the time variable is used as a regressor ($z_t = t$):

$$x_t = f(t;\theta^0) + u_t, \quad t \in \mathbb{Z}, \tag{3.5}$$

where $f(\cdot)$ is the *trend*—a fixed function of the time t and the parameter θ^0.

A linear trend model of a time series for $N = 1$ can be defined as follows:

$$x_t = \theta^{0'}\psi(t) + u_t, \quad t \in \mathbb{Z}, \tag{3.6}$$

where $\psi(t) = (\psi_i(t))$ is a column vector of m linearly independent functions and $\theta^0 = (\theta_i^0)$ is an m-column vector of trend coefficients:

$$f(t;\theta^0) = \theta^{0'}\psi(t).$$

The most commonly used special cases of the linear trend model (3.6) are the polynomial model given by

$$x_t = \sum_{i=1}^{m} \theta_i^0 t^{i-1} + u_t, \quad t \in \mathbb{Z},$$

and the harmonic trend model:

$$x_t = \sum_{i=1}^{m} \theta_i^0 \cos(\lambda_i t + \phi_i) + u_t, \quad t \in \mathbb{Z},$$

where $\{\lambda_i\}, \{\phi_i\}$ are, respectively, the frequencies and the initial phases of harmonic oscillations that define the trend.

Regression models of time series have numerous applications in economics, environmental studies, medicine, and technology [13, 16–18, 35, 39].

3.2 Stationary Time Series Models

Weak and strict stationarity of time series is defined as follows.

Definition 3.3. A time series $x_t \in \mathbb{R}$, $t \in \mathbb{Z}$, defined on a probability space $(\Omega, \mathcal{F}, \mathbb{P})$ is called *strictly stationary* if for all $n \in \mathbb{N}$ and $\tau \in \mathbb{Z}$ the n-dimensional probability distribution of the time series x_t is invariant to time shifts $t' = t + \tau$.

Definition 3.4. A time series x_t, $t \in \mathbb{Z}$, is called *stationary* or *weakly stationary* if its second order moments are bounded:

$$\mathbb{E}\{x_t^2\} < +\infty, \quad t \in \mathbb{Z}, \tag{3.7}$$

the expectation is independent of time:

$$\mathbb{E}\{x_t\} = \mu = \text{const}, \quad t \in \mathbb{Z},$$

and the covariance functions depend only on the time intervals between measurements:

$$\mathbf{Cov}\{x_t, x_{t'}\} = \mathbb{E}\{(x_t - \mu)(x_{t'} - \mu)\} = \sigma(t' - t), \quad t, t' \in \mathbb{Z}.$$

The relation between these two definitions of stationarity follows from the well-known property [8,31]: if x_t is strictly stationary and its second order moments are bounded (i.e., (3.7) is satisfied), then x_t is also weakly stationary.

The weak stationarity property is the most relevant to applications since establishing strict stationarity based on random observations is, usually, unfeasible. A weakly stationary time series can be characterized by its expectation μ and covariance function $\sigma(\tau) = \sigma(-\tau)$, $\tau \in \mathbb{Z}$. If the time series x_t has a "sufficiently short memory," i.e., if its covariance function tends to zero sufficiently quickly:

$$\sigma(\tau) \to 0, \quad \tau \to \pm\infty,$$

and the series

$$\sum_{\tau=1}^{+\infty} |\sigma(\tau)| < +\infty$$

converges, then we can define another important characteristic of the stationary time series x_t:

$$S(\lambda) = \frac{1}{2\pi} \sum_{\tau=-\infty}^{+\infty} \sigma(\tau)\cos(\lambda\tau), \quad \lambda \in [-\pi, +\pi]. \tag{3.8}$$

The quantity $S(\lambda)$ is called the *spectral density*. It can be seen from (3.8) that the spectral density is the cosine Fourier transform of the covariance function. The inverse relation also holds:

$$\sigma(\tau) = \int_{-\pi}^{+\pi} S(\lambda)\cos(\lambda\tau)d\lambda, \quad \tau \in \mathbb{Z}.$$

In applied forecasting, stationary time series models are used more often than any other type of statistical model [6]. The first theoretical results in statistical forecasting obtained by Andrey Kolmogorov [29, 30] were derived from stationary models.

3.3 ARIMA(p, d, q) Time Series Model

Let $(\Omega, \mathcal{F}, \mathbb{P})$ be a probability space. Suppose that a stationary time series $u_t, t \in \mathbb{Z}$, satisfying the very simple *discrete white noise* model WN$(0, \sigma^2)$ has been defined on this probability space:

$$\mathbb{E}\{u_t\} = 0, \quad \mathbb{E}\{u_t u_{t'}\} = \sigma^2 \delta_{t,t'}, \quad t, t' \in \mathbb{Z}. \tag{3.9}$$

The relation (3.9) defines $\{u_t\}$ as a sequence of uncorrelated random variables that have zero means and variances equal to σ^2. Note that sometimes the condition (3.9) is specialized by demanding the independence of $\{u_t\}$ and the normality of the joint probability distribution of $\{u_t\}$. Under these additional assumptions, (3.9) yields that $u_t, t \in \mathbb{Z}$, is a strictly stationary Gaussian time series.

To simplify the notation, time series models are defined using the *lag operator* (or *backshift operator*) B (cf. [5]):

$$Bu_t = u_{t-1}, \quad t \in \mathbb{Z}. \tag{3.10}$$

For arbitrary sequences $\{u_t\}$, $\{v_t\}$ and constants $a \in \mathbb{R}$, $d \in \mathbb{N}$, the operator B has the following properties:

$$\begin{gathered} Ba = a, \qquad B(au_t) = aBu_t, \qquad B(u_t + v_t) = Bu_t + Bv_t, \\ B^d u_t = u_{t-d}, \qquad (1 - aB)^{-1} = 1 + \sum_{i=1}^{\infty} a^i B^i. \end{gathered} \tag{3.11}$$

Let us also define the forward difference operator Δ:

$$\Delta u_t = u_t - u_{t-1}, \quad t \in \mathbb{Z},$$

and the forward difference operator of order $d \in \mathbb{N}$:

$$\Delta^d u_t = \Delta(\Delta^{d-1} u_t) = \cdots = \underbrace{\Delta\Delta \ldots \Delta}_{d} u_t, \quad t \in \mathbb{Z}. \tag{3.12}$$

The relations (3.10)–(3.12) imply that the operator Δ and the operator B are connected by the following identity:

$$\Delta^d u_t = (1 - B)^d u_t.$$

Definition 3.5. A time series x_t is said to follow an autoregressive moving average time series model of order (p, q), denoted as ARMA(p, q), if it satisfies the following stochastic difference equation:

$$x_t - \alpha_1 x_{t-1} - \cdots - \alpha_p x_{t-p} = u_t - \gamma_1 u_{t-1} - \cdots - \gamma_q u_{t-q}, \quad t \in \mathbb{Z}, \tag{3.13}$$

which can be rewritten in the shortened operator form as

$$\alpha(B)x_t = \gamma(B)u_t, \quad t \in \mathbb{Z}, \tag{3.14}$$

where

$$\alpha(B) = 1 - \sum_{i=1}^{p} \alpha_i B^i, \quad \gamma(B) = 1 - \sum_{j=1}^{q} \gamma_j B^j$$

are polynomials w.r.t. B, and $\{\alpha_i\}$, $\{\beta_j\}$ are the coefficients of these polynomials.

The ARMA(p, q) model utilizes the following parameters:

$$p, q \in \mathbb{N} \cup \{0\}, \quad \{\alpha_1, \ldots, \alpha_p\}, \{\beta_1, \ldots, \beta_q\} \subset \mathbb{R}^1, \quad \sigma^2 > 0.$$

If the roots $\{\lambda_1, \ldots, \lambda_p\} \subset \mathbb{C}$ of the characteristic equation (cf. [3])

$$\lambda^p - \sum_{i=1}^{p} \alpha_i \lambda^{p-i} = 0$$

corresponding to the polynomial $\alpha(B)$ lie within the unit circle in $\mathbb{C}$,

$$|\lambda_i| < 1, \quad i = 1, 2, \ldots, p,$$

then the time series x_t is weakly stationary. Observe that this condition is equivalent to demanding that all roots $\{\mu_1, \ldots, \mu_p\}$ of the polynomial $\alpha(\mu) = 1 - \sum_{i=1}^{p} \alpha_i \mu^i$ lie outside the unit circle: $|\mu_i| > 1$, $i = 1, \ldots, p$.

Definition 3.6. If $q = 0$, then the time series x_t is said to follow an order p autoregression model of time series, denoted as AR(p):

$$\alpha(B)x_t = x_t - \alpha_1 x_{t-1} - \cdots - \alpha_p x_{t-p} = u_t, \quad t \in \mathbb{Z}. \tag{3.15}$$

Definition 3.7. If $p = 0$, then the time series x_t is said to follow a moving average model of order q, denoted as MA(q):

$$x_t = \gamma(B)u_t = u_t - \gamma_1 u_{t-1} - \cdots - \gamma_q u_{t-q}, \quad t \in \mathbb{Z}. \tag{3.16}$$

For a detailed treatment of the models introduced by Definitions 3.5–3.7, see [5].

Thus, AR(p) and MA(q) are special cases of the ARMA(p, q) model. Let us establish the relation between the AR(p) and MA(q) models. If the roots of the polynomial $\alpha(\mu)$ lie outside the unit circle, then we can use a Taylor expansion w.r.t. powers of B to define the operator $\alpha^{-1}(B)$. Applying this operator to both sides of (3.15) and comparing the result to (3.16), we obtain a representation of the AR(p) model in the form of an MA($+\infty$) model ($q = +\infty$):

$$x_t = \alpha^{-1}(B)u_t, \quad t \in \mathbb{Z}.$$

If the roots of the polynomial $\gamma(\nu)$ lie outside the unit circle, then applying the operator $\gamma^{-1}(B)$ to both sides of (3.16) and comparing the result to (3.15) yields a representation of the MA(q) model in the form of an AR($+\infty$) model ($q = +\infty$):

$$\gamma^{-1}(B)x_t = u_t, \quad t \in \mathbb{Z}.$$

From the above argument, under the mentioned conditions on the roots of the polynomials $\alpha(\cdot)$ and $\gamma(\cdot)$, we can establish the following symbolic relations between the models [37]:

$$\begin{aligned} \text{AR}(p) &\equiv \text{MA}(+\infty), \\ \text{MA}(q) &\equiv \text{AR}(+\infty), \\ \text{ARMA}(p,q) &\equiv \text{MA}(+\infty), \\ \text{ARMA}(p,q) &\equiv \text{AR}(+\infty). \end{aligned} \tag{3.17}$$

It should also be noted that, as implied by (3.17), the ARMA(p, q) model is a special case of a so-called *general linear process* [37]:

$$x_t = \sum_{j=0}^{+\infty} \beta_j u_{t-j}, \quad t \in \mathbb{Z}, \tag{3.18}$$

where $\{\beta_j\}$ is a given sequence of coefficients such that the following series converges:

$$\sum_{j=0}^{+\infty} \beta_j^2 < +\infty. \tag{3.19}$$

Given (3.19), the general linear process (3.18) is a weakly stationary time series with a zero mean and the covariance function

$$\sigma(\tau) = \mathbb{E}\{x_t x_{t+\tau}\} = \sigma^2 \sum_{j=0}^{+\infty} \beta_j \beta_{j+\tau}, \quad \tau \in \mathbb{Z}.$$

By (3.16) and (3.18), for an MA(q) time series we can write $\beta_j = \gamma_j \mathbf{1}_{\{0,1,\dots,q\}}(j)$, and therefore

$$\sigma(\tau) = \sigma^2 \sum_{j=0}^{q-\tau} \gamma_j \gamma_{j+\tau} \mathbf{1}_{[-q,q]}(\tau), \quad \tau \in \mathbb{Z}.$$

For an AR(p) time series, the coefficients $\{\beta_j\}$ can be derived from $\{\alpha_i\}$ using well-known recurrence relations [3], and the covariances $\{\sigma(\tau) : \tau \in \mathbb{Z}\}$ satisfy the Yule–Walker equations [3]:

$$\begin{gathered} -\sum_{i=0}^{p} \alpha_i \sigma(i) = \sigma^2; \\ \sum_{i=0}^{p} \alpha_i \sigma(\tau - i) = 0, \quad \sigma(-\tau) = \sigma(\tau), \quad \tau \in \mathbb{N}. \end{gathered} \tag{3.20}$$

The ARMA(p, q) model serves as the basis for constructing a non-stationary ARIMA(p, d, q) time series model.

Definition 3.8. A non-stationary time series $\{x_t\}$ is said to be integrated of order $d \in \mathbb{N}$ if the time series $y_t = \Delta^d x_t$, $t \in \mathbb{Z}$, consisting of finite differences of the minimal order d is stationary.

An integrated time series x_t can be expressed from the time series of its finite differences y_t by using the summation operator:

$$S = \Delta^{-1} = (1 - B)^{-1} = 1 + \sum_{j=1}^{+\infty} B^j, \quad S y_t = \sum_{j=0}^{+\infty} y_{t-j},$$

$$S^d = \Delta^{-d} = (1 - B)^{-d} = 1 + \sum_{j=1}^{+\infty} C_{d-1+j}^{d-1} B^j, \quad S^d y_t = \sum_{j=0}^{+\infty} C_{d-1+j}^{d-1} y_{t-j}.$$

Definition 3.9. A non-stationary time series x_t is said to satisfy an integrated of order d autoregressive of orders (p, q) moving average model, which is denoted as ARIMA(p, d, q) [5], if it satisfies the following stochastic difference equation (written in the operator form):

$$\alpha(B) \Delta^d x_t = \gamma(B) u_t, \quad t \in \mathbb{Z}. \tag{3.21}$$

The time series $y_t = \Delta^d x_t$, $t \in \mathbb{Z}$, follows an ARMA(p, q) time series model considered above:

$$\alpha(B) y_t = \gamma(B) u_t, \quad t \in \mathbb{Z}.$$

To conclude the section, let us present the following three equivalent representations of the ARIMA(p, d, q) model defined by (3.21) [4, 37]:

1. ARMA$(p + d, q)$:

$$\beta(B)x_t = \gamma(B)u_t, \quad t \in \mathbb{Z},$$

$$\beta(B) = \alpha(B)\Delta^d = \left(1 - \sum_{i=1}^{p} \alpha_i B^i\right)(1 - B^d).$$

2. MA$(+\infty)$:

$$x_t = \alpha^{-1}(B)(1 - B)^d \gamma(B)u_t, \quad t \in \mathbb{Z}.$$

3. AR$(+\infty)$:

$$\gamma^{-1}(B)\alpha(B)(1 - B)^d x_t = u_t, \quad t \in \mathbb{Z}.$$

These equivalent representations are useful in constructing practical approximations to the ARIMA(p, d, q) model.

3.4 Nonlinear Time Series

3.4.1 A General Nonlinear Model

In the previous section, we have discussed ARIMA time series models—a classical family of models with a long history of statistical applications. ARIMA time series are an example of *linear time series models*—the generating stochastic difference equation (3.21) is linear in both the lag values $\{x_{t-i}\}$, $\{u_{t-j}\}$ and the parameters (coefficients) $\{\alpha_i\}$, $\{\beta_j\}$.

Due to both increasingly demanding applications and an exponential increase in computational power over the last years, modern statistical forecasting has become increasingly reliant on *nonlinear time series models* [9, 14, 41]. In the most general form, a nonlinear time series model can be represented as the following nonlinear stochastic difference equation:

$$x_t = F(x_{t-1}, \ldots, x_{t-p}, u_t, u_{t-1}, \ldots, u_{t-q}, v_t; \theta), \quad t \in \mathbb{Z}, \tag{3.22}$$

where u_t is discrete white noise WN$(0, \sigma^2)$ as defined in (3.9); $v_t \in V \subseteq \mathbb{R}^1$ is some auxiliary random process defined on $(\Omega, \mathcal{F}, \mathbb{P})$; integers $p, q \in \mathbb{N} \cup \{0\}$ are the maximal lags of the time series x_t and the innovation process u_t, respectively; $\theta \in \Theta \subseteq \mathbb{R}^m$ is the parameter vector; $F(\cdot) : \mathbb{R}^{p+q+1} \times \Theta \to \mathbb{R}^1$ is a certain nonlinear function. Observe that if $F(\cdot)$ in (3.22) is linear, then we obtain the previously considered ARIMA model. Also note that, in the most general case, stationarity conditions for the time series defined by (3.22) are not known.

Let us consider several well-studied special cases of (3.22) which are defined by a particular form of $F(\cdot)$.

3.4.2 *Bilinear Model BL(p, q, P, Q)*

Definition 3.10. A bilinear time series model $\mathrm{BL}(p, q, P, Q)$ of orders p, q, P, Q is defined by the following bilinear stochastic difference equation [19]:

$$x_t = \sum_{i=1}^{p} a_i x_{t-i} + \sum_{j=0}^{q} b_j u_{t-j} + \sum_{j=1}^{P} \sum_{k=1}^{Q} c_{jk} x_{t-j} u_{t-k}, \quad t \in \mathbb{Z}. \tag{3.23}$$

The parameters of this model are σ^2, $p, q, P, Q, \{a_i\}, \{b_j\}, \{c_{jk}\}$.

As an example of this model, let us consider the special case $\mathrm{BL}(1, 0, 1, 1)$ which is often used in econometric simulations [14, 24]:

$$x_t = \theta_1 x_{t-1} + u_t + \theta_2 x_{t-1} u_{t-1}, \quad t \in \mathbb{Z}. \tag{3.24}$$

It is known [14] that if

$$\mathbb{E}\{u_t^4\} < +\infty, \quad \theta_1^2 + \sigma^2 \theta_2^2 < 1, \tag{3.25}$$

then the model (3.24) defines a strictly stationary random process with a bounded fourth moment: $\mathbb{E}\{x_t^4\} < +\infty$.

3.4.3 *Functional-Coefficient Autoregression Model FAR(p, d)*

Definition 3.11. The functional-coefficient autoregression model $\mathrm{FAR}(p, d)$ is defined by the following stochastic difference equation [14]:

$$x_t = \theta_1(x_{t-d}) x_{t-1} + \dots + \theta_p(x_{t-d}) + \sigma(x_{t-d}) \varepsilon_t, \quad t \in \mathbb{Z}, \tag{3.26}$$

where p, d are positive integers, and $\theta_1(\cdot), \dots, \theta_p(\cdot), \sigma(\cdot) : \mathbb{R}^1 \to \mathbb{R}^1$ are some fixed functions.

If these functions are constant, we obtain the AR(p) model. In applications, the FAR(p, d) model can be characterized by either (a) parametric prior uncertainty, if the above functions are known up to certain parameters; or (b) nonparametric prior uncertainty, if the above functions are unknown, and only certain smoothness assumptions can be made (in the form of Lipschitz conditions).

3.4.4 *Generalized Exponential Autoregression Model EXPAR*(p, d)

Definition 3.12. The EXPAR(p, d) model is a special case of the FAR(p, d) model defined by the following nonlinear stochastic difference equation [36]:

$$x_t = \sum_{i=1}^{p} \left(\alpha_i + (\beta_i + \gamma_i x_{t-d}) \exp(-\theta_i x_{t-d}^2)\right) x_{t-i} + u_t, \quad t \in \mathbb{Z},$$

where $p, d \in \mathbb{N}$, σ^2, $\{\alpha_i, \beta_i, \gamma_i, \theta_i\}$ are model parameters.

3.4.5 *Threshold Autoregression Model TAR*(k)

Definition 3.13. It is said [14] that a time series x_t follows the threshold autoregression model with $2 \leq k < +\infty$ regimes, denoted as TAR(k), if it satisfies the following nonlinear stochastic difference equation:

$$x_t = \sum_{j=1}^{k} (\theta_{j0} + \theta_{j1} x_{t-1} + \cdots + \theta_{j,p_j} x_{t-p_j} + \sigma_j u_t) \mathbf{1}_{A_j}(x_{t-d}), \quad t \in \mathbb{Z}, \quad (3.27)$$

where $\{A_j\}$ is a partition of the real line: $\bigcup_{j=1}^{k} A_j = \mathbb{R}^1$, $A_i \cap A_j = \emptyset$ for $i \neq j$, $d, p_1, \ldots, p_k \in \mathbb{N}$ and $\sigma_j, \theta_{j0}, \theta_{j1}, \ldots, \theta_{j,p_j} \in \mathbb{R}^1$ for $j = 1, \ldots, k$.

The parameters of the TAR(k) model are k, $\{A_j\}$, d, $\{p, \sigma_j, \theta_j, p_j\}$.

TAR models have numerous applications. As an example, let us present a TAR(2) model that was used in [14] for logarithmic scale modeling of the American GDP growth index:

$$x_t = \begin{cases} -0.7x_{t-1} + u_t, & x_{t-1} \geq -0.5, \\ 0.7x_{t-1} + u_t, & x_{t-1} < -0.5. \end{cases}$$

3.4.6 ARCH(p) Model

Definition 3.14. The autoregressive conditional heteroscedasticity time series model of order p, or ARCH(p), is defined by the following equation [14, 40]:

$$x_t = \sigma_t u_t, \quad \sigma_t^2 = \theta_0 + \theta_1 x_{t-1}^2 + \cdots + \theta_p x_{t-p}^2, \quad t \in \mathbb{Z}, \tag{3.28}$$

where $\theta_0 > 0,\ \theta_1, \ldots, \theta_p \geq 0$, and u_t is the standard discrete white noise WN(0, 1).

It is known [14] that the time series (3.28) is strictly stationary if and only if

$$\sum_{j=1}^{p} \theta_j < 1,$$

and in that case we also have

$$\mathbb{E}\{x_t\} = 0, \quad \mathbb{D}\{x_t\} = \theta_0 \left(1 - \sum_{j=1}^{p} \theta_j\right)^{-1}.$$

3.4.7 GARCH(p, q) Model

Definition 3.15. Generalized autoregressive conditional heteroscedasticity model of orders p, q, denoted as GARCH(p, q), is defined by generalizing the second equation of (3.28) as follows [14]:

$$\sigma_t^2 = \theta_0 + \sum_{i=1}^{p} \theta_i x_{t-i}^2 + \sum_{j=1}^{q} \alpha_j \sigma_{t-j}^2.$$

Note that for $q = 0$ we obtain (3.28), implying that GARCH($p, 0$) =ARCH(p). The parameters of the GARCH(p, q) model are $p, q, \theta_0, \theta_1, \ldots, \theta_p, \alpha_1, \ldots, \alpha_q$.

3.5 Multivariate Time Series Models

3.5.1 Multivariate Stationary Time Series

The models presented in Sect. 3.2 for univariate time series $x_t \in \mathbb{R}^1$ can be easily generalized for the multivariate case $x_t \in \mathbb{R}^N$ [7, 9, 11, 20, 33, 34, 37].

An *N-variate stationary time series* $x_t = (x_{ti}) \in \mathbb{R}^N$ has the expectation

$$\mu = (\mu_i) = \mathbb{E}\{x_t\} \in \mathbb{R}^N$$

that doesn't depend on time t and the $(N \times N)$-matrix covariance function which is only dependent on the time difference:

$$\sigma(t-s) = (\sigma_{jk}(t-s)) = \mathbb{E}\{(x_t - \mu)(x_s - \mu)'\} = \mathbf{Cov}\{x_t, x_s\}, \quad t, s \in \mathbb{Z},$$

where the prime symbol denotes transposition, and

$$\sigma_{jk}(t-s) = \mathbf{Cov}\{x_{tk}, x_{sj}\}, \quad j, k \in \{1, \dots, N\},$$

is the covariance function of the kth and the jth coordinates of the multivariate (vector) time series x_t.

Similarly to (3.8), we introduce the $(N \times N)$-matrix spectral density:

$$S(\lambda) = (S_{jk}(\lambda)) = \frac{1}{2\pi} \sum_{\tau=-\infty}^{+\infty} \sigma_{jk}(\tau) e^{-i\lambda\tau}, \quad \lambda \in [-\pi, +\pi],$$

where i is the imaginary unit, and $S_{jk}(\lambda)$ is the joint spectral density of the univariate (scalar) time series $x_{t,j}, x_{t,k} \in \mathbb{R}^1$.

3.5.2 Vector Autoregression Model VAR(p)

Definition 3.16. The vector autoregression model of order p, or VAR(p), is a multivariate generalization of the AR(p) model defined in Sect. 3.3. It is defined by the following system of N stochastic difference equations (written in the matrix form):

$$x_t = \theta_0 + \theta_1 x_{t-1} + \dots + \theta_p x_{t-p} + u_t, \quad t \in \mathbb{Z}, \tag{3.29}$$

where $x_t = (x_{ti}) \in \mathbb{R}^N$ is the column vector of observations at time t, $x_{tj} \in \mathbb{R}^1$ is the jth coordinate at time t for $j = 1, \dots, N$; $\theta_0 = (\theta_{0i}) \in \mathbb{R}^N$ is a column vector of constants; $\theta_s = (\theta_{sij}) \in \mathbb{R}^{N \times N}$ is a matrix of autoregression coefficients corresponding to a lag $s \in \{1, 2, \dots, p\}$; $\{u_t\} \subset \mathbb{R}^N$ is the innovation process, which is a sequence of uncorrelated identically distributed random N-vectors with the expectation equal to zero, $\mathbb{E}\{u_t\} = \mathbf{0}_N$, and a nonsingular covariance matrix $\Sigma = (\sigma_{ij}) \in \mathbb{R}^{N \times N}$:

$$\mathbf{Cov}\{u_t, u_s\} = \mathbb{E}\{u_t u_s'\} = \delta_{ts}\Sigma, \quad t, s \in \mathbb{Z}.$$

Using the lag operator B defined in Sect. 3.3 and introducing a matrix polynomial of degree p,

$$P(B) = \mathbf{I}_N - \sum_{s=1}^{p} \theta_s B^s,$$

yields an equivalent representation of the VAR(p) model:

$$P(B)x_t = \theta_0 + u_t, \quad t \in \mathbb{Z}. \tag{3.30}$$

By increasing the number of variables N, the VAR(p) model can be represented as a first order vector autoregression VAR(1):

$$X_t = \Theta_0 + \Theta X_{t-1} + U_t, \quad t \in \mathbb{Z},$$

with the following block-matrix notation:

$$X_t = \begin{pmatrix} x_t \\ \hline x_{t-1} \\ \hline \vdots \\ \hline x_{t-p+1} \end{pmatrix}, \quad U_t = \begin{pmatrix} u_t \\ \hline u_{t-1} \\ \hline \vdots \\ \hline u_{t-p+1} \end{pmatrix}, \quad \Theta_0 = \begin{pmatrix} \theta_0 \\ \hline \mathbf{0}_N \\ \hline \vdots \\ \hline \mathbf{0}_N \end{pmatrix} \in \mathbb{R}^{Np},$$

$$\Theta = \left(\begin{array}{c|c|c|c|c|c} \theta_1 & \theta_2 & \theta_3 & \cdots & \theta_{p-1} & \theta_p \\ \hline \mathbf{I}_N & \mathbf{0}_{N\times N} & \mathbf{0}_{N\times N} & \cdots & \mathbf{0}_{N\times N} & \mathbf{0}_{N\times N} \\ \hline \mathbf{0}_{N\times N} & \mathbf{I}_N & \mathbf{0}_{N\times N} & \cdots & \mathbf{0}_{N\times N} & \mathbf{0}_{N\times N} \\ \hline \vdots & \vdots & \vdots & \cdots & \vdots & \vdots \\ \hline \mathbf{0}_{N\times N} & \mathbf{0}_{N\times N} & \mathbf{0}_{N\times N} & \cdots & \mathbf{I}_N & \mathbf{0}_{N\times N} \end{array}\right) \in \mathbb{R}^{Np\times Np},$$

$$Q = \left(\begin{array}{c|c|c|c} \Sigma & \mathbf{0}_{N\times N} & \cdots & \mathbf{0}_{N\times N} \\ \hline \mathbf{0}_{N\times N} & \mathbf{0}_{N\times N} & \cdots & \mathbf{0}_{N\times N} \\ \hline \vdots & \vdots & \cdots & \vdots \\ \hline \mathbf{0}_{N\times N} & \mathbf{0}_{N\times N} & \cdots & \mathbf{0}_{N\times N} \end{array}\right) \in \mathbb{R}^{Np\times Np}.$$

We also have

$$\mathbb{E}\{U_t\} = \mathbf{0}_{Np}, \quad \mathbf{Cov}\{U_t, U_s\} = Q \cdot \delta_{ts}, \quad t, s \in \mathbb{Z}.$$

It follows from (3.29) that the VAR(p) model has the following parameters: p, Σ, $\theta_0, \theta_1, \ldots, \theta_p$, and the total number of unknown parameters is

$$r_{VAR} = (p + 1/2)N^2 + 3N/2 + 1.$$

If we assume that all eigenvalues $\{\lambda_k\}$ of the matrix Θ, which are the roots of the following algebraic equation of order Np:

$$|\mathbf{I}_N \lambda^p - \theta_1 \lambda^{p-1} - \theta_2 \lambda^{p-2} - \cdots - \theta_p| = 0,$$

lie within the unit circle, $|\lambda_j| < 1$ for $j = 1, \ldots, Np$, then the time series (3.29) is stationary, and its expectation equals

$$\mu = \mathbb{E}\{x_t\} = (\mathbf{I}_N - \theta_1 - \theta_2 - \cdots - \theta_p)^{-1}\theta_0.$$

The covariance function $R(\tau) = (R_{jk}(\tau)) \in \mathbb{R}^{N \times N}$ is an $(N \times N)$-matrix function satisfying the Yule–Walker matrix equation [33]:

$$R(\tau) = \theta_1 R(\tau - 1) + \cdots + \theta_p R(\tau - p), \quad \tau = p, p+1, \ldots,$$

and the initial p matrices $R(0), \ldots, R(p-1)$ can be found from the following $(Np)^2 \times (Np)^2$-matrix equation:

$$\operatorname{vec}\begin{pmatrix} R(0) & R(1) & \cdots & R(p-1) \\ R'(1) & R(0) & \cdots & R(p-2) \\ \vdots & \vdots & \cdots & \vdots \\ R'(p-1) & R'(p-2) & \cdots & R(0) \end{pmatrix} = \left(\mathbf{I}_{(Np)^2} - \Theta \otimes \Theta\right)^{-1} \operatorname{vec}(Q),$$

where the symbol $\otimes$ denotes the Kronecker product of two matrices and vec(Q) denotes a vector representation of a matrix.

3.5.3 Vector Moving Average Model VMA(q)

Definition 3.17. The vector moving average model of order $q \in \mathbb{N} \cup \{0\}$, denoted as VMA($q$), is a multivariate generalization of the MA(q) model defined in Sect. 3.3 and is defined by the following matrix representation:

$$x_t = u_t + M_1 u_{t-1} + \cdots + M_q u_{t-q}, \quad t \in \mathbb{Z}, \tag{3.31}$$

where $M_1, \ldots, M_q \in \mathbb{R}^{N \times N}$ are matrix coefficients and the rest of the notation is the same as in (3.29).

Similarly to (3.30), the following operator representation of the model (3.31) can be written [33]:

$$x_t = M(B)u_t, \quad M(B) = \mathbf{I}_N + M_1 B + \cdots + M_q B^q, \quad t \in \mathbb{Z}. \tag{3.32}$$

By (3.31), the parameters of the VMA(q) model are q, Σ, $M_1, \ldots, M_q$, and the total number of unknown parameters is

$$r_{VMA} = (q + 1/2)N^2 + N/2 + 1.$$

Assuming that the roots of the algebraic equation of order Nq given by

$$|\mathbf{I}_N + M_1 z + \cdots + M_q z^q| = 0$$

lie outside the unit circle, the *time-reversibility* condition is satisfied for the time series (3.32):

$$M^{-1}(B)x_t = u_t, \quad t \in \mathbb{Z},$$

meaning that a VMA(q) model can be represented as a VAR($+\infty$) model:

$$x_t = \sum_{i=1}^{\infty} \theta_i x_{t-i} + u_t, \quad t \in \mathbb{Z},$$

where the matrix coefficients $\{\theta_i\}$ satisfy the following recurrence relation:

$$\theta_1 = M_1,$$
$$\theta_i = M_i - \sum_{j=1}^{i-1} \theta_{i-j} M_j, \quad i = 2, 3, \ldots .$$

A VMA(q) process is strictly stationary, has a zero expectation, and its matrix covariance function can be written as follows [33]:

$$R(\tau) = \mathbb{E}\{x_t x'_{t-\tau}\} = \sum_{i=0}^{q-\tau} M_{i+\tau} \Sigma M'_i \mathbf{1}_{[0,q]}(\tau); \quad R(-\tau) = R(\tau).$$

3.5.4 *VARMA*(p, q) *Model*

The VARMA(p, q) model is a generalization of VAR(p) and MA(q) models:

$$x_t - \theta_1 x_{t-1} - \cdots - \theta_p x_{t-p} = u_t + M_1 u_{t-1} + \cdots + M_q u_{t-q}, \quad t \in \mathbb{Z}, \tag{3.33}$$

or in the operator form:

$$P(B)x_t = M(B)u_t, \quad t \in \mathbb{Z},$$

where the polynomials $P(B)$, $M(B)$ are defined by (3.30), (3.32).

Under the condition formulated in Sect. 3.5.2, the time series (3.33) is strictly stationary and can be described by an MA$(+\infty)$ model:

$$x_t = \sum_{i=0}^{\infty} \Phi_i u_{t-i},$$

where the matrix coefficients $\{\Phi_i\}$ can be computed from the following recursion:

$$\Phi_0 = \mathbf{I}_N; \quad \theta_j = \mathbf{0}_{N\times N}, \ j > p; \quad M_i = \mathbf{0}_{N\times N}, \ i > q;$$

$$\Phi_i = M_i - \sum_{j=1}^{i} \theta_j \Phi_{i-j}, \quad i = 1, 2, \ldots .$$

It should be noted that the model (3.33) can be represented as a VAR$(+\infty)$ model if the time-reversibility condition is satisfied; there also exists a representation [33] of the N-dimensional VARMA(p, q) model as an $N(p+q)$-dimensional VAR(1) model.

3.5.5 System of Simultaneous Equations (SSE) Model

Definition 3.18. It is said that an N-variate time series $x_t \in \mathbb{R}^N$ follows the system of simultaneous equations model SSE(N, N_0, K, p) if it satisfies the following system of stochastic difference equations (written in the matrix form) [11, 25, 28]:

$$\Theta_0 x_t = \Theta_1 x_{t-1} + \cdots + \Theta_p x_{t-p} + A z_t + u_t, \quad t \in \mathbb{Z}, \tag{3.34}$$

where $x_t = (x_{ti}) \in \mathbb{R}^N$ is the column vector of N endogenous (interior) variables; $\Theta_0, \Theta_1, \ldots, \Theta_p \in \mathbb{R}^{N\times N}$, $A \in \mathbb{R}^{N\times N}$ are coefficient matrices; $z_t = (z_{tj}) \in \mathbb{R}^K$ is the column vector of K exogenous (exterior) variables; $\{u_t = (u_{ti}) \in \mathbb{R}^N : t \in \mathbb{Z}\}$ is a sequence of uncorrelated random vectors with zero mean vectors, $\mathbb{E}\{u_t\} = \mathbf{0}_N$, and covariance matrices equal to $\mathbb{E}\{u_t u_t'\} = \Sigma$.

Comparing (3.34) and (3.29), we can observe that the SSE model can be considered as a special case of the VAR model with exogenous variables (VARX) [33]. However, there are two significant differences: (1) since we usually have $\Theta_0 \neq \mathbf{I}_N$, the left-hand side of (3.34) does not explicitly contain x_t; (2) the system

(3.34) contains $0 \leq N_0 < N$ identities, therefore Σ is singular, contains N_0 zero rows and columns, and its rank is incomplete, $\operatorname{rank}(\Sigma) = N - N_0$.

The representation (3.34) is called the *structural form* of the SSE model. The structural form is used in econometrics to model the dynamics of various economic systems [25, 28].

As a rule, the SSE model satisfies the completeness condition (by construction): $|\Theta_0| \neq 0$. Thus, multiplying (3.34) by Θ_0^{-1} from the left, we can obtain the *reduced form* of the SSE model:

$$x_t = B_1 x_{t-1} + \cdots + B_p x_{t-p} + C z_t + v_t, \quad t \in \mathbb{Z},$$

where $B_1 = \Theta_0^{-1}\Theta_1, \ldots, B_p = \Theta_0^{-1}\Theta_p$, $C = \Theta_0^{-1}A$ are the "new" matrix coefficients, and $\{v_t\}$ is a sequence of uncorrelated random N-vectors with expectations equal to zero and a covariance matrix given by

$$\Sigma = \mathbb{E}\{v_t v_t'\} = \Theta_0^{-1}\Sigma(\Theta_0^{-1})'.$$

3.6 Discrete Time Series

3.6.1 *Markov Chains*

In many applied forecasting problems, the state space $\mathcal{A}$ where the process is observed ($x_t \in \mathcal{A}$) is discrete, and without loss of generality we can assume that

$$\mathcal{A} = \{0, 1, \ldots, N-1\}, \quad 2 \leq N \leq +\infty.$$

This setting becomes more and more common since modern measuring and recording equipment stores the data digitally [2, 10, 22, 42].

A time series with a discrete state space $\mathcal{A}$ is called a *discrete time series.* In other words, a discrete time series $x_t \in \mathcal{A}$, $t \in \mathbb{N}$, is a random process defined on a probability space $(\Omega, \mathcal{F}, \mathbb{P})$ with discrete time t and a discrete (finite or countable) state space $\mathcal{A}$. It should be noted that statistical analysis of discrete time series is significantly less investigated compared to the case of continuous time series where $\mathcal{A} = \mathbb{R}^N$ [15].

The simplest model of a discrete time series is a sequence of random variables x_t, $t \in \mathbb{N}$, which are mutually independent and have a certain discrete probability distribution:

$$\mathbb{P}\{x_t = i\} = p_i(t), \quad i \in \mathcal{A}, \quad t \in \mathbb{N}. \tag{3.35}$$

If the N-column vector of elementary probabilities $p(t) = (p_0(t), \dots, p_{N-1}(t))'$ doesn't depend on time t, then x_t is a homogeneous time series, and otherwise—an inhomogeneous time series. If $N < +\infty$, and

$$p(t) = (1/N, \dots, 1/N),$$

then we have a *uniformly distributed* sequence, which is sometimes called a purely random sequence [10].

Note that the distribution law $p(t)$ can be periodic (when describing cyclic or seasonal phenomena) with a period T_0:

$$\exists T_0 \in \mathbb{N}: \quad p(t + T_0) = p(t), \quad t \in \mathbb{N}.$$

Parametric distribution laws $p(t; \theta^0)$ present another important special case. The model (3.35) is, essentially, a discrete analogue of the trend model discussed in Sect. 3.1.

Definition 3.19. A discrete time series x_t is called a Markov chain if the Markov dependence property is satisfied:

$$\mathbb{P}\{x_{t+1} = i_{t+1} \mid x_t = i_t, \dots, x_1 = i_1\} = \mathbb{P}\{x_{t+1} = i_{t+1} \mid x_t = i_t\} = p_{i_t, i_{t+1}}(t) \tag{3.36}$$

for all $i_1, \dots, i_{t+1} \in \mathcal{A}$, $t \in \mathbb{N}$, where the matrix $P(t) = (p_{ij}(t)) \in [0, 1]^{N \times N}$ known as the matrix of one-step transition probabilities is stochastic; the N-column vector

$$\pi(1) = (\pi_i(1)), \quad \pi_i(1) = \mathbb{P}\{x_1 = i\}, \quad i \in \mathcal{A},$$

is the initial probability distribution.

A classification of Markov chains and their states can be found in [22]. Let us mention one of the important classes: *time-homogeneous* Markov chains are defined as satisfying the condition that $P(t) = P = \text{const}$ doesn't depend on t. Observe that the model of independent trials (3.35) is a special case of a Markov chain, where all N rows of the matrix $P(t)$ are identical and equal to $p'(t)$.

From now on, we are going to consider finite Markov chains with finite numbers of states, $2 \le N < +\infty$.

Note that the number of independent parameters defining a Markov chain (taking into account the normalization condition) is equal to $r_1 = N^2 - 1$, and there are only $r_0 = N - 1$ parameters in the model of independent trials.

3.6.2 Markov Chains of Order s

Definition 3.20. A Markov chain of order $s \in \mathbb{N}$, denoted as MC(s), is defined by the generalized Markov property [12]:

$$\begin{aligned}&\mathbb{P}\{x_{t+1} = i_{t+1} \mid x_t = i_t, \ldots, x_1 = i_1\} = \\&= \mathbb{P}\{x_{t+1} = i_{t+1} \mid x_t = i_t, \ldots, x_{t-s+1} = i_{t-s+1}\} = p_{i_{t-s+1},\ldots,i_t,i_{t+1}}(t) \quad (3.37)\end{aligned}$$

for all $i_1, \ldots, i_{t+1} \in \mathcal{A}$, $t \geq s$.

For $s = 1$, the relation (3.37) is identical to (3.36). A Markov chain of order s is characterized by an s-dimensional initial probability distribution

$$\pi_{i_1,\ldots,i_s} = \mathbb{P}\{x_1 = i_1, \ \ldots, \ x_s = i_s\}, \quad i_1, \ldots, i_s \in \mathcal{A},$$

and an $(s + 1)$-dimensional matrix of one-step transition probabilities at time t:

$$\begin{aligned}&P(t) = \left(p_{i_1,\ldots,i_{s+1}}(t)\right),\\&p_{i_1,\ldots,i_{s+1}}(t) = \mathbb{P}\{x_{t+1} = i_{s+1} \mid x_t = i_s, \ \ldots, \ x_{t-s+1} = i_1\}, \quad i_1, \ldots, i_{s+1} \in \mathcal{A}.\end{aligned}$$

If $P(t)$ doesn't depend on t, we have a *homogeneous Markov chain of order* s.

Markov chains of order s can be used to model processes and dynamic systems with a length of the memory that doesn't exceed s. A Markov chain MC(s) has $r_s = N^{s+1} - 1$ parameters, and the number of parameters grows exponentially as the order s increases. Model identification (estimation of the model parameters) requires a sample $x_1, x_2, \ldots, x_n$ of size $n > r_s$, which is often impractical. This motivates the development of so-called *parsimonious (small-parametric) models of higher order Markov chains* [23, 27].

Raftery's MTD model [38] defines a small-parametric form of the matrix P:

$$p_{i_1,\ldots,i_{s+1}} = \sum_{j=1}^{s} \lambda_j \, q_{i_j,i_{s+1}}, \quad i_1, \ldots, i_{s+1} \in \mathcal{A},$$

where $Q = (q_{ik})$ is a stochastic $(N \times N)$-matrix; $\lambda = (\lambda_1, \ldots, \lambda_s)'$ is an s-column vector of elementary probabilities, $\lambda_1 + \cdots + \lambda_s = 1$. The number of parameters in the MTD model is equal to $r_{\text{MTD}} = N(N-1)/2 + s - 1$ and grows linearly as s increases.

The papers [23, 26] present another small-parametric model, the *Markov chain of order s with r partial connections*, denoted as MC(s, r):

$$p_{i_1,\ldots,i_{s+1}} = q_{i_{m_1},\ldots,i_{m_r},i_{s+1}}, \quad i_1, \ldots, i_{s+1} \in \mathcal{A},$$

where $r \in \{1, 2, \ldots, s\}$ is the number of connections; $M = (m_1, \ldots, m_r)$ is the connection pattern—an arbitrary integer ordered r-vector, $1 = m_1 < m_2 < \cdots < m_r \leq s$; and $Q = (q_{j_1,\ldots,j_r})$ is an arbitrary stochastic matrix. Observe that if $r = s$, and thus $M = (1, 2, \ldots, s)$, then $P = Q$, and MC(s, s) is a fully connected Markov chain $MC(s)$ of order s.

3.6.3 *DAR*(s) *Model of Jacobs and Lewis*

Definition 3.21. A discrete autoregression model of order s, denoted as DAR(s), is defined by the following stochastic difference equation [21]:

$$x_t = \mu_t x_{t-\eta_t} + (1-\mu_t)\xi_t, \quad t > s,$$

where $s \geq 2$, $\{x_1, \ldots, x_s\}, \{\xi_t, \eta_t, \mu_t : t > s\}$ are mutually independent discrete random variables with probability distributions defined as follows:

$$\mathbb{P}\{\xi_t = i\} = \pi_i, \quad i \in \mathcal{A}; \qquad \mathbb{P}\{\eta_t = j\} = \lambda_j, \quad j \in \{1,\ldots,s\}, \ \lambda_s \neq 0;$$
$$\mathbb{P}\{\mu_t = 1\} = 1 - \mathbb{P}\{\mu_t = 0\} = \varrho; \quad \mathbb{P}\{x_t = i\} = \pi_i, \quad i \in \mathcal{A}, \ k \in \{1,\ldots,s\}.$$

This model was successfully used to simulate air pollution [32].

3.6.4 *DMA*(q) *Model*

Definition 3.22. A discrete moving average model of order $q \in \mathbb{N}$, denoted as DMA(q), is defined by the following stochastic difference equation [32]:

$$x_t = \xi_{t-\eta_t}, \quad t > q,$$

where $q \geq 2$, and $\{\xi_t, \eta_t\}$ are mutually independent random variables with probability distributions defined as follows:

$$\mathbb{P}\{\xi_t = i\} = \pi_i, \qquad i \in \mathcal{A};$$
$$\mathbb{P}\{\eta_t = j\} = \lambda_j, \qquad j \in \{1,\ldots,q\}.$$

3.6.5 *INAR*(m) *Model*

Definition 3.23. An integer autoregression model of order m, denoted as INAR(m), is defined by the following stochastic difference equation [1]:

$$x_t = \sum_{j=1}^{m} \sum_{i=1}^{x_{t-j}} \xi_{ti}^{(j)} + \eta_t, \quad t \in \mathbb{N},$$

where

$$\{\xi_{ti}^{(j)} : t = m+1, m+2, \ldots\ ; \ i, j \in \mathbb{N}\}$$

are mutually independent Bernoulli random variables,

$$\mathbb{P}\{\xi_{ti}^{(j)} = 1\} = 1 - \mathbb{P}\{\xi_{ti}^{(j)} = 0\} = p_j,$$

and $\{\eta_t\}$ are integer random variables which are independent of $\{\xi_{ti}^{(j)}\}$.

This model was proposed in [1] to study time series in economics.

References

1. Alzaid, A., Al-Osh, M.: An integer-valued pth-order autoregressive structure (INAR(p)) process. J. Appl. Probab. **27**, 314–324 (1990)
2. Amagor, H.: A Markov analysis of DNA sequences. J. Theor. Biol. **104**, 633–642 (1983)
3. Anderson, T.: The Statistical Analysis of Time Series. Wiley, New York (1994)
4. Borovkov, A.: Mathematical Statistics. Gordon & Breach, Amsterdam (1998)
5. Box, G., Jenkins, G., Reinsel, G.: Time Series Analysis: Forecasting and Control. Wiley-Blackwell, New York (2008)
6. Brillinger, D.: Time Series: Data Analysis and Theory. Holt, Rinehart and Winston, New York (1975)
7. Brockwell, P., Davis, R.: Introduction to Time Series and Forecasting. Springer, New York (2002)
8. Bulinsky, A., Shyryaev, A.: Theory of Random Processes (in Russian). Fizmatlit, Moscow (2003)
9. Chatfield, C.: Time Series Forecasting. Chapman & Hall/CRC, Boca Raton (2001)
10. Collet, D.: Modeling Binary Data. Chapman & Hall, London (2002)
11. Davison, A.: Statistical Models. Cambridge University Press, Cambridge (2009)
12. Doob, J.: Stochastic Processes. Wiley, New York (1953)
13. Draper, N., Smith, H.: Applied Regression Analysis. Wiley, New York (1998)
14. Fan, J., Yao, Q.: Nonlinear Time Series: Nonparametric and Parametric Methods. Springer, New York (2003)
15. Fokianos, K., Kedem, B.: Regression theory for categorical time series. Stat. Sci. **18**, 357–376 (2003)
16. Fried, R., Gather, U.: Fast and robust filtering of time series with trends. In: Härdle, W., RönzComp, B. (eds.) COMPSTAT 2002, pp. 367–372. Physica, Heidelberg (2002)
17. Gather, U., Schettlinger, K., Fried, R.: Online signal extraction by robust linear regression. Comput. Stat. **21**, 33–51 (2006)
18. Gelper, S., Fried, R., Croux, C.: Robust forecasting with exponential and Holt–Winters smoothing. J. Forecast. **29**, 285–300 (2010)
19. Granger, C., Andersen, A.: An Introduction to Bilinear Time Series Models. Vandenhoeck and Ruprecht, Gottingen (1978)
20. Hannan, E.: Multiple Time Series. Wiley, New York (1970)
21. Jacobs, P., Lewis, A.: Discrete time series generated by mixtures. J. Roy. Stat. Soc. B **40**(1), 94–105 (1978)
22. Kemeny, J., Snell, J.: Finite Markov Chains. Springer, New York (1976)
23. Kharin, Y.: Markov chains with r-partial connections and their statistical estimation (in Russian). Trans. Natl. Acad. Sci. Belarus **48**(1), 40–44 (2004)
24. Kharin, Y.: Robustness of autoregressive forecasting under bilinear distortions. In: Computer Data Analysis and Modeling, vol. 1, pp. 124–128. BSU, Minsk (2007)
25. Kharin, Y.: Robustness of the mean square risk in forecasting of regression time series. Comm. Stat. Theor. Meth. **40**(16), 2893–2906 (2011)

26. Kharin, Y., Piatlitski, A.: A Markov chain of order s with r partial connections and statistical inference on its parameters. Discrete Math. Appl. **17**(3), 295–317 (2007)
27. Kharin, Y., Piatlitski, A.: Statistical analysis of discrete time series based on the MC(s, r)-model. Aust. J. Stat. **40**(1-2), 75–84 (2011)
28. Kharin, Y., Staleuskaya, S.: Robustification of "approximating approach" in simultaneous equation models. In: New Trends in Probability and Statistics, vol. 5, pp. 143–150. TEV, Vilnius (2000)
29. Kolmogorov, A.: On the use of statistically estimated forecasting formulae (in Russian). Geophys. J. **3**, 78–82 (1933)
30. Kolmogorov, A.: Interpolation and extrapolation of stationary stochastic series (in Russian). Izv. Akad. Nauk SSSR Ser. Mat **5**(1), 3–14 (1941)
31. Koroljuk, V.: Handbook on Probability Theory and Mathematical Statistics (in Russian). Nauka, Moscow (1985)
32. Lewis, P.: Simple models for positive-valued and discrete-valued time series with ARMA correlation structure. Multivariate Anal. **5**, 151–166 (1980)
33. Lutkepohl, H.: Introduction to Multiple Time Series Analysis. Springer, Berlin (1993)
34. Makridakis, S., Hyndman, R., Wheelwright, S.: Forecasting: Methods and Applications. Wiley, New York (1998)
35. Mosteller, F., Tukey, J.: Data Analysis and Regression: A Second Course in Statistics. Addison-Wesley, Reading (1977)
36. Ozaki, T.: Non-linear time series models and dynamical systems. In: Time Series in the Time Domain. Handbook of Statistics, vol. 5, pp. 25–83. Elsevier, Amsterdam (1985)
37. Priestley, M.: Spectral Analysis and Time Series. Academic Press, New York (1999)
38. Raftery, A.: A model for high-order Markov chains. J. Roy. Stat. Soc. B **47**(3), 528–539 (1985)
39. Seber, G., Lee, A.: Linear Regression Analysis. Wiley-Interscience, Hoboken (2003)
40. Shephard, N.: Statistical aspects of ARCH and stochastic volatility. In: Cox, D., Hinkley, D. (eds.) Time Series Models in Econometrics, Finance and Other Fields, pp. 1–67. Chapman & Hall, London (1996)
41. Tong, H.: Non-linear Time Series. Clarendon Press, Oxford (1993)
42. Waterman, M.: Mathematical Methods for DNA Sequences. CRC Press, Boca Raton (1989)

Chapter 4
Performance and Robustness Characteristics in Statistical Forecasting

Abstract In this chapter, we define optimality and robustness under distortions for statistical forecasting. The problem of statistical forecasting is stated in the most general form and then specialized for point or interval forecasting and different levels of prior uncertainty. The chapter introduces performance characteristics of forecasting statistics based on loss functions and risk functionals. In order to define mathematically rigorous robustness characteristics, a classification of the types of distortions common to applications is made, and the relevant mathematical distortion models are constructed. Robustness of statistical forecasting techniques is defined in terms of the following robustness characteristics: the guaranteed (upper) risk, the risk instability coefficient, the δ-admissible distortion level.

4.1 A General Formulation of the Statistical Forecasting Problem

Based on the approaches formulated in the previous chapters, let us state the statistical forecasting problem in the most general form.

Assume that a d-variate time series $x_t \in \mathbb{R}^d$, $t \in \mathbb{Z}$, defined in the probability space $(\Omega, \mathcal{F}, \mathbb{P})$ follows a certain stochastic model—for instance, one of the models described in Chap. 3. Furthermore, assume that the two following sets of ordered time points have been chosen in the time region $\mathbb{Z}$:

$$\mathcal{T}_0 = \{t_1^0, \ldots, t_{T_0}^0\}, \quad \mathcal{T}_+ = \{t_1^+, \ldots, t_{T_+}^+\},$$
$$t_1^0 < t_2^0 < \cdots < t_{T_0}^0 < t_1^+ < t_2^+ < \cdots < t_{T_+}^+,$$

where $T_0 \in \mathbb{N}$ is the number of recorded observations, $\mathcal{T}_0 \subset \mathbb{Z}$ is a set of T_0 time points where the time series was observed (recorded), $\mathcal{T}_+ \subset \mathbb{Z}$ is a set of T_+ future time points where the values of the time series are unknown and should be forecast.

Y. Kharin, *Robustness in Statistical Forecasting*, DOI 10.1007/978-3-319-00840-0_4,

In econometric applications, $\mathcal{T}_0$ is often called the basic period or the base of the forecast, and $\mathcal{T}_1$—the forecast period [7].

Observations at time points that belong to $\mathcal{T}_0$ will be represented as a composite random T_0-vector:

$$X^0 = \begin{pmatrix} x_{t_1^0} \\ \hdashline \vdots \\ \hdashline x_{t_{T_0}^0} \end{pmatrix} \in \mathbb{R}^{T_0 d},$$

and the subset of the time series that is to be forecast (corresponding to time points from $\mathcal{T}_+$) as a composite random T_+-vector:

$$X^+ = \begin{pmatrix} x_{t_1^+} \\ \hdashline \vdots \\ \hdashline x_{t_{T_+}^+} \end{pmatrix} \in \mathbb{R}^{T_+ d}.$$

Further, let $g(\cdot) : \mathbb{R}^{T_+ d} \to \mathbb{R}^\nu$, $1 \le \nu \le T_+ d$, be a given Borel function expressing certain properties of the future values X^+ that are of interest to the researcher. Stated the most generally, the problem of forecasting is to construct a statistical estimator for the observed random ν-vector $Y = g(X^+) \in \mathbb{R}^\nu$ based on the recorded observations $X^0 \in \mathbb{R}^{T_0 d}$. We will immediately note that in most applications, either $g(X^+) = X^+$ is an identity function ($\nu = T_+ d$), and a forecast is made of the future segment $Y = X^+$ of the time series, or $g(X^+) = x_{t_{T_+}^+}$, and only the value $x_{t_{T_+}^+}$ is being predicted.

Let us introduce several special cases of the general forecasting problem.

Depending on the type of the forecast, the following two cases are distinguished:

1. *Point forecasting*, where a Borel function $f(\cdot) : \mathbb{R}^{T_0 d} \to \mathbb{R}^\nu$ is constructed, defining a *forecasting statistic*

$$\hat{Y} = f(X^0) \in \mathbb{R}^\nu;$$

2. *Set (interval) forecasting*, where based on X^0 a Borel set $\mathcal{Y}_\gamma \in \mathcal{B}^\nu$ ($\mathcal{B}^\nu$ is a Borel σ-algebra of subsets of $\mathbb{R}^\nu$) is constructed such that with a high probability $\gamma \in (0, 1)$ we have $Y \in \mathcal{Y}_\gamma$.

Depending on the number $T_+ \in \mathbb{N}$ of the future time series elements that are being predicted, the following two cases are distinguished:

1. The most common case of *single-point forecasting*, where $T_+ = 1$, $\mathcal{T}_+ = \{t_1^+\}$, $t_1^+ = t_{T_0}^0 + \tau$, and a forecast should be made of the vector $Y = g(x_{t_{T_0}^0+\tau})$

depending on a single future value $x_{t^0_{T_0}+\tau}$, where $\tau \in \mathbb{N}$ is the so-called *forecast horizon*;
2. The case of *interval forecasting*, where a forecast is made of the vector

$$Y = g(x_{t_1^+}, \ldots, x_{t_{T_+}^+})$$

depending on an interval of $T_+ > 1$ future values of the time series.

It is a common mistake to make interval forecasts by combining several consecutive single-point forecasts $\hat{x}_{t_1^+}, \ldots, \hat{x}_{t_{T_+}^+}$ and constructing a plug-in forecasting statistic as

$$\tilde{Y} = g(\hat{x}_{t_1^+}, \ldots, \hat{x}_{t_{T_+}^+}).$$

The *stochastic model* of the general forecasting problem formulated above is characterized by the joint probability distribution of the observed segment X^0 and the predicted segment X^+:

$$P = P_{X^0,X^+}(B), \quad B \in \mathcal{B}^{(T_0+T_+)d},$$

the marginal probability distributions of each of the segments:

$$P = P_{X^0}(A), \quad A \in \mathcal{B}^{T_0 d}; \quad P = P_{X^+}(C), \quad C \in \mathcal{B}^{T_+ d},$$

as well as the probability distribution of the random vector $Y = g(X^+)$ that is being forecast:

$$P = P_Y(D) = P_{X^+}(g^{-1}(D)), \quad D \in \mathcal{B}^{\nu}.$$

Depending on the level of prior uncertainty (PU) of the probability distribution $P_{X^0,X^+}(\cdot)$, the following three cases are distinguished:

- PU-C is *the level of complete prior information*, where the probability distribution $P_{X^0,X^+}(\cdot)$ is completely known;
- PU-P is *the level of parametric prior uncertainty*, where $P_{X^0,X^+}(\cdot)$ is known up to a parameter $\theta^0 \in \Theta \subseteq \mathbb{R}^m$:

$$P = P_{X^0,X^+}(B;\theta), \quad B \in \mathcal{B}^{(T_0+T_+)d}, \quad \theta \in \Theta;$$

- PU-NP is *the level of nonparametric prior uncertainty*, where the probability distribution $P_{X^0,X^+}(\cdot)$ lies in a non-parametrizable function space.

Finally, let us consider the *informational aspect* of statistical forecasting. We are going to assume that the above probability measures $P_{X^0,X^+}(\cdot)$, $P_{X^0}(\cdot)$, $P_{X^+}(\cdot)$ are absolutely continuous w.r.t. a certain measure $\mu(\cdot)$, and thus we can write their respective Radon–Nikodym derivatives as $p_{X^0,X^+}(\cdot)$, $p_{X^0}(\cdot)$, $p_{X^+}(\cdot)$. Let us introduce the functional that evaluates the Shannon information on the predicted random vector X^+ which is contained in the observed random vector X^0:

$$\mathfrak{I}_{X^0,X^+} = \int\limits_{\mathbb{R}^{(T_0+T_+)d}} \ln \frac{p_{X^0,X^+}(X^0,X^+)}{p_{X^0}(X^0)p_{X^+}(X^+)} p_{X^0,X^+}(X^0,X^+)\mu(d(X^0,X^+)) \geq 0.$$

If $\mathfrak{I}_{X^0,X^+} = 0$, i.e., X^0 and X^+ are independent, and

$$P_{X^0,X^+}(B_1 \times B_2) = P_{X^0}(B_1)P_{X^+}(B_2), \quad B_1 \in \mathcal{B}^{T_0 d}, \ B_2 \in \mathcal{B}^{T_+ d},$$

then the observation X^0 provides zero information on X^+, and thus the forecast is trivial. The quantity $\mathfrak{I}_{X^0,X^+}$ can serve as the measure of potential forecasting accuracy.

Evaluation of the robustness of point forecasts isn't limited to the approaches described in this section. For example, a new approach to evaluation of point forecasts was proposed very recently by T. Gneiting [6].

4.2 The Risk Functional and Optimality of Forecasting Statistics

Consider the general problem of statistical forecasting which was formulated in Sect. 4.1 and consists of predicting the values of a "future" T_+-segment of the time series $X^+ \in \mathbb{R}^{T_+}$ based on the T_0-segment of recorded observations $X^0 \in \mathbb{R}^{T_0}$ by a certain Borel function (we assume $d = 1$ to simplify the notation):

$$\hat{X}^+ = f(X^0) = (f_i(X^0)) : \mathbb{R}^{T_0} \to \mathbb{R}^{T_+}, \tag{4.1}$$

which is called a forecasting statistic. For simplicity, we are considering the case described in Sect. 4.1, where $g(\cdot)$ is a functional identity, and a random vector formed by T_+ futures values, $Y = X^+$, is being predicted.

Forecast error of the forecasting statistic (4.1), defined as

$$\xi = (\xi_i) = \hat{X}^+ - X^+ = f(X^0) - X^+ \in \mathbb{R}^{T_+},$$

is a random T_+-vector. As in Sect. 2.1, let us introduce a matrix loss function

$$w = w(X', X'') : \mathbb{R}^{T_+} \times \mathbb{R}^{T_+} \to \mathbb{R}^M,$$

where w is the matrix loss if the actual "future" state is $X^+ = X'$ and its forecast is $\hat{X}^+ = X''$. The *risk functional* is defined as the expectation of the loss function:

$$R = R(f(\cdot)) = \mathbb{E}\{w(X^+, f(X^0))\} = \int w(X^+, f(X^0)) P_{X^0,X^+}(d(X^0, X^+)), \tag{4.2}$$

which will exist provided that $\mathbb{E}\{\|w(X^+, f(X^0))\|\} < +\infty$.

Let us consider a few special definitions of the loss functions $w(\cdot)$ which are most frequently used in applications.

1. For a $(0-1)$ *loss function*

$$w(X', X'') = 1 - \delta_{X',X''} \in \{0, 1\},$$

which is an indicator of inequality between the forecasting statistic and the actual value, the risk (4.2) is a scalar functional—the *probability of an incorrect forecast*:

$$r = r(f(\cdot)) = \mathbb{P}\{\hat{X}^+ \neq X^+\} = \int\limits_{f(X^0) \neq X^+} P_{X^0,X^+}(d(X^0, X^+)) \in [0, 1]. \tag{4.3}$$

In applications, this benchmark of forecasting performance proves itself to be extremely rigid. In particular, for the commonly encountered setting where the probability distribution $P_{X^0,X^+}(\cdot)$ is absolutely continuous w.r.t. the Lebesgue measure, and continuous statistics $f(\cdot)$ are used to construct a forecast, the risk attains the maximum possible value $r(f(\cdot)) \equiv 1$ for any given $f(\cdot)$. Thus, the risk functional (4.3) makes it impossible to draw a comparison between different forecasting statistics $f(\cdot)$. Usefulness of the risk functional (4.3) is therefore restricted to certain discrete time series models (see Sect. 3.6).

2. If

$$w(X^+, \hat{X}^+) = \sum_{i=1}^{T_+} |\hat{x}_{t_i^+} - x_{t_i^+}|^p \geq 0,$$

where $p > 0$ is a parameter, then the risk (4.2) is also a scalar functional:

$$r = r(f(\cdot)) = \mathbb{E}\left\{\sum_{i=1}^{T_+} |\hat{x}_{t_i^+} - x_{t_i^+}|^p\right\} \geq 0. \tag{4.4}$$

We can also define $(r(f(\cdot)))^{1/p}$, which is called the *expected forecast risk w.r.t. the L_p-norm*. If $p = 2$, then the definition (4.4) yields the *mean square risk of forecasting*:

$$r = r(f(\cdot)) = \mathbb{E}\{(\hat{X}^+ - X^+)'(\hat{X}^+ - X^+)\} \geq 0, \tag{4.5}$$

which is the most commonly used performance metric of forecasting algorithms.

3. If

$$w(X^+, \hat{X}^+) = (\hat{X}^+ - X^+)(\hat{X}^+ - X^+)' \in \mathbb{R}^{T_+ \times T_+},$$

then the risk (4.2) is a $(T_+ \times T_+)$-matrix functional:

$$R = R(f(\cdot)) = (r_{ij}(f(\cdot))) = \mathbb{E}\{(\hat{X}^+ - X^+)(\hat{X}^+ - X^+)'\} \in \mathbb{R}^{T_+ \times T_+}. \quad (4.6)$$

A diagonal element $r_{i,i}$ of this *matrix risk* is called the i*th scalar risk*:

$$r_{ii}(f(\cdot)) = \mathbb{E}\{(\hat{x}_{t_i^+} - x_{t_i^+})^2\} \geq 0, \quad (4.7)$$

which is the mean square error of forecasting a future value $x_{t_i^+}$ using the statistic $\hat{x}_{t_i^+} = f_i(X^0)$. A non-diagonal element

$$r_{ij}(f(\cdot)) = \mathbb{E}\{(\hat{x}_{t_i^+} - x_{t_i^+})(\hat{x}_{t_j^+} - x_{t_j^+})\}$$

characterizes the correlation dependence of the random forecast errors of the ith and jth components, $i, j \in \{1, \ldots, T_+\}$.

Note that there exists a simple relation between the mean square risk of forecasting (4.5) and the risks defined by (4.7):

$$r(f(\cdot)) = \sum_{i=1}^{T_+} r_{ii}(f(\cdot)). \quad (4.8)$$

Definition 4.1. The forecasting statistic $f(\cdot)$ and the forecast $\hat{X}^+ = f(X^0)$ obtained through this statistic are called unbiased if

$$\mathbb{E}\{|X^+|\} < +\infty, \quad \mathbb{E}\{|f(X^0)|\} < +\infty,$$

and the bias vector of the forecast is zero:

$$b(T_0, T_+) = \mathbb{E}\{\hat{X}^+ - X^+\} = \mathbb{E}\{f(X^0) - X^+\} = \mathbf{0}_{T_+}, \quad (4.9)$$

otherwise the forecast (and the forecasting statistic) are called biased. If we have

$$\lim_{T_0 \to \infty} b(T_0, T_+) = \mathbf{0}_{T_+},$$

then it is said that the forecasting statistic and the forecast are asymptotically unbiased.

Corollary 4.1. *The matrix risk* (4.6) *is related to the bias vector* (4.9) *and the covariance matrix of the random vector* $\xi = \hat{X}^+ - X^+$*:*

$$R = \mathbf{Cov}\{\hat{X}^+ - X^+, \hat{X}^+ - X^+\} + b\,b',$$

and in the case of an unbiased forecast it coincides with that matrix:

$$R = \mathbf{Cov}\{\hat{X}^+ - X^+, \hat{X}^+ - X^+\}.$$

Proof. Let us apply (4.6), (4.9) and the well-known properties of expectations:

$$\begin{aligned} R &= \mathbb{E}\{(\hat{X}^+ - X^+)(\hat{X}^+ - X^+)'\} \equiv \\ &\equiv \mathbb{E}\{((\hat{X}^+ - X^+) - \mathbb{E}\{\hat{X}^+ - X^+\} + b)(\cdot)'\} = \mathbf{Cov}\{\xi, \xi\} + b\,b'. \end{aligned}$$ □

From now on, unless stated otherwise, two special cases of the risk functional (4.2) will be considered: the scalar mean square risk (4.5) and the matrix mean square risk (4.6), which coincide for $T_+ = 1$.

Definition 4.2. The forecasting statistic $\hat{X}^+ = f_0(X^0)$ is called mean square optimal if it yields the minimum value of the mean square risk (4.5):

$$r(f_0(\cdot)) = \inf_{f(\cdot) \in F} r(f(\cdot)), \tag{4.10}$$

where the infimum is taken over the set F of all possible forecasting statistics. Here $r_0 = r(f_0(\cdot)) \geq 0$ is the minimum risk, which is attained if the forecasting statistic $f_0(\cdot)$ is used.

Definition 4.3. A forecasting statistic $\hat{X}^+ = f_0(X^0)$ is called optimal w.r.t. the matrix mean square risk (4.6) if

$$\forall f(\cdot) \in F,\ f(\cdot) \neq f^0(\cdot) \quad \Rightarrow \quad \Delta = R(f) - R(f_0) \succeq \mathbf{0}, \tag{4.11}$$

i.e., the matrix Δ is positive-semidefinite.

Let us consider the problem of finding an optimal forecasting statistic, i.e., a solution to one of the optimization problems (4.10) and (4.11), in the most general setting defined in Sect. 4.1.

First, note that the solutions depend on the set F of the possible forecasting statistics. This set is defined by the researcher based on prior information and the available computational resources. As mentioned in Sect. 4.1, three levels of prior uncertainty of the probability distribution $P_{X^0,X^+}(B)$, i.e., the stochastic model of the studied phenomenon, are possible when solving the problems (4.10) and (4.11). These levels are PU-C (complete prior information), where the probability distribution of $P_{X^0,X^+}(B)$ is completely known, PU-P (parametric prior uncertainty), where the distribution is known up to a parameter $\theta^0 \in \Theta \subseteq \mathbb{R}^m$, and PU-NP (nonparametric prior uncertainty), where the probability distribution $P_{X^0,X^+}(\cdot)$ is an unknown element of a function space $\mathcal{P}$ which doesn't allow for a finite-dimensional parameterization.

In the case of complete prior information PU-C, solving the optimization problems (4.10) and (4.11) yields the optimal forecasting statistic

$$\hat{X}^+ = f_0(X^0; P_{X^0,X^+}(\cdot)), \tag{4.12}$$

which depends on the true probability distribution $P_{X^0,X^+}(\cdot)$.

Let us write an explicit expression for the optimal forecasting statistic (4.12).

Theorem 4.1. *If* $\mathbb{E}\{|X^+|^2\} < +\infty$, *then the mean square optimal forecasting statistic defined by the criterion* (4.11) *can be written as the following matrix form conditional expectation:*

$$\hat{X}^+ = f_0(X^0; P_{X^0,X^+}(\cdot)) ::= \mathbb{E}\{X^+ \mid X^0\} = \int\limits_{\mathbb{R}^{dT_+}} X^+ P_{X^+|X^0}(dX^+). \tag{4.13}$$

This forecast is unbiased, and its matrix risk can be expressed from the conditional covariance matrix:

$$R_0 = R(f_0) = \mathbb{E}\{\mathbf{Cov}\{X^+, X^+ \mid X^0\}\}. \tag{4.14}$$

Proof. Using the notation of (4.13), let us rewrite the matrix risk functional (4.6) as the following expansion:

$$\begin{aligned} R(f) &= \mathbb{E}\left\{(f(X^0) - X^+)(\cdot)'\right\} \equiv \\ &\equiv \mathbb{E}\left\{((f(X^0) - f_0(X^0)) - (X^+ - \mathbb{E}\{X^+ \mid X^0\}))\,(\cdot)'\right\} = \\ &= \mathbb{E}\left\{(f(X^0) - f_0(X^0))\,(\cdot)'\right\} + \mathbb{E}\left\{(X^+ - \mathbb{E}\{X^+ \mid X^0\})\,(\cdot)'\right\} - \\ &\quad - \mathbb{E}\left\{(f(X^0) - f_0(X^0))\,(X^+ - \mathbb{E}\{X^+ \mid X^0\})'\right\} - (\cdot)'. \end{aligned}$$

Using the total expectation formula yields

$$\begin{aligned} R(f) &= \mathbb{E}\left\{(f(X^0) - f_0(X^0))\,(\cdot)'\right\} + \mathbb{E}\left\{\mathbb{E}\left\{(X^+ - \mathbb{E}\{X^+ \mid X^0\})\,(\cdot)' \mid X^0\right\}\right\} - \\ &\quad - \mathbb{E}\left\{(f(X^0) - f_0(X^0))\,\mathbb{E}\left\{(X^+ - \mathbb{E}\{X^+ \mid X^0\})' \mid X^0\right\}\right\} - (\cdot)' = \\ &= \mathbb{E}\left\{(f(X^0) - f_0(X^0))\,(\cdot)'\right\} + \mathbb{E}\left\{\mathbf{Cov}\{X^+, X^+ \mid X^0\}\right\}; \end{aligned} \tag{4.15}$$

here the final equality is based on the fact that two of the summands in the initial expression for $R(f)$ are equal to zero. Thus,

$$\Delta = R(f) - R(f_0) = \mathbb{E}\{(f(X^0) - f_0(X^0))(\cdot)'\} \succeq \mathbf{0},$$

and for $f(\cdot) = f_0(\cdot)$ the minimal risk (4.14) is attained. It is quite easy to see that the forecast (4.13) is unbiased:

$$\mathbb{E}\{\hat{X}^+ - X^+\} = \mathbb{E}\{\mathbb{E}\{X^+ \mid X^0\} - X^+\} = \mathbb{E}\{X^+\} - \mathbb{E}\{X^+\} = \mathbf{0}_{dT_+}. \quad \square$$

Corollary 4.2. *The optimal forecasting statistic* (4.13) *is defined uniquely up to a set* $B_0 \in \mathcal{B}^{dT_0}$ *of measure zero,* $P_{X^0}(B_0) = 0$.

Proof. It is sufficient to apply the representation (4.15) to prove this corollary. □

Corollary 4.3. *The forecasting statistic* (4.13) *is also optimal w.r.t. the criterion* (4.10)*; the minimal mean square risk* (4.5) *equals*

$$r_0 = r(f_0) = \sum_{i=1}^{T_+} r_{ii}(f_0), \quad r_{ii}(f_0) = \mathbb{E}\left\{\mathbb{D}\{x_{t_i^+} \mid X^0\}\right\}. \tag{4.16}$$

Proof. The proof easily follows from (4.5) and an expansion of the forecast risk similar to (4.15). □

Lemma 4.1. *If* $X \in \mathbb{R}^m$, $Y \in \mathbb{R}^n$ *are arbitrary random vectors in* $(\Omega, \mathcal{F}, \mathbb{P})$*, and* $\mathbb{E}\{|Y|^2\} < +\infty$*, then we have the following identity:*

$$\mathbf{Cov}\{Y, Y\} = \mathbb{E}\{\mathbf{Cov}\{Y, Y \mid X\}\} + \mathbf{Cov}\{\mathbb{E}\{Y \mid X\}, \mathbb{E}\{Y \mid X\}\}. \tag{4.17}$$

Proof. Using the total expectation formula yields

$$\begin{aligned}
\mathbf{Cov}\{Y, Y\} &= \mathbb{E}\{(Y - \mathbb{E}Y)(\cdot)'\} \equiv \\
&\equiv \mathbb{E}\{\mathbb{E}\{((Y - \mathbb{E}\{Y \mid X\}) + (\mathbb{E}\{Y \mid X\} - \mathbb{E}Y))(\cdot)' \mid X\}\} = \\
&= \mathbf{Cov}\{Y, Y \mid X\} + \mathbb{E}\{\mathbb{E}\{(\mathbb{E}\{Y \mid X\} - \mathbb{E}Y)(\cdot)' \mid X\}\} + \\
&\quad + \mathbb{E}\{\mathbb{E}\{(Y - \mathbb{E}\{Y \mid X\})(\mathbb{E}\{Y \mid X\} - \mathbb{E}Y)' \mid X\}\} + (\cdot)',
\end{aligned}$$

which coincides with the right-hand side of (4.17) since the last two summands are equal to zero. □

Corollary 4.4. *For any random variables* $X, Y \in \mathbb{R}$ *such that* $\mathbb{E}\{Y^2\} < +\infty$*, the following identity is satisfied:*

$$\mathbb{D}\{Y\} = \mathbb{E}\{\mathbb{D}\{Y \mid X\}\} + \mathbb{D}\{\mathbb{E}\{Y \mid X\}\}. \tag{4.18}$$

The relation (4.18) is quite well known [7].

Taking into account Lemma 4.1 and its corollary, we can obtain more accurate expressions for the minimal risks (4.14), (4.16).

Corollary 4.5. *Under the conditions of Theorem 4.1, the matrix risk of the optimal forecasting statistic* (4.13) *can be written as*

$$R_0 = R(f_0) = \mathbf{Cov}\{X^+, X^+\} - \mathbf{Cov}\left\{\mathbb{E}\{X^+ \mid X^0\}, \mathbb{E}\{X^+ \mid X^0\}\right\}.$$

Corollary 4.6. *Under the conditions of Corollary 4.4, the following formula can be written for the scalar risks:*

$$r_{ii}(f_0) = \mathbb{D}\{x_{t_i^+}\} - \mathbb{D}\left\{\mathbb{E}\{x_{t_i^+} \mid X^0\}\right\} \geq 0, \quad i = 1, \ldots, T_+. \tag{4.19}$$

From (4.19) we can see that if $x_{t_i^+}$ doesn't depend on X^0, then

$$\mathbb{E}\left\{x_{t_i^+} \mid X^0\right\} = \mathbb{E}\left\{x_{t_i^+}\right\} = \text{const}, \quad \mathbb{D}\left\{\mathbb{E}\left\{x_{t_i^+} \mid X^0\right\}\right\} = 0, \quad r_{ii}(f_0) = \mathbb{D}\left\{x_{t_i^+}\right\}.$$

If a dependence is found between $x_{t_i^+}$ and X^0, then the forecast risk is reduced by

$$\mathbb{D}\left\{\mathbb{E}\{x_{t_i^+} \mid X^0\}\right\} > 0.$$

In the case of PU-P parametric prior uncertainty, the optimal forecast (4.12) is undefined since it depends on an unknown parameter $\theta^0 \in \Theta \subseteq \mathbb{R}^m$:

$$\hat{X}^+ = f_0\left(X^0; P_{X^0,X^+}(\cdot; \theta^0)\right). \tag{4.20}$$

The uncertainty of θ^0 in (4.20) can be eliminated by applying the plug-in principle. This means replacing the unknown actual value of θ^0 by some consistent statistical estimator $\tilde{\theta} \in \mathbb{R}^m$ based on a length $\tilde{T}$ realization $\tilde{X} = (\tilde{x}_{\tilde{t}_1}, \ldots, \tilde{x}_{\tilde{t}_{\tilde{T}}})' \in \mathbb{R}^{T+d}$ of the investigated time series $\{x_t\}$:

$$\tilde{\theta} \xrightarrow[\tilde{T}\to\infty]{\mathbf{P}} \theta^0,$$

$$\tilde{X}^+ = f^0\left(X^0; P_{X^0,X^+}\left(\cdot; \tilde{\theta}\right)\right) =:: g(X_0, \tilde{X}). \tag{4.21}$$

Definition 4.4. The forecasting statistic (4.21) and the forecast $\tilde{X}^+$ are called consistent and asymptotically optimal if the following convergence holds in probability:

$$\tilde{X}^+ \xrightarrow[\tilde{T}\to\infty]{\mathbf{P}} \hat{X}^+. \tag{4.22}$$

A *strongly consistent forecast* is defined by requiring an almost sure convergence in (4.22).

Construction of a forecasting statistic $\tilde{X}^+$ defined by (4.21) based on an auxiliary realization $\tilde{X}$ is usually called the *training stage of the forecasting algorithm*, and $\tilde{X}$ is called a training realization. Often, the observed realization X^0 is also used as the training realization, $\tilde{X} = X^0$. It should also be noted that some approaches to construction of forecasting statistics in a PU-P setting are based on simultaneous estimation of X^+ and θ^0 (see, e.g., Sect. 8.2).

The plug-in approach summarized in (4.21) can also be used in the PU-NP (nonparametric prior uncertainty) case:

$$\tilde{X}^+ = f_0\left(X^0, \tilde{P}_{X^0,X^+}(\cdot)\right), \tag{4.23}$$

where $\tilde{P}_{X^0,X^+}(\cdot)$ is a nonparametric estimator of the unknown probability distribution $P_{X^0,X^+}(\cdot)$ based on a training sample $\tilde{X}$. Consistent asymptotically optimal forecasts are then defined similarly to (4.22).

A more detailed review of the methods used to construct forecasting statistics (4.12), (4.21), (4.23) will be given in the following chapters.

4.3 Classification of Model Distortions

As mentioned in the previous section, a statistical forecast

$$\hat{X}^+ = f_0(X^0) : \mathbb{R}^{dT_0} \to \mathbb{R}^{dT_+} \tag{4.24}$$

depends only partially on the observation X^0; an important basis of the forecast is the hypothetical model M_0 of the experimental data, i.e., the system of prior assumptions on the investigated process or phenomenon. Unfortunately, these model assumptions are distorted (or disturbed) in most applications [16]. In the presence of distortion, the optimal forecasting statistic, which minimizes the risk under the hypothetical model M_0, often proves unstable: the forecast risk can be much higher than the value calculated for the undistorted model.

Instability of traditional forecasting statistics in real-world applications motivates construction of robust statistics. A statistic is said to be robust if its forecast risk is "weakly affected by small distortions of the model M_0" [8, 9, 17]. A mathematical evaluation of robustness requires a formal description and classification of the most commonly encountered types of distortions.

Figure 4.1 presents a classification diagram of model distortion types common to applied statistical forecasting problems. The distortion types shown in the diagram are briefly described below, and more detailed descriptions can be found in the following chapters.

Following the structure of Chap. 3, let us start by assigning distortion types to one of the two classes based on the manner in which the hypothetical model M_0 is defined:

D.1. *A hypothetical model is defined explicitly* as a probability distribution $P_{X^0,X^+}(B)$, $B \in \mathcal{B}^{d(T_0+T_+)}$, where $X^0 = (x_1', \ldots, x_{T_0}')' \in \mathbb{R}^{dT_0}$ is a vector of T_0 observations, and $X^+ = (x_{T_0+1}', \ldots, x_{T_0+T_+}')' \in \mathbb{R}^{dT_+}$ is a composite vector of T_+ future random variables that are being predicted. To simplify the

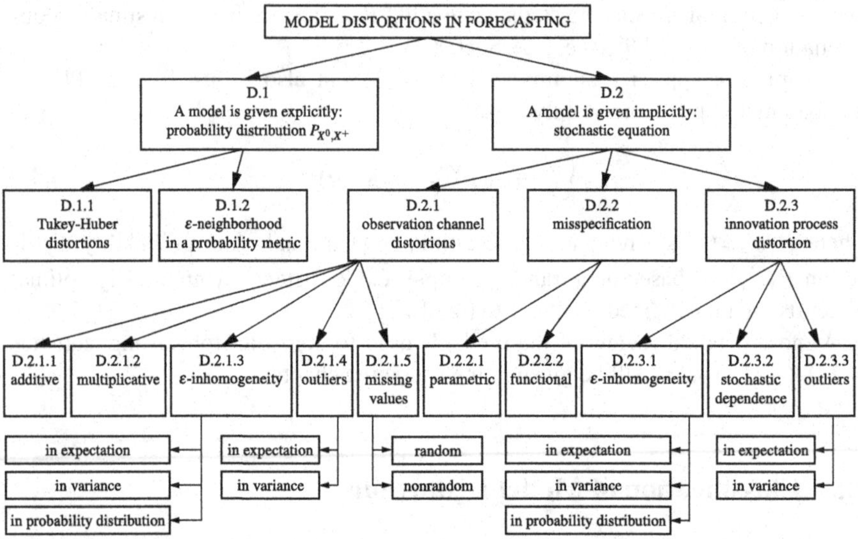

Fig. 4.1 Classification of model distortions in forecasting

notation introduced in Sect. 4.1, we assume that $\mathcal{T}_0 = \{1, \ldots, T_0\}$, $\mathcal{T}_+ = \{T_0 + 1, \ldots, T_0 + T_+\}$; also, for simplicity, we will assume that the random vectors X^0, X^+ have a joint absolutely continuous (w.r.t. the Lebesgue measure in $\mathbb{R}^{d(T_0+T_+)}$) distribution with the probability density $p_{X^0,X^+}(X^0, X^+)$, where $X^0 \in \mathbb{R}^{dT_0}$, $X^+ \in \mathbb{R}^{dT_+}$. If the probability distribution is discrete, the results are similar.

D.2. A *hypothetical model is defined implicitly* by a stochastic equation

$$x_t = G(x_{t-1}, \ldots, x_{t-s}, u_t, u_{t-1}, \ldots, x_{t-L}; \theta^0), \quad t \in \mathbb{Z}, \tag{4.25}$$

where $u_t \in \mathbb{R}^\nu$ is a random process called the *innovation process* defined on $(\Omega, \mathcal{F}, \mathbb{P})$; positive integers s, L define, respectively, the lengths of the memories of the predicted and the innovation processes; $\theta^0 \in \Theta \subseteq \mathbb{R}^m$ is a parameter vector;

$$G(\cdot) : \mathbb{R}^{ds} \times \mathbb{R}^{\nu(L+1)} \times \Theta \to \mathbb{R}^d$$

is a Borel function. Examples of time series models defined in the form (4.25) can be found in Chap. 3.

In the class *D.1*, the most common distortion types are Tukey–Huber distortions *D.1.1* and distortions defined by ε-neighborhoods in probability metrics *D.1.2*.

Tukey–Huber distortions [1, 4, 5, 10, 13–15, 18] of the observation vector X^0 are defined as mixtures of two probability distributions:

$$p_{X^0}(X^0) = (1 - \varepsilon) p^0_{X^0}(X^0) + \varepsilon\, h(X^0),$$

where $p^0(\cdot)$ is the undistorted (hypothetical) probability density, $h(\cdot)$ is the so-called contaminating probability density, $\varepsilon \in [0,1)$ is the distortion level. If $\varepsilon = 0$, then $p_{X^0}(\cdot) = p^0_{X^0}(\cdot)$, and no distortion is present.

To define a distortion of type *D.1.2*, an *ε-neighborhood in a probability metric* is constructed:

$$0 \leq \varrho\left(p_{X^0}(\cdot), p^0_{X^0}(\cdot)\right) \leq \varepsilon,$$

where $\varrho(\cdot)$ is a certain probability metric. Let us give some examples of probability metrics used to define this type of distortions:

- Kolmogorov variational distance [3, 11, 12]:

$$\varrho\left(p, p^0\right) = \frac{1}{2} \int\limits_{\mathbb{R}^{dT_0}} \left|p(X) - p^0(X)\right| dX \in [0,1];$$

- Hellinger distance [12]:

$$\varrho\left(p, p^0\right) = \frac{1}{2} \int\limits_{\mathbb{R}^{dT_0}} \left(\sqrt{p(X)} - \sqrt{p^0(X)}\right)^2 dX \in [0,1];$$

- Kullback–Leibler distance [2]:

$$\varrho\left(p, p^0\right) = \int \left(p(X) - p^0(X)\right) \ln \frac{p(X)}{p^0(X)} dX \in [0, +\infty].$$

As in the previous case, $\varepsilon = 0$ corresponds to absence of distortion.

Distortion class *D.2* consists of three subclasses. The subclass *D.2.1* is characterized by observation channel distortions:

$$X = H(X^0, \Xi), \tag{4.26}$$

where $X^0 \in \mathbb{R}^{dT_0}$ is the "unobservable history" and $X \in \mathbb{R}^{dT_0}$ are the observations, or the "observable history" of the random process; $\Xi = (\xi_1', \dots, \xi_{T_0}')' \in \mathbb{R}^{dT_0}$ is the unobservable random distortion vector (vector of observation channel errors); and $H(\cdot)$ is the function that describes recording of observations under distortion of level $\varepsilon \geq 0$. For example, an *observation channel with additive distortions (D.2.1.1)* is defined as follows:

$$x_t = x_t^0 + \varepsilon \xi_t, \quad t \in \mathbb{Z}, \tag{4.27}$$

where $\{\xi_t\}$ is a sequence of jointly independent random variables with zero expectations and identical finite variances,

$$\mathbb{E}\{\xi_t\} = 0, \quad \mathbb{D}\{\xi_t\} = \sigma^2 < +\infty.$$

An *observation channel with multiplicative distortions* (*D.2.1.2*) is described by the relation

$$x_t = (1 + \varepsilon \xi_t) x_t^0, \quad t \in \mathbb{Z}. \tag{4.28}$$

It is said that observation channel distortions are *ε-inhomogeneous* (*D.2.1.3*) if the random distortion vectors $\xi_1, \ldots, \xi_{T_0}$ aren't identically distributed, but their probability distributions differ by no more than ε w.r.t. some probability metric. Another possible definition of ε-inhomogeneity is based on ε-deviations of expectations or variances.

The subclass *D.2.1.4* is characterized by the presence of *outliers*:

$$x_t = (1 - \eta_t) x_t^0 + \eta_t \, \upsilon_t, \quad t \in \mathbb{Z}, \tag{4.29}$$

where $\eta_t \in \{0, 1\}$ is a Bernoulli random variable distributed as

$$\mathbb{P}\{\eta_t = 1\} = 1 - \mathbb{P}\{\eta_t = 0\} = \varepsilon \in [0, 1),$$

and $\{\eta_t\}$ are jointly independent; $\{\upsilon_t\}$ is a sequence of independent random variables describing the outliers (in expectation or variance). If $\eta_t = 0$, then the true value is observed:

$$x_t = x_t^0;$$

and if $\eta_t = 1$, then the observation x_t is an outlier:

$$x_t = x_t^0 + \upsilon_t.$$

The quantity ε is the outlier probability.

The subclass *D.2.1.5* is defined by the presence of missing values (gaps) in the observed statistical data $X^0 \in \mathbb{R}^{dT_0}$. To formalize these observation gaps, let us introduce a binary $(T_0 \times d)$-matrix $O = (o_{ti})$, where

$$o_{ti} = \begin{cases} 1, & \text{if } x_{ti}^0 \text{ is recorded,} \\ 0, & \text{if } x_{ti}^0 \text{ is missing.} \end{cases}$$

Depending on randomness of the matrix O, two types of missing values are distinguished, random and nonrandom. The settings where the matrix O depends or doesn't depend on the matrix of observations X^0 are classified separately.

The subclass *D.2.2* is described by *misspecification errors* in (4.25) and includes two distortion types:

- *Parametric* distortion means that in (4.25), instead of the true value of the parameter θ^0, we have a priori specified (or found by statistical estimation) a different value $\tilde{\theta}$ with $|\tilde{\theta} - \theta^0| \leq \varepsilon$; the quantity ε is the distortion level;

- *Functional* distortion, meaning that in (4.25), instead of the true function $G(\cdot)$, we have specified a different function $\tilde{G}(\cdot)$, and $\|\tilde{G}(\cdot) - G(\cdot)\| \leq \varepsilon$ holds in a certain metric.

The subclass *D.2.3* includes *innovation process distortions*, i.e., deviations of the innovation process $u_t \in \mathbb{R}^\nu$, $t \in \mathbb{Z}$, in (4.25) from the model assumptions. Three distortion types are included in this subclass:

- *ε-inhomogeneity* (*D.2.3.1*), defined similarly to *D.2.1.3*;
- *Stochastic dependence* (*D.2.3.2*): usually, the innovation process is assumed to be a white noise, or a sequence of independent random vectors $\{u_t\}$; dependence of these vectors is a deviation from the hypothetical stochastic model M_0;
- *Outliers* (*D.2.3.3*) in the observations $\{u_t\}$, defined similarly to *D.2.1.4*.

4.4 Robustness Characteristics

In Sect. 4.1, we have formulated the general problem of statistical forecasting, and Sect. 4.2 presents optimality criteria and defines mean square optimal forecasting statistics for a fixed hypothetical model M_0. However, in practice the hypothetical model M_0 is subject to distortions. In this section, we are going to introduce functionals that characterize the robustness (stability) of forecasting statistics under distortions of the model M_0 introduced in the previous section.

Assume that the future values $X^+ \in \mathbb{R}^{dT_+}$ of a time series $\{x_t\}$ are predicted based on observations $X^0 \in \mathbb{R}^{dT_0}$. In the absence of distortion, the investigated time series is accurately described by the hypothetical model M_0 defined by the joint probability distribution $P_{X^0,X^+}(\cdot)$, and a mean square optimal forecasting statistic belonging to a function family F can be written as

$$\hat{X}^+ = f_0(X^0), \tag{4.30}$$

where $f_0(\cdot)$ minimizes the mean square risk:

$$r_0 = r(f_0) = \inf_{f(\cdot)\in F} r(f), \tag{4.31}$$

$$r(f) = \mathbb{E}\left\{\left(f(X^0) - X^+\right)'\left(f(X^0) - X^+\right)\right\} \geq 0.$$

The stochastic model of the observed time series under distortions described in Sect. 4.3 will be represented by the following family of probability measures:

$$\left\{P^{\varepsilon}_{X,X^0}(B), \quad B \in \mathcal{B}^{d(T_0+T_+)}: \quad \varepsilon \in [0, \varepsilon_+]\right\}, \tag{4.32}$$

where ε is the model distortion level and $\varepsilon_+ \geq 0$ is its maximal value. If $\varepsilon_+ = 0$, then no distortion is present, and the time series fully conforms to the hypothetical model M_0. A more specific definition of the family (4.32) follows from the choice of one of the distortion models presented in Sect. 4.3.

Under distortion (4.32), the mean square forecast risk of the forecasting statistic $f(\cdot) \in F$, written as

$$\tilde{X}^+ = f(X^0),$$

depends on the distortion level:

$$r_\varepsilon = r_\varepsilon(f) = \mathbb{E}_\varepsilon \left\{ \left(f(X^0) - X^+\right)' \left(f(X^0) - X^+\right) \right\} \geq 0, \tag{4.33}$$

where $\mathbb{E}_\varepsilon\{\cdot\}$ is the expectation w.r.t. the probability measure $P^\varepsilon_{X,X^0}(\cdot)$.

Definition 4.5. The *guaranteed upper risk* is defined as the exact upper bound for the set of the risk values obtained under every allowed distortion of the model M_0:

$$r_+ = r_+(f) = \sup_{0 \leq \varepsilon \leq \varepsilon_+} r_\varepsilon(f). \tag{4.34}$$

Similarly to (4.34), we can introduce the *guaranteed lower risk*:

$$r_- = r_-(f) = \inf_{0 \leq \varepsilon \leq \varepsilon_+} r_\varepsilon(f). \tag{4.35}$$

It follows from (4.33)–(4.35) that for any distortion of the hypothetical model M_0 allowed by (4.32), the mean square risk is bounded by its guaranteed lower and upper values:

$$r_-(f) \leq r_\varepsilon(f) \leq r_+(f), \quad f(\cdot) \in F.$$

Definition 4.6. The *risk instability coefficient* of the forecasting statistic $f(\cdot)$ under distortion (4.32) of the hypothetical model M_0 is defined as the relative increment of the guaranteed risk (4.34) w.r.t. the hypothetical risk $r_0 > 0$ defined by (4.31):

$$\kappa = \kappa(f) = \frac{r_+(f) - r_0}{r_0} \geq 0. \tag{4.36}$$

Definition 4.7. For a fixed $\delta > 0$, the quantity

$$\varepsilon_* = \varepsilon_*(\delta) = \sup\{\varepsilon : \kappa(f) \leq \delta\} \geq 0 \tag{4.37}$$

is called the *δ-admissible distortion level.*

The quantity ε_* is the maximum distortion level such that the risk instability coefficient doesn't exceed a given constant δ.

Higher robustness (stability) of a forecasting statistic $f(\cdot) \in F$ under distortions (4.32) corresponds to smaller values of κ and larger values of ε_*.

Definition 4.8. A forecasting statistic $f_*(\cdot)$ defined as

$$\tilde{X}^+ = f_*(X^0)$$

is called *minimax robust* if it minimizes the risk instability coefficient:

$$\kappa(f_*) = \inf_{f(\cdot)\in F} \kappa(f). \tag{4.38}$$

Let us also define the *Hampel breakdown point* ε^* in the setting of statistical forecasting as the maximum proportion of outliers in the sample X^0 such that for arbitrary changes in their values the forecasting statistic remains bounded. This characteristic of "qualitative robustness" was introduced in [8] for parameter estimation problems.

To conclude the chapter, let us give a more intuitive explanation of the introduced robustness characteristics.

The guaranteed upper risk $r_+(f)$ defined by (4.34) estimates the maximal mean square forecast error of the forecasting statistic $f(\cdot) \in F$ under every possible distortion of a given type and distortion level not exceeding some fixed value ε_+.

The interval

$$\Delta(f) = (r_-(f), r_+(f))$$

is defined by (4.34) and (4.35) as the range of the mean square forecast error of the forecasting statistic $f(\cdot) \in F$ if distortion level varies from the minimum $\varepsilon = 0$ to the maximal value $\varepsilon = \varepsilon_+$.

The risk instability coefficient $\kappa(f)$ defined by (4.36) has the following meaning: for the forecasting statistic $f(\cdot)$ under distortion of level $\varepsilon \in [0, \varepsilon_+]$, the forecast risk (i.e., the mean square forecast error) increases by no more than $r_0\kappa(f) \times 100\,\%$ compared to the hypothetical risk r_0 (i.e., the minimum forecast risk which is attained in the absence of distortion, $\varepsilon = 0$):

$$r(f) \leq r_0 + r_0\kappa(f). \tag{4.39}$$

The robust forecasting statistic $f_*(\cdot) \in F$ defined by the minimax condition (4.38) minimizes the upper bound for the risk in (4.39).

The δ-admissible distortion level $\varepsilon_*(\delta)$ defined by (4.37) indicates the critical distortion level for the forecasting statistic $f(\cdot) \in F$. If the distortion level is higher than this critical value, $\varepsilon_+ > \varepsilon_*$, then the risk instability coefficient is unacceptably high:

$$\kappa(f) > \delta,$$

and use of the statistic $f(\cdot)$ is no longer justified. Instead, a robust forecasting statistic $f_*(\cdot)$ should be constructed from (4.38).

References

1. Atkinson, A., Riani, M.: Robust Diagnostic Regression Analysis. Springer, New York (2000)
2. Borovkov, A.: Mathematical Statistics. Gordon & Breach, Amsterdam (1998)
3. Borovkov, A.: Probability Theory. Gordon & Breach, Amsterdam (1998)
4. Croux, C., Filzmoser, P.: Discussion of "A Survey of Robust Statistics" by S. Morgenthaler. Stat. Methods Appl. **15**(3), 271–293 (2007)
5. Francq, C., Zakoian, J.: GARCH Models: Structure, Statistical Inference and Financial Applications. Wiley, Chichester (2010)
6. Gneiting, T.: Making and evaluating point forecasts. J. Am. Stat. Assoc. **106**(494), 746–761 (2011)
7. Greene, W.: Econometric Analysis. Macmillan, New York (2000)
8. Hampel, F., Ronchetti, E., Rousseeuw, P., Stahel, W.: Robust Statistics: The Approach Based on Influence Functions. Wiley, New York (1986)
9. Huber, P.: Some mathematical problems arising in robust statistics. In: Proceedings of the International Congress of Mathematicians, pp. 821–824. University of Vancouver, Vancouver (1974)
10. Huber, P.: Robust Statistics. Wiley, New York (1981)
11. Kharin, A.: Minimax robustness of Bayesian forecasting under functional distortions of probability densities. Austrian J. Stat. **31**, 177–188 (2002)
12. Koroljuk, V.: Handbook on Probability Theory and Mathematical Statistics (in Russian). Nauka, Moscow (1985)
13. Liese, F., Miescke, K.: Statistical Decision Theory: Estimation, Testing, and Selection. Springer, New York (2008)
14. Mosteller, F., Tukey, J.: Data Analysis and Regression: A Second Course in Statistics. Addison-Wesley Pub. Co., Reading (1977)
15. Rieder, H.: Robust Asymptotic Statistics. Springer, New York (1994)
16. Tsay, R.: Outliers, level shifts and variance changes in time series. J. Forecasting **7**, 1–20 (1988)
17. Tukey, J.: A survey of sampling from contaminated distributions. In: Contributions to Probability and Statistics: Essays in Honor of Harold Hotelling, pp. 448–485. Stanford University Press, Stanford (1960)
18. Varmuza, K., Filzmoser, P.: Introduction to Multivariate Statistical Analysis in Chemometrics. Taylor & Francis/CRC Press, Boca Raton (2009)

Chapter 5
Forecasting Under Regression Models of Time Series

Abstract This chapter is devoted to statistical forecasting under regression models of time series. These models are defined as additive mixtures of regression components and random observation errors, where the regression components are determined by regression functions. In the case of complete prior information, the optimal forecasting statistic is constructed and its mean square risk is evaluated. Consistent forecasting statistics are constructed for the different levels of parametric and nonparametric uncertainty introduced in Chap. 3, and explicit expressions are obtained for the forecast risk. A special case of regression—the logistic regression—is considered.

5.1 Optimal Forecasting Under Complete Prior Information

The defining feature of regression models is the dependence of the expectation of the observed random vector $x_t \in \mathbb{R}^d$ on some other known nonrandom variables $z_t \in \mathbb{R}^M$ called regressors. Knowing this dependence allows us to use the values $\{z_t : t > T_0\}$ at future time points to forecast future elements of the time series $\{x_t : t > T_0\}$. Let us start with some real-world examples where it is natural to apply regression analysis:

(1) for a certain product, establishing the dependence of the amount of sales x_t on the price of the product, prices of competing products, and the marketing budget (these quantities form the regressor vector z_t); using the established dependence to predict future values of x_t;
(2) predicting future GDP values by finding the dependence between the size x_t of the GDP and the variables z_t, which in this case define the state economy regulation policy; this prediction can then be used to evaluate the various proposed economic policies;
(3) forecasting the efficiency x_t of a chemical reaction chain based on the vector z_t composed of the relevant characteristics of raw materials and the reaction environment.

Y. Kharin, *Robustness in Statistical Forecasting*, DOI 10.1007/978-3-319-00840-0_5,

Consider a general (nonlinear) regression model of time series defined in Sect. 3.1:

$$x_t = F(z_t; \theta^0) + u_t, \quad t \in \mathbb{Z}. \tag{5.1}$$

Here $F(z; \theta^0) : \mathbb{R}^M \times \Theta \to \mathbb{R}^d$ is the regression function, $\{z_t\} \subset \mathbb{R}^M$ is a known nonrandom sequence, $\{u_t\} \subset \mathbb{R}^d$ is an unobservable sequence of jointly independent random vectors representing random observation errors. The vectors $\{u_t\}$ have zero expectations and a finite covariance matrix $\Sigma = (\sigma_{jk}) \in \mathbb{R}^{d \times d}$:

$$\mathbb{E}\{u_t\} = \mathbf{0}_d, \quad \mathbf{Cov}\{u_t, u_{t'}\} = \delta_{tt'}\Sigma, \quad t, t' \in \mathbb{Z}. \tag{5.2}$$

First, let us consider the case of complete prior information PU-C, where $F(\cdot)$, Σ, $\{z_1, \ldots, z_{T_0}, z_{T_0+1}\}$ are known. Let the recorded observations be $\{x_1, \ldots, x_{T_0}\}$, and let x_{T_0+1} be the predicted future observation.

Theorem 5.1. *Under the regression model* (5.1), (5.2) *above, assuming prior knowledge of the stochastic characteristics* $F(\cdot)$, Σ, *the following two statements hold:*

(1) a regression mean square optimal point forecast defined by (4.6) *and* (4.11) *can be written as*

$$\widehat{x}_{T_0+1} = F(z_{T_0+1}; \theta^0); \tag{5.3}$$

this forecast is unbiased,

$$\mathbb{E}\{\widehat{x}_{T_0+1} - x_{T_0+1}\} = \mathbf{0}_d, \tag{5.4}$$

and has the minimal risk

$$R_0 = \Sigma; \quad r_0 = \mathrm{tr}(\Sigma); \tag{5.5}$$

(2) if, in addition, the random errors are normally distributed, $\mathcal{L}\{u_t\} = \mathcal{N}_d(\mathbf{0}_d, \Sigma)$ *with a nonsingular covariance matrix* Σ, *then for a confidence level* $\varepsilon \in (0, 1)$ *a* $(1-\varepsilon)$*-confidence region for* x_{T_0+1} *can be defined as an ellipsoid in* $\mathbb{R}^d$:

$$\mathcal{X}_{1-\varepsilon} = \left\{ x \in \mathbb{R}^d : (x - F(z_{T_0+1}; \theta^0))' \, \Sigma^{-1}(\cdot) < G_d^{-1}(1-\varepsilon) \right\}, \tag{5.6}$$

$$\mathbb{P}\{x_{T_0+1} \in \mathcal{X}_{1-\varepsilon}\} = 1 - \varepsilon, \tag{5.7}$$

where $G_d^{-1}(1-\varepsilon)$ *is the* $(1-\varepsilon)$*-quantile of the standard* χ^2 *distribution with* d *degrees of freedom.*

Proof. Applying Theorem 4.1 with $T_+ = 1$ (one-step-ahead forecasting) to the regression model (5.1), (5.2) yields

$$\widehat{x}_{T_0+1} = \mathbb{E}\{x_{T_0+1} \mid x_{T_0}, \ldots, x_1\} = \mathbb{E}\left\{F(z_{T_0+1}; \theta^0) + u_{T_0+1} \mid x_{T_0}, \ldots, x_1\right\}.$$

The nonrandomness of z_{T_0+1} together with the independence of $\{x_t : t \leq T_0\}$ and u_{T_0+1} now imply (5.3). Unbiasedness of the estimator (5.3) directly follows from Theorem 4.1, and the forecast risk then coincides with (5.5):

$$R_0 = \mathbb{E}\{\mathbf{Cov}\{x_{T_0+1}, x_{T_0+1} \mid x_{T_0}, \ldots, x_1\}\} =$$
$$= \mathbb{E}\{\mathbf{Cov}\{u_{T_0+1}, u_{T_0+1} \mid x_{T_0}, \ldots, x_1\}\} = \Sigma.$$

The second statement will be proved by directly verifying (5.7). Let us define an auxiliary nonnegative random variable

$$\eta = \left(x_{T_0+1} - F(z_{T_0+1}; \theta^0)\right)' \Sigma^{-1} \left(x_{T_0+1} - F(z_{T_0+1}; \theta^0)\right) \geq 0.$$

By (5.1) and the assumption that u_{T_0+1} is normal, we have

$$\eta = u'_{T_0+1} \Sigma^{-1} u_{T_0+1} \equiv v'_{T_0+1} v_{T_0+1},$$

where $v_{T_0+1} = \Sigma^{-1/2} u_{T_0+1}$. By the properties of linear transformations of normal vectors [1], we have $\mathcal{L}\{v_t\} = \mathcal{N}_d(\mathbf{0}_d, \Sigma^{-1/2}\Sigma(\Sigma^{-1/2})') = \mathcal{N}_d(\mathbf{0}_d, \mathbf{I}_d)$, i.e., v_t has the standard normal distribution. Then, by the definition of the χ^2 distribution [1], η has a χ^2 distribution with d degrees of freedom and the probability distribution function

$$\mathbb{P}\{\eta < y\} = G_d(y), \quad y \geq 0.$$

Thus, by (5.6), we have

$$\mathbb{P}\{x_{T_0+1} \in \mathcal{X}_{1-\varepsilon}\} = \mathbb{P}\{\eta < G_d^{-1}(1-\varepsilon)\} = G_d(G_d^{-1}(1-\varepsilon)) = 1 - \varepsilon,$$

proving (5.7). □

Corollary 5.1. *If $d = 1$ and $\Sigma = \sigma^2$ or, in other words, if the predicted time series is scalar, then a $(1-\varepsilon)$-confidence interval forecast of the future value $x_{T_0+1} \in \mathbb{R}^1$ can be written as follows: $x_{T_0+1} \in \mathcal{X}_{1-\varepsilon} = (x^-_{T_0+1}, x^+_{T_0+1})$ with probability $1-\varepsilon$, where*

$$x^{\pm}_{T_0+1} = F\left(z_{T_0+1}; \theta^0\right) \pm \sigma \Phi^{-1}(1 - \varepsilon/2),$$

and $\Phi^{-1}(\gamma)$ is a γ-quantile of the standard normal distribution.

Note that the optimal forecast (5.3) doesn't depend on the past data $\{x_1, \ldots, x_{T_0}\}$, $\{z_1, \ldots, z_{T_0}\}$, which is due to the lack of prior uncertainty (remember that we are assuming complete prior knowledge).

Also note that an optimal simultaneous forecast of $T_+ > 1$ future values of the time series can be made using the same formula (5.3):

$$\widehat{x}_{T_0+\tau} = F(z_{T_0+\tau}; \theta^0), \quad \tau = 1, \ldots, T_+.$$

In other words, the independence of $\{u_t\}$ implies the independence of the forecast vector elements. The minimal mean square risk of forecasting a composite vector $X^+ = (x_{T_0+1}, \ldots, x_{T_0+T_+})$ can be written as $r_0 = T_+ \mathrm{tr}\,(\Sigma)$.

5.2 Regression Forecasting Under Parametric Prior Uncertainty

In most applications, complete prior knowledge of the regression model (5.1), (5.2) is unavailable, or, in other words, the forecasting problem includes prior uncertainty (parametric or nonparametric). In this section, we are going to discuss the PU-P (parametric prior uncertainty) setting where the true value of the parameter vector $\theta^0 \in \Theta \subseteq \mathbb{R}^m$ defining the regression function $F(z; \theta^0)$ is a priori unknown, and the same is true of the covariance matrix $\Sigma^0 = (\sigma^0_{jk}) \in \mathbb{R}^{d\times d}$. The following three main directions have been proposed to solve this type of forecasting problems:

(1) Bayesian approach;
(2) joint estimation of the parameters θ^0, Σ^0 and the future value X^+ by applying the maximum likelihood principle;
(3) use of the plug-in principle.

These directions will be discussed in more detail in the following subsections for the general setting of the forecasting problem introduced in Sect. 3.1.

5.2.1 Bayesian Approach in the PU-P Setting

The Bayesian approach is based on the assumption that the unknown model parameters θ^0, Σ^0 are random and follow a probability distribution with a certain a priori known probability density $p_{\theta^0,\Sigma^0}(\theta, \Sigma)$ (to simplify the argument, we are going to restrict ourselves to the absolutely continuous case). We are also going to assume that the probability density $q_{\Sigma^0}(u)$ of the random observation error vector u_t is known up to a parameter Σ^0 and satisfies the conditions (5.2).

From (5.1), (5.2), and well-known properties of joint probability densities, we can obtain the conditional probability density of X^+ given fixed past values X^0:

$$p_{X^+|X^0}(X^+ \mid X^0) = \frac{p_{X^+,X^0}(X^+, X^0)}{\int\limits_{\mathbb{R}^{dT_+}} p_{X^+,X^0}(X^+, X^0) dX^+}, \tag{5.8}$$

$$p_{X^+,X^0}(X^+, X^0) = \int p_{X^+,X^0|\theta^0,\Sigma^0}(X^+, X^0 \mid \theta, \Sigma) p_{\theta^0,\Sigma^0}(\theta, \Sigma) d\theta d\Sigma, \tag{5.9}$$

$$p_{X^+,X^0|\theta^0,\Sigma^0}(X^+, X^0 \mid \theta, \Sigma) = \left(\prod_{i=1}^{T_0} q_\Sigma \left(x_{t_i^0} - F(z_{t_i^0}; \theta) \right) \right) \times$$
$$\times \left(\prod_{j=1}^{T_+} q_\Sigma \left(x_{t_j^+} - F(z_{t_j^+}; \theta) \right) \right). \tag{5.10}$$

By Theorem 4.1, the *Bayesian regression forecasting statistic* which is optimal w.r.t. matrix mean square risk (4.6) can be written in the following integral form:

$$\widehat{X}^+ = f_B(X^0) ::= \frac{\int_{\mathbb{R}^{dT_+}} X^+ p_{X^+,X^0}(X^+, X^0) dX^+}{\int_{\mathbb{R}^{dT_+}} p_{X^+,X^0}(X^+, X^0) dX^+}. \tag{5.11}$$

The Bayesian forecasting statistic (5.11) is universal and possesses all of the optimal properties mentioned in Theorem 4.1. However, in practice this statistic is hard to compute due to the following two reasons:

(1) the prior probability density $p_{\theta^0,\Sigma^0}(\cdot)$ is usually unknown;
(2) the formulas (5.8)–(5.11) require multiple integration in many dimensions, which often has an enormous computational cost.

5.2.2 *Joint Estimation Using the Maximum Likelihood (ML) Principle*

Assuming the parameters θ^0, Σ^0 to be unknown, let us use the representation (5.10) to construct a logarithmic likelihood function (LLF) for X^0, X^+, i.e., a logarithm of the joint probability density of X^0, X^+:

$$l(X^+, \theta, \Sigma) = l_{X^0}(\theta, \Sigma) + l_{X^+}(\theta, \Sigma), \tag{5.12}$$

$$l_{X^0}(\theta, \Sigma) = \ln p_{X^0}(X^0, \theta, \Sigma) = \sum_{i=1}^{T_0} \ln q_\Sigma \left(x_{t_i^0} - F(z_{t_i^0}; \theta) \right), \tag{5.13}$$

$$l_{X^+}(\theta, \Sigma) = \ln p_{X^+}(X^+; \theta, \Sigma) = \sum_{j=1}^{T_+} \ln q_\Sigma \left(x_{t_j^+} - F(z_{t_j^+}; \theta) \right). \tag{5.14}$$

Theorem 5.2. *Under PU-P uncertainty in the regression model* (5.1), (5.2), *let the probability density* $q_\Sigma(u)$ *have a single mode at* $u = \mathbf{0}_d$. *Then the ML forecasting statistic has the form*

$$\widehat{x}_{t_j^+} = F(z_{t_j^+}; \widetilde{\theta}), \quad j = 1, \ldots, T_+, \tag{5.15}$$

where $\widetilde{\theta} = \widetilde{\theta}(X^0)$ *is a statistical estimator for* θ^0 *defined as a solution of the following optimization problem*

$$l_{X^0}(\theta, \Sigma) + T_+ \ln q_\Sigma(\mathbf{0}_d) \to \max_{\theta, \Sigma}. \tag{5.16}$$

Proof. Following the ML principle, joint estimators for θ^0, Σ^0, X^+ are found by maximizing the log likelihood function (5.12):

$$l(X^+, \theta, \Sigma) \to \max_{X^+, \theta, \Sigma}. \tag{5.17}$$

By (5.14) and the unimodality of $q_\Sigma(\cdot)$, this maximum is attained at $\widehat{X}^+ = (\widehat{x}_{t_j^+})$, where $\widehat{x}_{t_j^+}$ is defined by (5.15). Substituting (5.15) into (5.17) yields the optimization problem (5.16). □

From (5.15), (5.16) it follows that under the assumptions of Theorem 5.2, the ML principle leads to plug-in forecasting statistics introduced in Sect. 4.2. In particular, if Σ is known, or if $q_\Sigma(\mathbf{0}_d)$ doesn't depend on Σ, then $\widetilde{\theta}$ defined by (5.16) is the maximum likelihood estimator (MLE).

5.2.3 Using the Plug-In Principle

As discussed in Sect. 4.2, a *plug-in regression forecasting statistic* has the following general form, which is similar to (5.15):

$$\widehat{x}_{t_j^+} = F(z_{t_j^+}; \widehat{\theta}), \quad j = 1, \ldots, T_+, \tag{5.18}$$

where $\widehat{\theta} = \widehat{\theta}(X^0) : \mathbb{R}^{dT_0} \to \Theta \subseteq \mathbb{R}^m$ is a consistent statistical estimator of the parameter vector θ^0 of the regression function based on observations X^0. Construction of the estimator $\widehat{\theta}$ can be based on numerous classical estimators: MLEs (or their generalization—minimal contrast estimators [2]), least squares estimators (LSE), or moment estimators. The main difficulty is the estimation of the forecast risk associated with the use of different estimators. The most common choices in regression analysis—MLEs and LSEs—are discussed below [11].

Theorem 5.3. *Under the conditions of Theorem 5.2, if $q_\Sigma(u) = n_d(u \mid \mathbf{0}_d, \Sigma)$ is the d-dimensional normal probability density, then the plug-in regression forecasting statistic based on MLEs $\widehat{\theta}$, $\widehat{\Sigma}$ can be written in the form* (5.18), *where $\widehat{\theta}$, $\widehat{\Sigma}$ are found by solving the following optimization problem:*

$$\ln|\Sigma| + \frac{1}{T_0}\sum_{i=1}^{T_0}\left(x_{t_i^0} - F(z_{t_i^0};\theta)\right)'\Sigma^{-1}\left(x_{t_i^0} - F(z_{t_i^0};\theta)\right) \to \min_{\theta,\Sigma}. \tag{5.19}$$

Proof. Due to the normal distribution of $\{u_t\}$, by (5.13) we have the following optimization problem:

$$l_{X^0}(\theta,\Sigma) = \sum_{i=1}^{T_0}\ln\left((2\pi)^{-\frac{d}{2}}|\Sigma|^{-\frac{1}{2}}\exp\left(-\frac{1}{2}\left(x_{t_i^0} - F(z_{t_i^0};\theta)\right)'\Sigma^{-1}(\cdot)\right)\right) =$$

$$= -\frac{T_0 d}{2}\ln(2\pi) - \frac{T_0}{2}\ln|\Sigma| - \frac{1}{2}\sum_{i=1}^{T_0}\left(x_{t_i^0} - F(z_{t_i^0};\theta)\right)'\Sigma^{-1}(\cdot) \to \max_{\theta,\Sigma},$$

which is equivalent to (5.19). □

Note that the estimator $\widehat{\theta}$ defined by (5.19) is, assuming nonlinear dependence between $F(\cdot)$ and θ, a nonlinear LSE (a generalized nonlinear LSE for $\Sigma \neq c\mathbf{I}_d$). The most extensive theoretical results have been obtained in the case of linear dependence between $F(\cdot)$ and θ:

$$F(z;\theta^0) = \theta^{0\prime}\psi(z), \quad z \in \mathbb{R}^M, \tag{5.20}$$

where $\psi(z) = (\psi_i(z)) \in \mathbb{R}^m$ is an m-column vector of known linearly independent (basis) functions,

$$\theta^0 = \left(\theta^0_{(1)} \vdots \theta^0_{(2)} \vdots \ldots \vdots \theta^0_{(d)}\right) = (\theta^0_{ij})$$

is an $(m \times d)$-matrix of regression coefficients, and

$$\theta^0_{(j)} = (\theta^0_{ij}) \in \mathbb{R}^m$$

is the jth m-column vector of this matrix. The multivariate regression function (5.20) is linear w.r.t. its parameters, and thus the regression model (5.1), (5.20) is called the *multivariate linear regression model* [1]. Under the assumptions (5.2), LSEs of the parameters of this model θ^0, Σ based on $T_0 > m$ observations $\{x_t : t \in \mathcal{T}_0\}$ and a priori given regressor values $\{z_t : t \in \mathcal{T}_0\}$ can be written in the following form [1]:

$$\widehat{\theta} = A^{-1}C', \tag{5.21}$$

where $\mathcal{T}_0 = \{t_1^0, \ldots, t_{T_0}^0\}$,

$$A = \sum_{t\in\mathcal{T}_0} \psi(z_t)\psi'(z_t) \quad \text{is an } (m \times m)\text{-matrix},$$

$$C = \sum_{t\in\mathcal{T}_0} x_t \psi'(z_t) \quad \text{is a } (d \times m)\text{-matrix},$$

$$\widehat{\Sigma} = \frac{1}{T_0 - m} \sum_{t\in\mathcal{T}_0} \left(x_t - \widehat{\theta}'\psi(z_t)\right)\left(x_t - \widehat{\theta}'\psi(z_t)\right)'. \tag{5.22}$$

The LSEs defined by (5.21) require an additional assumption on the *design of the experiment*—the dynamics of regressor vectors z_t:

$$|A| \neq 0. \tag{5.23}$$

The estimators (5.21), (5.22) coincide with the LSE and are obtained by substituting (5.20) into (5.19).

By Theorem 5.3, under the multivariate linear regression model (5.20) the plug-in forecasting statistic (5.18) has the form

$$\widehat{x}_{t_j^+} = \widehat{\theta}'\psi(z_{t_j^+}), \quad j = 1, \ldots, T_+. \tag{5.24}$$

Theorem 5.4. *Under PU-P uncertainty in the multivariate linear regression model* (5.1), (5.2), (5.20), *the plug-in forecasting statistic* (5.24), (5.21) *is unbiased:*

$$\mathbb{E}\{\widehat{x}_{t_j^+} - x_{t_j^+}\} = \mathbf{0}_d, \quad j = 1, \ldots, T_+, \tag{5.25}$$

and the respective matrix mean square forecast risk equals

$$R = \mathbb{E}\left\{(\widehat{x}_{t_j^+} - x_{t_j^+})(\widehat{x}_{t_j^+} - x_{t_j^+})'\right\} = \left(1 + \psi'(z_{t_j^+})A^{-1}\psi(z_{t_j^+})\right)\Sigma. \tag{5.26}$$

Proof. It is known [1] that, under the conditions of the theorem, the MLE (5.21) is unbiased. Thus, (5.24) yields

$$\mathbb{E}\{\widehat{x}_{t_j^+} - x_{t_j^+}\} = \mathbb{E}\left\{\theta^{0\prime}\psi(z_{t_j^+}) - x_{t_j^+}\right\} = \mathbb{E}\{-u_{t_j^+}\} = \mathbf{0}_d,$$

which coincides with (5.25). Let us use a similar argument to calculate the forecast risk.

Due to (5.1), (5.20), and (5.24), we have

$$\widehat{x}_{t_j^+} - x_{t_j^+} = (\widehat{\theta} - \theta^0)'\psi(z_{t_j^+}) - u_{t_j^+} = \left((\widehat{\theta}_{(i)} - \theta^0_{(i)})'\psi(z_{t_j^+})\right)_{i=1,\ldots,d} - u_{t_j^+},$$

and thus by (5.2) we can write

$$R = \Sigma + \left(\psi'(z_{t_j^+})\mathbb{E}\left\{(\widehat{\theta}_{(i)} - \theta^0_{(i)})(\widehat{\theta}_{(k)} - \theta^0_{(k)})'\right\}\psi(z_{t_j^+})\right)_{i,k=1,\dots,d}.$$

By applying well-known results from multivariate linear regression theory [1], we have:

$$\mathbb{E}\left\{(\widehat{\theta}_{(i)} - \theta^0_{(i)})(\widehat{\theta}_{(k)} - \theta^0_{(k)})'\right\} = \sigma_{ik}A^{-1}, \quad i,k = 1,\dots,d.$$

Substituting the above equality into the previous formula leads to (5.26). □

Theorem 5.5. *Under the conditions of Theorem 5.4, if, additionally, the random observation errors are normal with the probability density* $q_\Sigma(u) = n_d(u \mid \mathbf{0}_d, \Sigma)$*, and the covariance matrix* Σ *is known, then a* $(1-\varepsilon)$*-confidence forecast region for* $x_{t_j^+}$, $j = 1,\dots,T_+$, *can be defined as the following ellipsoid:*

$$\mathcal{X}_{1-\varepsilon} = \left\{x \in \mathbb{R}^d : \frac{\left(x - \widehat{\theta}'\psi(z_{t_j^+})\right)' \Sigma^{-1}\left(x - \widehat{\theta}'\psi(z_{t_j^+})\right)}{1 + \psi'(z_{t_j^+})A^{-1}\psi(z_{t_j^+})} < G_d^{-1}(1-\varepsilon)\right\}, \tag{5.27}$$

where $G_d(\cdot)$ *is the standard* χ^2 *distribution function with* d *degrees of freedom.*

Proof. Using the representation (5.24), let us introduce an auxiliary random d-column vector of the "remainders:"

$$\eta = x_{t_j^+} - \widehat{\theta}'\psi(z_{t_j^+}) = x_{t_j^+} - \widehat{x}_{t_j^+}.$$

Due to linearity of the LSE (5.21) w.r.t. the observations and Theorem 5.4, we can write

$$\mathcal{L}\{\eta\} = \mathcal{N}_d\left(\mathbf{0}_d, \left(1 + \psi'(z_{t_j^+})A^{-1}\psi(z_{t_j^+})\right)\Sigma\right).$$

From the well-known properties of probability distributions, the quadratic form

$$Q = \frac{\eta'\Sigma^{-1}\eta}{1 + \psi'(z_{t_j^+})A^{-1}\psi(z_{t_j^+})}$$

follows the standard χ^2 distribution with d degrees of freedom, $\mathcal{L}\{Q\} = \chi^2_d$, and thus its probability distribution function is $G_d(y)$.

Therefore, in the notation of (5.27), we have:

$$\mathbb{P}\left\{x_{t_j^+} \in \mathcal{X}_{1-\varepsilon}\right\} = \mathbb{P}\left\{Q < G_d^{-1}(1-\varepsilon)\right\} = G_d\left(G_d^{-1}(1-\varepsilon)\right) = 1-\varepsilon,$$

proving the theorem. □

Corollary 5.2. *Interval forecasts of a scalar quantity* $x_{t_j^+} \in \mathbb{R}^1$ $(d = 1, \Sigma = \sigma^2)$ *have the following form: at* $(1-\varepsilon)$*-confidence,* $x_{t_j^+} \in (x^-, x^+)$*, where*

$$x^{\pm} = \widehat{\theta}'\psi(z_{t_j^+}) \pm g_{1-\varepsilon/2}\,\sigma\sqrt{1 + \psi'(z_{t_j^+})A^{-1}\psi(z_{t_j^+})}, \tag{5.28}$$

and $g_u = \Phi^{-1}(u)$ *is a u-quantile of the* $\mathcal{N}(0,1)$ *distribution,* $u \in (0,1)$.

Note that in applications the design of the experiment $\{z_{t_j^0}\}$ usually satisfies the Eicker condition: as the observation length tends to infinity, $T_0 \to +\infty$, the minimal characteristic value of the matrix A satisfies $\lambda_{min}(A) \to +\infty$, and thus

$$\psi'(z_{t_j^+})A^{-1}\psi(z_{t_j^+}) \to 0.$$

In this case the length of the confidence interval defined by (5.28) tends to the length of the confidence interval in the PU-C (complete prior information) setting:

$$\Delta = 2\sigma g_{1-\varepsilon/2}.$$

Theorem 5.6. *If, under the assumptions of Theorem 5.5, the covariance matrix of observation errors is unknown, and* $T_0 \geq m + d$*, then the following d-dimensional ellipsoid is a* $(1-\varepsilon)$*-confidence forecast region for* $x_{t_j^+}$*:*

$$\widetilde{X}_{1-\varepsilon} = \left\{ x \in \mathbb{R}^d : \frac{\left(x - \widehat{\theta}'\psi(z_{t_j^+})\right)' \widehat{\Sigma}^{-1} \left(x - \widehat{\theta}'\psi(z_{t_j^+})\right)}{1 + \psi'(z_{t_j^+})A^{-1}\psi(z_{t_j^+})} < \right.$$

$$\left. < \frac{d(T_0 - m)}{T_0 + 1 - (d+m)} F^{-1}_{d, T_0+1-(d+m)}(1-\varepsilon) \right\},$$

where $j = 1, \ldots, T_+$*;* $F^{-1}_{m,n}(p)$ *is a p-quantile of the standard central F-distribution with* (m,n) *degrees of freedom,* $p \in (0,1)$.

Proof. Consider an auxiliary random d-column vector:

$$\widetilde{\eta} = \left(1 + \psi'(z_{t_j^+})A^{-1}\psi(z_{t_j^+})\right)^{-1/2} \eta \in \mathbb{R}^d.$$

Let us also define a quadratic form $\widetilde{Q} = \widetilde{\eta}'\widehat{\Sigma}^{-1}\widetilde{\eta} \geq 0$, where Σ is defined by (5.22).

We are going to use the following results from multivariate linear regression theory [1]:

(1) $\mathcal{L}\{\widetilde{\eta}\} = \mathcal{N}_d(\mathbf{0}_d, \Sigma)$;
(2) $\mathcal{L}\{(T_0 - m)\widetilde{\Sigma}\} = W_d(\Sigma, T_0 - m)$ is the Wishart probability distribution with the matrix Σ and $T_0 - m$ degrees of freedom;
(3) $\widetilde{\eta}$ and $\widetilde{\Sigma}$ are independent random elements.

Applying the theorem on probability distributions of the generalized Hotelling's T^2-statistic [1], we obtain

$$\mathcal{L}\left\{\frac{\widetilde{Q}}{T_0 - m}\frac{T_0 + 1 - (d + m)}{d}\right\} = F_{d,T_0+1-(d+m)},$$

and thus

$$\mathbb{P}\left\{x_{t_j^+} \in \widetilde{X}_{1-\varepsilon}\right\} = 1 - \varepsilon. \qquad \square$$

Corollary 5.3. *In the scalar case ($d = 1$), let $\widehat{\Sigma} = s^2$ be the sample variance of the remainders. Then the limits of a $(1 - \varepsilon)$-confidence interval $\widetilde{X}_{1-\varepsilon} = (\widetilde{x}^-, \widetilde{x}^+)$ can be defined as*

$$\widetilde{x}^{\pm} = \widehat{\theta}\,'\psi(z_{t_j^+}) \pm t^{-1}_{T_0-m}\left(1 - \frac{\varepsilon}{2}\right) s\sqrt{1 + \psi'(z_{t_j^+})A^{-1}\psi(z_{t_j^+})},$$

where $t_f^{-1}(p)$ is a p-quantile of Student's t-distribution with f degrees of freedom.

Proof. As in the proof of Theorem 5.6, we are going to rely on the following facts:

(1) $\mathcal{L}\{\widetilde{\eta}/\sigma\} = \mathcal{N}(0, 1)$;
(2) $\mathcal{L}\{(T_0 - m)s^2/\sigma^2\} = \chi^2_{T_0-m}$;
(3) $\widetilde{\eta}$, s^2 are independent.

By well-known properties of probability distributions [3], we have that

$$\mathcal{L}\left\{\frac{\widetilde{\eta}/\sigma}{\sqrt{s^2/\sigma^2}}\right\} = \mathcal{L}\left\{\frac{\widetilde{\eta}}{s}\right\} = t_{T_0-m},$$

where t_{T_0-m} denotes Student's t-distribution with $T_0 - m$ degrees of freedom. Thus,

$$\mathbb{P}\left\{\widetilde{x}^- < x_{t_j^+} < \widetilde{x}^+\right\} = 1 - \varepsilon. \qquad \square$$

5.3 Logistic Regression Forecasting

Let us consider a commonly encountered applied problem which leads to the logistic regression model. Assume that a complex technological process is being studied to establish the dependence between the average ratio of defective items

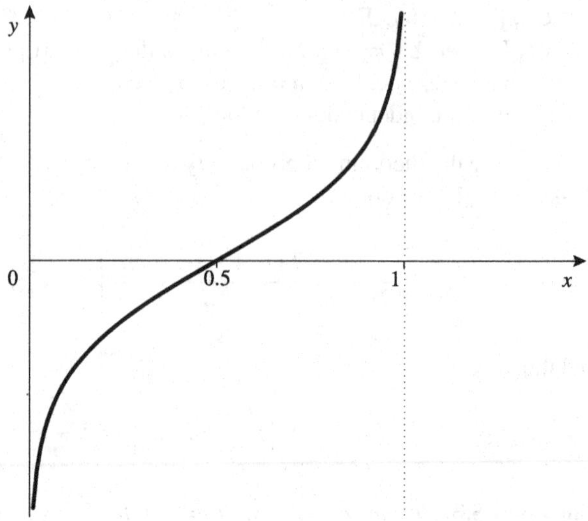

Fig. 5.1 The logit function

(e.g., computer chips) $x \in [0,1]$ on M factors (or characteristics) of the underlying technological process $z = (z_j) \in \mathbb{R}^M$. Statistical data on the process has been obtained by sampling and is a set of T_0 series of experimental data. In the tth series, $t = 1, 2, \ldots, T_0$, the vector of technological factors assumed the values $z_t = (z_{t,j}) \in \mathbb{R}^M$, and N_t^0 defective items have been found in N_t items that have been subjected to quality control; thus, the recorded defect ratio was equal to $x = x_t = N_t^0/N_t$. We would like to construct a forecast $\widehat{x}_{T_0+1} \in [0,1]$ for the unknown true defect ratio x_{T_0+1} for a given (expected or future) collection of technological factors $z_{T_0+1} = (z_{T_0+1,j}) \in \mathbb{R}^M$. The logistic regression model proves adequate for this type of forecasting problems [8].

Definition 5.1. The logit transformation is defined as the following $[0,1] \leftrightarrow \mathbb{R}^1$ bijection (Fig. 5.1):

$$y = \operatorname{logit}(x) ::= \ln \frac{x}{1-x}. \tag{5.29}$$

From (5.29) we can find the inverse function:

$$x = F(y) = \frac{1}{1+e^{-y}} = 1 - \frac{1}{1+e^y} = \frac{e^y}{1+e^y}, \tag{5.30}$$

which is known as the standard logistic distribution function.

The probability density of the standard logistic distribution function is

$$f(y) = F'(y) = \frac{e^y}{(1+e^y)^2},$$

the expectation is zero, $\mu = 0$, and the variance equals $D = \pi^2/3$ [7,8].

Thus,

$$y = \text{logit}(x) = F^{-1}(x). \tag{5.31}$$

Definition 5.2. It is said that a random variable $x \in [0,1]$ has a logistic regression dependence on $z = (z_k) \in \mathbb{R}^M$ with a coefficient vector $\theta = (\theta_0, \theta_1, \ldots, \theta_M) \in \mathbb{R}^{M+1}$ if the following conditions are satisfied:

(1) $\mathbb{E}\{x\} = p_\theta(z) : \mathbb{R}^M \to [0,1]$;
(2) there exists a positive integer $N = N(x)$ such that $N\,x \in \{0, 1, \ldots, N\}$ is a discrete binomially distributed random variable, $\mathcal{L}\{Nx\} = \text{Bi}(N, p_\theta(z))$;
(3) the function $p_\theta(z)$ is chosen so that its logit transformation is a linear function w.r.t. the factors z and the parameters θ:

$$\text{logit}(p_\theta(z)) = \ln \frac{p_\theta(z)}{1 - p_\theta(z)} = \theta_0 + \sum_{j=1}^{M} \theta_j z_j. \tag{5.32}$$

In the above example, the function $p_\theta(z)$ represents the average defect ratio, i.e., the probability of a defect for technological factors z and coefficients θ. Definition 5.2 and (5.30) imply the following nonlinear dependence of the probability $p = p_\theta(z)$ on the factors z and the coefficients θ:

$$p = p_\theta(z) = F\left(\theta_0 + \sum_{j=1}^{M} \theta_j z_j\right) = \left(1 + \exp\left(-\theta_0 - \sum_{j=1}^{M} \theta_j z_j\right)\right)^{-1}. \tag{5.33}$$

Due to the conditions (1), (2), and equality (5.33), we have the following representation:

$$x = F\left(\theta_0 + \sum_{j=1}^{M} \theta_j z_j\right) + \xi, \quad \mathbb{E}\{\xi\} = 0, \quad \mathbb{D}\{\xi\} < +\infty, \tag{5.34}$$

and thus the logistic model is a special case of the general regression model discussed in Sect. 3.1. Thus, a suboptimal plug-in forecast for the vector of factors $z = (z_j)$ defined by (5.18) can be written as follows:

$$\widehat{x} = F\left(\widehat{\theta}_0 + \sum_{j=1}^{M} \widehat{\theta}_j z_j\right), \tag{5.35}$$

where $\widehat{\theta} = (\widehat{\theta}_0, \ldots, \widehat{\theta}_M)' \in \mathbb{R}^{M+1}$ is a consistent estimator of the logistic regression coefficients based on the collected statistical data.

Let us consider the problem of constructing a consistent statistical estimator $\widehat{\theta}$. Under the logistic regression model defined by conditions (1)–(3), the collected statistical data

$$\left\{(x_t, z_t) \in \mathbb{R}^{M+1} : t = 1, 2, \ldots, T_0\right\}$$

has the following properties:

(A_1) $\{x_1, x_2, \ldots, x_{T_0}\} \subset [0, 1]$ is a sample of T_0 jointly independent random variables;

(A_2) x_t is a random variable derived from $N_t \geq 1$ independent Bernoulli trials with a binomial (after normalization) probability distribution, $t = 1, \ldots, T_0$:

$$\mathcal{L}\{N_t x_t\} = \mathrm{Bi}\left(N_t, p_{\theta^0}(z_t)\right),$$

$$p_{\theta^0}(z_t) = F\left(\theta^{0\prime}\overline{z}_t\right),$$

where

$$\overline{z}_t = (1 \vdots z_t')' \in \mathbb{R}^{M+1}$$

is a composite column vector of factors, and $F(\cdot)$ is defined by (5.30).

Let us discuss two main methods for statistical estimation of θ^0.

Use of Maximum Likelihood

Theorem 5.7. *The MLE $\widehat{\theta} = (\widehat{\theta}_j) \in \mathbb{R}^{M+1}$ for the logistic regression model described above is the solution of $M+1$ vector nonlinear equations:*

$$\sum_{t=1}^{T_0} N_t\left(x_t - F(\theta'\overline{x}_t)\right)\overline{x}_t = \mathbf{0}_{M+1}. \tag{5.36}$$

Proof. The logistic regression model satisfies the conditions A_1, A_2. The condition A_2 implies that

$$\mathbb{P}_\theta\{x = x_t\} = C_{N_t}^{N_t x_t}\left(p_\theta(z_t)\right)^{N_t x_t}\left(1 - p_\theta(z_t)\right)^{N_t(1-x_t)},$$

and thus, taking into account the condition A_1, the LLF can be written as

$$l(\theta) = \ln\prod_{t=1}^{T_\theta}\mathbb{P}_0\{x = x_t\} = \sum_{t=1}^{T_0}\ln\left(C_{N_t}^{N_t x_t} + N_t x_t \ln p_\theta(z_t) + N_t(1-x_t)\ln\left(1 - p_\theta(z_t)\right)\right). \tag{5.37}$$

The MLE $\widehat{\theta}$ is found from the maximality condition

$$l(\theta) \to \max_{\theta}. \tag{5.38}$$

Let us rewrite (5.37) by applying (5.32), (5.33):

$$l(\theta) = \sum_{t=1}^{T_0} \left(\ln C_{N_t}^{N_t x_t} + N_t x_t \ln \frac{p_\theta(z_t)}{1 - p_\theta(z_t)} + N_t \ln\left(1 - p_\theta(z_t)\right) \right) =$$

$$= \text{const} + \sum_{t=1}^{T_0} \left(N_t x_t \theta' \overline{z}_t - N_t \ln\left(1 + \exp(\theta' \overline{z}_t)\right)\right).$$

Then the necessary extremum condition (5.38) leads to a system of equations

$$\nabla_\theta l(\theta) = \sum_{t=1}^{T_0} \left(N_t x_t \overline{z}_t - N_t \frac{\exp(\theta' \overline{z}_t)}{1 + \exp(\theta' \overline{z}_t)} \overline{z}_t \right) = \mathbf{0}_{M+1},$$

which by (5.30) is equivalent to (5.36). □

Unfortunately, (5.36) defines an analytically unsolvable system of transcendent equations. Thus, numerical methods must be applied, specifically linearization of $F(\theta' \overline{z}_t)$ over θ.

Use of the Generalized Least Squares Method

Let us define the logit transformed sample (5.29):

$$l_t = \text{logit}(x_t) = \ln \frac{x_t}{1 - x_t}, \quad t = 1, \ldots, T_0. \tag{5.39}$$

Theorem 5.8. *Under the conditions of Theorem 5.7, the random variables $l_1, \ldots, l_{T_0}$ defined by (5.39) are jointly independent, and for $N_t \to \infty$ the following asymptotic relations hold for $t = 1, \ldots, T_0$:*

$$l_i \xrightarrow{\mathbf{P}} \text{logit}\left(p_{\theta^0}(z_t)\right) = \theta^{0\prime} \overline{z}_t,$$

$$\mathcal{L}\left\{\sqrt{N_t}(l_t - \theta^{0\prime} \overline{z}_t)\right\} \longrightarrow \mathcal{N}_1\left(0, \left(p_{\theta^0}(z_t)\left(1 - p_{\theta^0}(z_t)\right)\right)^{-1}\right). \tag{5.40}$$

Proof. From the condition A_2 and certain well-known properties of the binomial distribution, we have

$$\mathbb{E}\{N_t x_t\} = N_t p_{\theta^0}(z_t), \quad \mathbb{D}\{N_t x_t\} = N_t p_{\theta^0}(z_t)\,(1 - p_{\theta^0}(z_t)),$$
$$\mathcal{L}\left\{\sqrt{N_t}\,(x_t - p_{\theta^0}(z_t))\right\} \underset{N_t\to\infty}{\longrightarrow} \mathcal{N}_1\left(0, p_{\theta^0}(z_t)(1 - p_{\theta^0}(z_t))\right),$$
$$x_t \overset{\mathbf{P}}{\underset{N_t\to\infty}{\longrightarrow}} p_{\theta^0}(z_t), \quad t = 1,\ldots,T_0. \tag{5.41}$$

Then (5.40) follows from (5.41) and the Anderson theorem on functional transformations of asymptotically normal random sequences [1] due to (5.32) and the following fact implied by (5.29):

$$\frac{d}{dx}(\text{logit}(x)) = \frac{1}{x(1-x)}. \qquad \square$$

Given a sufficiently long series of observations ($N_t \to \infty$, $t = 1,\ldots,T_0$), Theorem 5.8 allows for a representation of a logit-transformed sample (5.39) as a random sample satisfying a heteroscedastic (corresponding to a heterogeneous population) multiple linear regression model [8]:

$$l_t = \theta'\overline{z}_t + \eta_t, \quad t = 1,2,\ldots,T_0, \tag{5.42}$$

where $\{\eta_t\}$ are the jointly independent "induced" random errors:

$$\mathbb{E}\{\eta_t\} \approx 0, \quad \sigma_t^2 = \mathbb{D}\{\eta_t\} \approx \left(N_t p_\theta(z_t)(1 - p_\theta(z_t))\right)^{-1}. \tag{5.43}$$

The relations (5.42), (5.43) allow us to apply the generalized (weighted) least squares method using plug-in variance estimators ($\widehat{p_\theta(z_t)} = x_t$):

$$\widehat{\sigma}_t^2 = \left(N_t x_t (1 - x_t)\right)^{-1}, \quad t = 1,2,\ldots,T_0;$$
$$\varrho^2(\theta) = \sum_{t=1}^{T_0} \frac{(l_t - \theta'\overline{z}_t)^2}{\widehat{\sigma}_t^2} = \sum_{t=1}^{T_0} N_t x_t (1 - x_t)(l_t - \theta'\overline{z}_t)^2 \longrightarrow \min_\theta.$$

The necessary condition for this minimum can be written as

$$\nabla_\theta \varrho^2(\theta) = -2\sum_{t=1}^{T_0} N_t x_t (1 - x_t)(l_t - \theta'\overline{z}_t)\overline{z}_t = \mathbf{0},$$

or equivalently

$$\sum_{t=1}^{T_0} N_t x_t (1 - x_t) l_t \overline{z}_t = \left(\sum_{t=1}^{T_0} N_t x_t (1 - x_t)\overline{z}_t\overline{z}_t'\right)\theta.$$

Thus, if the design of the experiment is nonsingular, or, formally,

$$\left| \sum_{t=1}^{T_0} N_t x_t (1 - x_t) \bar{z}_t \bar{z}_t' \right| \neq 0, \tag{5.44}$$

we can find a generalized (weighted) LSE for the logistic regression coefficients:

$$\widehat{\theta} = \left(\sum_{t=1}^{T_0} N_t x_t (1 - x_t) \bar{z}_t \bar{z}_t' \right)^{-1} \sum_{t=1}^{T_0} N_t x_t (1 - x_t) l_t \bar{z}_t. \tag{5.45}$$

Note that to satisfy the nonsingularity condition (5.44), the number of series with different values of $\{z_t\}$ must be no smaller than the number of the parameters. In particular, we must have that $T_0 \geq M + 1$.

In conclusion, note that other nonlinear regression models can be constructed similarly to the logistic model (5.33) by taking another probability distribution function instead of the logistic function (5.30). For instance, we can base our argument on the Cauchy probability distribution function

$$F(y) = 1/2 + \pi^{-1} \text{arctg}(y),$$

which would lead to the following nonlinear regression model:

$$\text{tg}\left((p_\theta(z_t) - 1/2)\pi \right) = \theta' z_t.$$

5.4 Nonparametric Kernel Regression Forecasting

As in Sects. 5.2, 5.3, consider the problem of forecasting a random vector $x \in \mathbb{R}^d$ stochastically dependent on a vector $z \in \mathbb{R}^M$ of regressors (factors):

$$x = F(z) + u, \tag{5.46}$$

where $u \in \mathbb{R}^d$ is a random d-vector with a zero expectation and a finite variance, and $F(\cdot) : \mathbb{R}^M \to \mathbb{R}^d$ is the unknown regression function. In Sects. 5.2, 5.3 we have studied the case of PU-P parametric prior uncertainty where the regression function $F(\cdot)$ was assumed to be known up to an m-vector of parameters

$$\theta^0 = (\theta_j^0) \in \mathbb{R}^m : \quad F(z) ::= F(z; \theta^0).$$

Based on statistical data $\{(x_t' \vdots z_t')' \in \mathbb{R}^{d+M} : t = 1, 2, \dots, T_0\}$ collected over T_0 time units, a consistent statistical estimator $\widehat{\theta}$ and the respective parametric regression function estimator $\widehat{F}(z) ::= F(z;\widehat{\theta})$ were used to forecast future values $x_{T+\tau} \in \mathbb{R}^d$:

$$\widehat{x}_{T+\tau} = F\big(z_{T+\tau};\widehat{\theta}\big).$$

Unfortunately, a parametric representation of the regression function $F(\cdot)$ is usually unknown; in other words, we are in a PU-NP (nonparametric prior uncertainty) setting. This situation will be considered in the remaining part of Chap. 5. For brevity, let us consider the univariate case: $d = M = 1$; the results obtained for univariate distributions can be easily generalized for multivariate x, z. Without loss of generality, we assume that $z \in [0, 1]$; any different finite range for z can be obtained by shifting and scaling.

Let us define the *general nonparametric regression model* [6]:

$$x_t = F(z_t) + u_t, \quad t = 1, 2, \ldots, T_0, \ldots , \tag{5.47}$$

where $F(\cdot) \in \mathbb{C}(\mathbb{R}^1)$ is the unknown real-valued continuous regression function; $\{u_t\}$ are jointly independent identically distributed random observation errors,

$$\mathbb{E}\{u_t\} = 0, \quad \mathbb{D}\{u_t\} = \mathbb{E}\{u_t^2\} = \sigma^2 < +\infty; \tag{5.48}$$

the sequence $z_t \in [0, 1]$ represents the observed regressor (factor, predictor) values; $x_t \in \mathbb{R}^1$ are the observed values of the dependent variable x in (5.46); T_0 is the observation length. Statistical data used to forecast the future value of $x_{T_0+\tau} \in \mathbb{R}^1$ for a given predictor value $z_{T_0+\tau} \in [0, 1]$ will be represented as a $(T_0 \times 2)$-matrix:

$$Y_{T_0} = \begin{pmatrix} z_1 & x_1 \\ \vdots & \vdots \\ z_{T_0} & x_{T_0} \end{pmatrix} \in \mathbb{R}^{T_0 \times 2}.$$

Two predictor models are known, the *model of random predictors (MRP)* and the *model of deterministic predictors (MDP)*. In the *MRP model*, we assume that $\{z_t\}$ is an observed random sample from a probability distribution with a distribution function $G(z) = \mathbb{P}\{z_t < z\}$ and a probability density function $g(z) = G'(z)$, where $z \in [0, 1]$; in this model Y_{T^0} is a random sample of size T_0 from the joint bivariate probability distribution of the random vector (z_t, x_t) with a probability density

$$p(z, x) = g(z)p(x \mid z).$$

In the *MDP model* the predictor values $\{z_t\}$ are nonrandom, and randomness of $\{x_t\}$ is only due to random observation errors $\{u_t\}$.

In order to consider MDP as a special case of MRP, let us construct an empirical distribution function for the "sample" $\{z_t\}$ [6]:

$$G_{T_0}(z) = \frac{1}{T_0} \sum_{t=1}^{T_0} \mathbf{1}(z - z_t),$$

and assume that the following uniform convergence holds as $T_0 \to \infty$:

$$\sup_{z \in [0,1]} |G_{T_0}(z) - G(z)| \to 0,$$

where $G(\cdot)$ is some absolutely continuous probability distribution function. An example of such "pseudorandom" sample $\{z_t\}$ is

$$z_t = G^{-1}\left(\frac{t - 0.5}{T_0}\right), \quad t = 1, \ldots, T_0.$$

In order to construct a plug-in forecasting statistic under PU-NP prior uncertainty, we are going to use *nonparametric estimation of the regression function* $F(z)$ based on statistical data Y_{T^0}.

Definition 5.3. The Nadaraya–Watson nonparametric kernel estimator of the regression function $F(\cdot)$ is defined as the following statistic [6]

$$\widehat{F}(z) = F_0(z; Y_{T_0}) = \frac{\sum_{t=1}^{T_0} K_h(z - z_t)x_t}{\sum_{t=1}^{T_0} K_h(z - z_t)}, \quad z \in [0, 1], \tag{5.49}$$

$$K_h(u) = \frac{1}{h} K\left(\frac{u}{h}\right), \quad u \in \mathbb{R}, \tag{5.50}$$

where $K(\cdot)$ is the kernel—a continuous bounded even function satisfying the normalization condition

$$\int_{-\infty}^{+\infty} K(u)du = 1; \tag{5.51}$$

the number $h = h_{T_0} > 0$ is a scaling multiplier known as the smoothing parameter.

Let us define several kernel functions which are commonly used in applications:

- rectangular kernel

$$K(u) = \frac{1}{2} \mathbf{1}_{[-1,+1]}(u); \tag{5.52}$$

- Gauss kernel

$$K(u) = \frac{1}{\sqrt{2\pi}} e^{-u^2/2}; \tag{5.53}$$

- Epanechnikov kernel

$$K(u) = \frac{3}{4}\left(1 - u^2\right) \mathbf{1}_{[-1,+1]}(u). \tag{5.54}$$

The following asymptotic properties of the Nadaraya–Watson estimator follow directly from the definition (5.49):

(1) if $h \to 0$, the kernel $K(u)$ has a global maximum at $u = 0$, and all of the values $\{z_1, \ldots, z_{T_0}\}$ are different, then for $t \in \{1, \ldots, T_0\}$ we have

$$\widehat{F}(z_t) \longrightarrow \frac{K(0)x_t}{K(0)} = x_t,$$

i.e., for $h \to 0$ we achieve "perfect forecasting" of the observed training data Y_{T_0};

(2) if $h \to +\infty$, then $K\left((z - z_t)/h\right) \to K(0)$, and

$$\widehat{F}(z) \longrightarrow \frac{1}{T_0} \sum_{t=1}^{T_0} x_t = \overline{x},$$

i.e., as $h \to +\infty$, the regression function estimator tends to the sample mean.

Theorem 5.9. *Under a nonparametric regression model with a random predictor* (5.47), (5.48), *assume that a plug-in nonparametric kernel forecast has been constructed:*

$$\widehat{x}_{T_0+\tau} = \widehat{F}(z_{T_0+\tau}) = F_0(z_{T_0+\tau}; Y_{T_0}), \quad \tau \in \mathbb{N}, \tag{5.55}$$

where $F_0(\cdot)$ is defined by (5.49), *and the following conditions hold:*

(B$_1$) *the regression function $F_0(\cdot)$ is continuous and bounded;*

(B$_2$) *the probability density function $g(z)$ of the random predictor z_t is continuous and bounded, and $g(z_{T_0+\tau}) > 0$;*

(B$_3$) $K(-u) = K(u)$, $\sup |K(u)| \leq c_0 < +\infty$,

$$\int_{-\infty}^{+\infty} |K(u)| du < +\infty;$$

(B$_4$) *for $T_0 \to +\infty$, the smoothing parameter satisfies*

$$h_{T_0} \to 0, \quad T_0 h_{T_0} \to +\infty.$$

Then the forecast (5.55), (5.49) *is consistent:*

$$\widehat{x}_{T_0+\tau} - \widehat{F}(z_{T_0+\tau}) \xrightarrow[T_0\to+\infty]{\mathbf{P}} 0, \quad z_{T_0+\tau} \in [0,1]. \tag{5.56}$$

Proof. Denote

$$\psi(z) = F(z)g(z), \quad \widehat{g}_h(z) = \frac{1}{T_0}\sum_{t=1}^{T_0} K_h(z - z_t),$$

$$\widehat{\psi}_h(z) = \frac{1}{T_0}\sum_{t=1}^{T_0} K_h(z - z_t)x_t, \qquad z \in [0,1], \tag{5.57}$$

where $\widehat{g}_h(z)$ is the Rosenblatt–Parzen nonparametric kernel estimator [9] for the probability density function $g(z)$ of the random predictor based on the sample $\{z_t\}$. From (5.55), (5.57), and (5.49), we have the following equivalent representation of the forecasting statistic:

$$\widehat{x}_{T_0+\tau} = \widehat{F}(z_{T_0+\tau}) = \frac{\widehat{\psi}_{h_{T_0}}(z_{T_0+\tau})}{\widehat{g}_{h_{T_0}}(z_{T_0+\tau})}. \tag{5.58}$$

Let us consider the asymptotic behaviors of the numerator and the denominator in (5.58) as $T_0 \to +\infty$. It is known [9] that under the conditions B_2, B_3, B_4, the nonparametric probability density estimator $\widehat{g}_{h_{T_0}}(\cdot)$ is consistent:

$$\widehat{g}_{h_{T_0}}(z) \xrightarrow[T_0\to+\infty]{\mathbf{P}} g(z), \quad z \in C(g), \tag{5.59}$$

where $C(g) = [0,1]$ is the set of continuity points of $g(\cdot)$.

Now let us prove the convergence

$$\psi_{h_{T_0}}(z) \xrightarrow[T_0\to+\infty]{\mathbf{P}} \psi(z), \quad z \in C(F) = [0,1]. \tag{5.60}$$

By (5.57), (5.47), (5.48), and the total expectation formula, we have:

$$\mathbb{E}\{\psi_h(z)\} = \mathbb{E}\{K_h(z - z_t)x_t\} = \int_{-\infty}^{+\infty} K_h(z-u)F(u)g(u)du =$$

$$= \int_{-\infty}^{+\infty} K_h(z-u)\psi(u)du \equiv \int_{-\infty}^{+\infty} K_h(u)\psi(z-u)du. \tag{5.61}$$

Applying (5.61) and the condition B_3, let us evaluate the bias of the estimator $\widehat{\psi}_h(z)$ for an arbitrary $\delta > 0$ (by substituting $\upsilon = (z-u)/h$):

$$\left|\mathbb{E}\{\widehat{\psi}_h(z)\} - \psi(z)\right| \equiv \left| \int_{-\infty}^{+\infty} K(\upsilon)\left(\psi(z-\upsilon h) - \psi(z)\right) d\upsilon \right| \le$$

$$\le \int_{|\upsilon|\le\delta/h} (\cdot) + \int_{|\upsilon|>\delta/h} (\cdot) \le \sup_{|u|\le\delta} |\psi(z-u) - \psi(z)|\, c_0 +$$

$$+ |\psi(z)| \int_{|\upsilon|>\delta/h} |K(\upsilon)| d\upsilon + \frac{1}{\delta} \sup_{|\upsilon|>\delta/h} \left(|\upsilon|\,|K(\upsilon)|\right) \int_{-\infty}^{+\infty} |\psi(u)| du. \tag{5.62}$$

In the right-hand side of (5.62), the first summand tends to zero due to the free choice of $\delta > 0$; the second summand tends to zero due to boundedness of $\psi(z)$ (which follows from (5.57) and the conditions B_1, B_2); the third summand tends to zero due to condition B_3 and boundedness of the integral

$$\int_{-\infty}^{+\infty} |\psi(u)| du = \int_{-\infty}^{+\infty} |F(u)| g(u) du \le c_1,$$

since condition B_1 implies that $|F(u)| \le c_1 < +\infty$. Thus, we have

$$\mathbb{E}\left\{\widehat{\psi}_h(z) - \psi(z)\right\} = o(1). \tag{5.63}$$

Similarly to (5.62) and (5.63), the independence of $\{(x_t, z_t)\}$ allows us to write the variance as

$$\mathbb{D}\{\widehat{\psi}_h(z)\} = \frac{1}{T_0} \int_{-\infty}^{+\infty} K_h^2(z-u) \left(F^2(u) + \sigma^2\right) g(u) du - \frac{1}{T_0} \left(\int_{-\infty}^{+\infty} K_h(z-u) \psi(u) du \right)^2 =$$

$$= \frac{1}{T_0 h} \int_{-\infty}^{+\infty} K^2(u) \left(F^2(z+uh) + \sigma^2\right) g(z+uh) du + o\left(\frac{1}{T_0 h}\right).$$

Now, by applying the conditions B_1–B_4 and using the asymptotic analysis technique which led to (5.63), we obtain the following asymptotic expansion of the variance:

$$\mathbb{D}\left\{\widehat{\psi}_h(z)\right\} = \frac{1}{T_0 h} \left(F^2(z) + \sigma^2\right) g(z) \int_{-\infty}^{+\infty} K^2(u) du + o\left(\frac{1}{T_0 h}\right) \to 0. \tag{5.64}$$

The relations (5.63) and (5.64) imply that the variance of the estimator $\widehat{\psi}_h(z)$ tends to zero:

$$\mathbb{E}\left\{(\widehat{\psi}_h(z) - \psi_h(z))^2\right\} = \mathbb{D}\left\{\widehat{\psi}_h(z)\right\} + \mathbb{E}^2\left\{\widehat{\psi}_h(z) - \psi_h(z)\right\} \to 0,$$

i.e., the following mean square convergence holds:

$$\widehat{\psi}_h(z) \xrightarrow{\text{m.s.}} \psi(z), \quad z \in [0, 1].$$

Now the relations between different types of stochastic convergence [12] allow us to obtain (5.60).

From (5.59), (5.60), as well as continuity and boundedness of the functional transformation (5.57) (since by B_2 we have $g(z_{T_0+\tau}) > 0$), we can write:

$$\widehat{x}_{T_0+\tau} - F(z_{T_0+\tau}) \xrightarrow{\mathbf{P}} \frac{F(z_{T_0+\tau})g(z_{T_0+\tau})}{g(z_{T_0+\tau})} - F(z_{T_0+\tau}) = 0,$$

which coincides with (5.56). □

Let us rewrite the nonparametric estimator (5.49) in a generalized form:

$$\widehat{F}(z) = \sum_{t=1}^{T_0} w_{T_0 t} x_t, \quad w_{T_0 t} = \frac{K_{h_{T_0}}(z - z_t)}{\sum_{t=1}^{T_0} K_{h_{T_0}}(z - z_t)} = \frac{K_{h_{T_0}}(z - z_t)}{T_0 \widehat{g}_{h_{T_0}}(z)}, \tag{5.65}$$

where $\{w_{T_0 t}\}$ are weight coefficients satisfying the normalization condition

$$w_{T_0 1} + \cdots + w_{T_0 T_0} \equiv 1.$$

Like the Nadaraya–Watson estimator, certain other nonparametric kernel estimators can be represented in the "weighted" form (5.65):

- Greblicki estimator [5] (used if the probability density $g(\cdot)$ is a priori known):

$$w_{T_0 t} = K_{h_{T_0}}(z - z_t)/\big(T_0 g(z)\big); \tag{5.66}$$

- Priestley–Chao estimator [10]:

$$w_{T_0 t} = (z_t - z_{t-1}) K_{h_{T_0}}(z - z_t), \quad z_0 = 0; \tag{5.67}$$

- Gasser–Müller estimator [4]:

$$w_{T_0 t} = \int_{S_{t-1}}^{S_t} K_{h_{T_0}}(z - u) du, \quad S_t = \frac{z_{(t)} - z_{(t+1)}}{2}, \tag{5.68}$$

where $0 = z_{(0)} \le z_{(1)} \le \cdots \le z_{(T_0)} \le z_{(T_0+1)} = 1$ are order statistics of $\{z_t\}$.

As a rule, weight coefficients (5.66)–(5.68) are used in the MDP case [6]. The consistency of the forecasting statistic (5.65) can also be proved for weight coefficients (5.66)–(5.68).

To conclude the section, let us consider the mean square risk of the nonparametric forecasting statistic (5.65):

$$r = r(T_0, \tau) = \operatorname{Var}\left\{\widehat{x}_{T_0+\tau}\right\} = \mathbb{E}\left\{(\widehat{x}_{T_0+\tau} - x_{T_0+\tau})^2\right\} \geq 0. \tag{5.69}$$

We are going to use the following auxiliary theorem [4].

Theorem 5.10. *Assume that in a nonparametric regression model* (5.47), (5.48) *with a deterministic predictor, a nonparametric regression function estimator* (5.65) *is used and that the following conditions are satisfied:*

(C_1) *the regression function $F(\cdot)$ is doubly continuously differentiable;*
(C_2) *the kernel $K(u)$ is a bounded function with a finite $[-1, +1]$ support;*
(C_3) *for $T_0 \to +\infty$, we have*

$$\max_{2 \leq t \leq T_0} (z_{(t)} - z_{(t-1)}) = O(T_0^{-1});$$

(C_4) *for $T_0 \to +\infty$, we have*

$$h_{T_0} \to 0, \quad T_0 h_{T_0} \to +\infty.$$

Then the mean square error of the estimator for the regression function at any point $z \in [-1, +1]$ satisfies the asymptotic expansion

$$\mathbb{E}\left\{(\widehat{F}(z) - F(z))^2\right\} = \frac{\sigma^2 c_k}{T_0 h_{T_0}} + \frac{h_{T_0}^4}{4} d_k^2 \left(F''(z)\right)^2 + o\left(h_{T_0}^4 + \frac{1}{T_0 h_{T_0}}\right),$$

where c_k, d_k are constants dependent on the kernel:

$$c_k = \int_{-1}^{+1} K^2(u)du > 0, \quad d_k = \int_{-1}^{+1} u^2 K(u)du > 0.$$

Corollary 5.4. *The mean square forecast risk satisfies the asymptotic expansion*

$$r(T_0, \tau) = \sigma^2 \left(1 + \frac{c_k}{T_0 h_{T_0}}\right) + h_{T_0}^4 \frac{d_k^2 \left(F''(z_{T_0+\tau})\right)^2}{4} + o\left(h_{T_0}^4 + \frac{1}{T_0 h_{T_0}}\right). \tag{5.70}$$

Proof. By (5.47), (5.48), (5.55), and (5.65), taking into account the independence of $\{x_1, \ldots, x_{T_0}\}$ and $u_{T_{0+\tau}}$, the expression (5.69) can be rewritten as

$$r(T_0, \tau) = \mathbb{E}\left\{\left(\widehat{F}(z_{T_{0+\tau}}) - (F(z_{T_{0+\tau}}) + u_{T_{0+\tau}})\right)^2\right\} \equiv$$

$$\equiv \mathbb{E}\left\{\left((\widehat{F}(z_{T_{0+\tau}}) - F(z_{T_{0+\tau}}) - u_{T_{0+\tau}}\right)^2\right\} =$$

$$= \sigma^2 + \mathbb{E}\left\{\left(\widehat{F}(z_{T_{0+\tau}}) - F(z_{T_{0+\tau}})\right)^2\right\}.$$

Substituting the asymptotic expansion of Theorem 5.10 leads to (5.70). □

Corollary 5.5. *Assume that the smoothing parameter in the forecasting statistic* (5.58) *is chosen to satisfy the condition* C_4, *and that it has the following form:*

$$h_{T_0} = cT^{-\alpha}, \quad c > 0, \ 0 < \alpha < 1, \tag{5.71}$$

then the asymptotically fastest convergence of the risk $r(T_0, \tau)$ *as* $T_0 \to +\infty$ *is attained for* $\alpha^* = 1/5$:

$$r(T_0, \tau) = \sigma^2 + O(T_0^{-4/5}). \tag{5.72}$$

Proof. Substituting (5.71) in the expansion (5.70), we obtain

$$r(T_0, \tau) = \sigma^2 + O(T_0^{-1+\alpha}) + O(T_0^{-4\alpha}) + o(T_0^{-1+\alpha} + T_0^{-4\alpha}).$$

The asymptotically fastest convergence of the risk is obtained when the orders of decreasing are the same for the second and the third summand: $-1 + \alpha = 4\alpha$, and thus we obtain $\alpha^* = 1/5$ and (5.72). □

Note that for an M-variate predictor, the expansion (5.72) has the following general form:

$$r(T_0, \tau) = \sigma^2 + O\left(T_0^{-\frac{4}{M+4}}\right).$$

Thus, as the number of regressors M increases (i.e., as the model becomes more complex), the asymptotic order of convergence of the risk becomes lower. In the PU-P case, when parametric regression forecasting is used (see Sect. 5.2), the order of convergence is higher:

$$r(T_0, \tau) = \sigma^2 + O(T_0^{-1}).$$

In conclusion, let us note that the computational complexity of the nonparametric forecasting algorithm (5.55) can be estimated as $O(T_0)$. For large values of T_0, special measures must be taken to make the computations feasible:

(1) choosing a kernel $K(\cdot)$ with a bounded support;
(2) restricting the values of the predictor z to some finite grid;
(3) using FFT (Fast Fourier Transform) techniques.

5.5 Nonparametric $k - NN$-Regression Forecasting

Under the same nonparametric regression model (5.47), (5.48) as in Sect. 5.4, let us apply a different nonparametric technique to estimate the regression function in the forecasting algorithm (5.55).

Definition 5.4. Assume that $k \in \{1, \ldots, T_0\}$ is a positive integer, $z \in \mathbb{R}^M$ is an arbitrary value of the regressor (predictor) vector, $\varrho_t = \varrho(z_t, z) = |z_t - z|$ is the Euclidean distance between points z_t, $z \in \mathbb{R}^M$, the order statistics of $\{\varrho_t\}$ are denoted as $0 \leq \varrho_{(1)} \leq \varrho_{(2)} \leq \cdots \leq \varrho_{(T_0)}$, and the set

$$J_k(z) = \{t \in \{1, \ldots, T_0\} : \varrho(z_t, z) \leq \varrho_{(k)}\} \tag{5.73}$$

consists of time points corresponding to the k predictor values in $\{z_t\}$ that are closest to z, $|J_k(z)| = k$; then the $k - NN$*-estimator*, or the k nearest neighbors estimator of the regression function $F(\cdot)$ at the point z is defined as follows [6]:

$$\widehat{F}(z) = \frac{1}{k} \sum_{t \in J_k(z)} x_t. \tag{5.74}$$

To illustrate this definition, let us consider the following two special cases:

1. If $k = T_0$, then by (5.73), (5.74) we have $J_{T_0}(z) = \{1, \ldots, T_0\}$ and

$$\widehat{F}(z) = \frac{1}{T_0} \sum_{t=1}^{T_0} x_t = \overline{x}$$

is the sample mean of the entire observed sample. Thus, in that case we have "complete smoothing" of the data leading to a trivial forecast $\widehat{x}_{T_0+\tau} = \overline{x}$. Note that this case is equivalent to kernel forecasting for $h \to +\infty$ (see Sect. 5.3).
2. If $k = 1$, then by (5.73), (5.74) we have $\widehat{F} = x_{t_*(z)}$, where $t_*(z)$ is the number of the observation in the sample $\{z_1, \ldots, z_{T_0}\}$ that lies the closest to z, written formally as $t_*(z) = \arg\min_{1 \leq t \leq T_0} |z_t - z|$. The forecast is then made using the nearest neighbor criterion:

$$\widehat{x}_{T_0+\tau} = x_{t_*(z_{T_0+\tau})}.$$

In the general case, the forecast (5.73), (5.74) is averaged over k nearest neighbors of $z_{T_0+\tau}$.

We are going to use the following auxiliary result (see [6] for a detailed proof).

Theorem 5.11. *Under the nonparametric regression model* (5.47), (5.48) *with* $d = M = 1$ *and a random predictor with a differentiable probability distribution* $g(z)$, $z \in \mathbb{R}^1$, *assume that the regression function* $F(\cdot)$ *is twice differentiable. If we have* $T_0 \to +\infty$, $k = k_{T_0} \to +\infty$, *and also* $k_{T_0}/T_0 \to 0$, *then for all* $z : g(z) > 0$ *the bias and the variance of the* $k - NN$*-estimator* (5.73), (5.74) *satisfy the following asymptotic expansions:*

$$\mathbb{E}\left\{\widehat{F}(z) - F(z)\right\} = \frac{F''(z)g(z) + 2F'(z)g'(z)}{24g^3(z)}\left(\frac{k_{T_0}}{T_0}\right)^2 + o\left(\left(\frac{k_{T_0}}{T_0}\right)^2\right),$$

$$\mathbb{D}\left\{\widehat{F}(z)\right\} = \frac{\sigma^2}{k_{T_0}} + o\left(\frac{1}{k_{T_0}}\right). \tag{5.75}$$

From (5.75) we can see that under the conditions of this theorem, the $k - NN$-estimator is consistent.

Corollary 5.6. *If* $k_{T_0} = T_0^{\gamma}$, $0 < \gamma < 1$, *then the asymptotically smallest mean square error of the* $k - NN$*-estimator* (5.73), (5.74) *is attained for* $\gamma^* = 4/5$*:*

$$\mathbb{E}\left\{\left(\widehat{F}(z) - F(z)\right)^2\right\}\left\{\widehat{F}(z)\right\} = O(T_0^{-4/5}). \tag{5.76}$$

Proof. We have

$$\mathbb{E}\left\{(\widehat{F}(z) - F(z))^2\right\} = \mathbb{D}\left\{\widehat{F}(z)\right\} + \mathbb{E}^2\left\{\widehat{F}(z) - (F(z)\right\}.$$

Substituting (5.75) into this expression leads to

$$\mathbb{E}\left\{\left(\widehat{F}(z) - F(z)\right)^2\right\} = c_1\left(\frac{k_{T_0}}{T_0}\right)^4 + c_2\frac{1}{k_{T_0}} + o\left(\frac{1}{k_{T_0}} + \left(\frac{k_{T_0}}{T_0}\right)^4\right) =$$
$$= c_1 T_0^{4(\gamma-1)} + c_2 T^{-\gamma} + o\left(T^{4(\gamma-1)} + T^{-\gamma}\right),$$

where c_1, c_2 are constants independent of T_0. The asymptotically smallest sum is obtained when the exponents are equal: $4(\gamma - 1) = -\gamma$, which leads to $\gamma^* = 4/5$ and proves (5.76). □

Corollary 5.7. *Under the conditions of Corollary 5.6, the asymptotically smallest risk is obtained for* $\gamma^* = 4/5$*:*

$$r(T_0, \tau) = \mathbb{E}\left\{(\widehat{x}_{T_0+\tau} - x_{T_0+\tau})^2\right\} = \sigma^2 + O\left(T_0^{-4/5}\right). \tag{5.77}$$

By comparing (5.77) and (5.72), we can see that for $T_0 \to \infty$ the asymptotics of the minimum risk values $r(T_0, T)$ are the same for the kernel estimator and the $k - NN$-estimator.

In addition to (5.73), (5.74), let us introduce another $k - NN$-estimator [6], which is quite similar to the nonparametric kernel estimator described in Sect. 5.4:

$$\widehat{F}(z) = \frac{\sum\limits_{t \in J_k(z)} K\left((z - z_t)/\varrho_{(k)}\right) x_t}{\sum\limits_{t \in J_k(z)} K\left((z - z_t)/\varrho_{(k)}\right)}, \tag{5.78}$$

where the set of k nearest neighbors $J_k(z)$ and the distance $\varrho_{(k)}$ between the point z and the kth nearest neighbor are defined by relations (5.73); the kernel $K(u)$ of this estimator is a continuous bounded even function satisfying the normalization property (5.51). In particular, for a rectangular kernel $K(u)$ defined as

$$K(u) = \begin{cases} 0.5, & u \in [-1, 1], \\ 0, & \text{otherwise}, \end{cases}$$

the nonparametric estimator (5.78) becomes the estimator (5.74) considered earlier. Thus, the $k - NN$-estimator defined by (5.78) is:

(1) a generalization of the simplest $k - NN$-estimator (5.74) for the case of an arbitrary kernel;
(2) a generalization of the kernel estimator (5.49) for the case where the smoothing parameter h depends on statistical data, i.e., is a statistic: $h_{T_0} = \varrho_{(k)}$.

The paper [6] proves the consistency of the estimator (5.78) and presents asymptotic expansions of its bias and variance similar to (5.75).

Let us briefly mention another type of nonparametric $k - NN$-estimators [6]:

$$F(z) = \frac{1}{T_0 h_{T_0}} \sum_{t=1}^{T_0} K\left(\frac{G_{T_0}(z_t) - G_{T_0}(z)}{h_{T_0}}\right) x_t,$$

where $G_{T_0}(z)$ is the empirical distribution function of the "sample" $\{z_t\}$ defined in Sect. 5.4. Compared to (5.49), this definition uses a metric based on the empirical distribution function instead of the Euclidean metric $|z_t - z|$ to express the distance between the points z_t and z.

In conclusion, note that for $k < T_0$ the $k - NN$-estimator is less computationally intensive than the kernel estimator.

5.6 Some Other Nonparametric Regression Forecasting Methods

This section reviews some other methods of nonparametric regression forecasting not mentioned in Sects. 5.4, 5.5. A more detailed treatment of these methods can be found in [6].

5.6.1 *Functional Series Expansions of Regression Functions*

Assume that an orthonormal system of basis functions $\{\phi_0(z), \phi_1(z), \dots\}$ is defined on $[-1, +1]$, and that the regression function $F(\cdot)$ allows for a functional series expansion:

$$F(z) = \sum_{j=0}^{+\infty} \beta_j \phi_j(z), \quad z \in [-1, +1]. \tag{5.79}$$

As before, a nonparametric regression forecast at time $T_0 + \tau$ is a statistic

$$\widehat{x}_{T_0+\tau} = \widehat{F}(z_{T_0+\tau}), \tag{5.80}$$

where $\widehat{F}(\cdot)$ is a nonparametric estimator of the regression function $F(\cdot)$ defined as an L_{T^0} partial sum of the functional series (5.79):

$$\widehat{F}(z) = \sum_{j=0}^{L_{T_0}} \widehat{\beta}_j \phi_j(z), \quad \widehat{\beta}_j = \sum_{t=1}^{T} x_t \int_{A_t} \phi_j(u) du. \tag{5.81}$$

Here $A_1, \dots, A_{T_0}$ is some partition of the regressor domain into T_0 nonoverlapping intervals:

$$[-1, +1] = \bigcup_{t=1}^{T_0} A_t; \quad A_t \cap A_{t'} = \emptyset, \quad t \neq t',$$

with $z_t \in A_t$, $t = 1, \dots, T_0$. Let us give an example of such partition if $\{z_t\}$ are ordered:

$$A_1 = [-1, z_1], \; A_2 = (z_1, z_2], \; \dots, \; A_{T_0} = (z_{T_0-1}, +1].$$

The parameter L_{T_0} has the same meaning as the smoothing parameter h_{T^0} in the case of kernel estimators. The basis functions can be chosen to be polynomials of

Legendre, Hermite, Lagger, Fourier, etc. For instance, Legendre polynomials, which are written as

$$\phi_0(z) = \frac{1}{\sqrt{2}}, \quad \phi_1(z) = \sqrt{\frac{3}{2}}z, \quad \phi_2(z) = \sqrt{\frac{5}{2}}\frac{1}{2}\left(3z^2 - 1\right), \ldots ,$$

are defined by the following well-known recurrence relation:

$$(l+1)\phi_{l+1}(z) = (2l+1)z\phi_l(z) - l\phi_{l-1}(z), \quad l = 1, 2, \ldots .$$

The following result has been proved in [6].

Theorem 5.12. *If for some* $0 < s < 1$ *we have*

$$\mathbb{E}\left\{|u_t|^{1+\frac{1}{s}}\right\} < +\infty,$$

and the asymptotics $T_0 \to +\infty$, $L_{T_0} \to +\infty$ *are such that*

$$T_0^{s-1} \sum_{j=0}^{L_{T_0}} \sup_{-1 \le z \le +1} \phi_j^2(z) < +\infty, \tag{5.82}$$

then the nonparametric estimator (5.81) *is consistent:*

$$\widehat{F}(z) \xrightarrow{\mathbf{P}} F(z), \quad z \in [-1, +1].$$

The following asymptotic relation is satisfied as $j \to +\infty$ for many practical orthonormal systems of functions $\{\phi_j(\cdot)\}$:

$$\sup_{-1 \le z \le +1} \phi_j^2(z) = O(j^{\varrho}),$$

$$\varrho = \begin{cases} -1/4 & \text{for Hermite and Lagger polynomials,} \\ 0 & \text{for Fourier polynomials,} \\ 1/2 & \text{for Legendre polynomials.} \end{cases}$$

The condition (5.82) becomes an asymptotic upper limit on L_{T_0} as $T_0 \to +\infty$:

$$L_{T_0} = O\left(T_0^{\frac{1-s}{1+2\varrho}}\right).$$

For example, a system of Legendre polynomials leads to the condition

$$L_{T_0} = O\left(T_0^{\frac{1-s}{2}}\right),$$

and for $s = 1/2$ we can write

$$L_{T_0} = O\left(T_0^{1/4}\right).$$

5.6.2 Spline Smoothing

If $\widetilde{F}(\cdot)$ is a nonparametric estimator of the regression function $F(\cdot)$, then the measure of fitting between the collected statistical data $\{(x_t, z_t) : t = 1, \ldots, T_0\}$ and the estimator $\widetilde{F}(\cdot)$ is the so-called residual sum of squares:

$$S^2\left(\widetilde{F}(\cdot)\right) = \sum_{t=1}^{T_0} \left(x_t - \widetilde{F}(z_t)\right)^2 \geq 0. \quad (5.83)$$

The smaller $S^2(\widetilde{F}(\cdot))$ is, the better the fitting. Taking (5.83) as the objective functional and minimizing it over $\widetilde{F}(\cdot)$ lying in the set of all real functions, we are obviously going to find infinitely many estimators $\widetilde{F}(\cdot)$ with $S^2 = 0$. However, these functions $\widetilde{F}(\cdot)$ turn out to be extremely nonsmooth and result in inefficient forecasts (the forecast risk is too high). To overcome this difficulty, Härdle [6] proposed to impose a "roughness penalty" on $\widetilde{F}(\cdot)$:

$$\int_{-1}^{+1} \left(\widetilde{F}''(z)\right)^2 dz \geq 0. \quad (5.84)$$

In the same monograph, Härdle has also proposed to use a different optimality criterion for $\widetilde{F}(\cdot)$ by introducing weights in the functionals (5.83), (5.84):

$$S_\lambda\left(\widetilde{F}(\cdot)\right) = \sum_{t=1}^{T_0} \left(x_t - \widetilde{F}(z_t)\right)^2 + \lambda \int_{-1}^{+1} \left(\widetilde{F}''(z)\right)^2 dz \to \min_{\widetilde{F}(\cdot)}, \quad (5.85)$$

where $\lambda = \lambda_{T_0} \geq 0$ is a smoothing parameter.

In [6] it is proved that the optimization problem (5.85) has a unique solution in the class of doubly continuously differentiable functions $\widetilde{F}(\cdot)$—the cubic spline $\widetilde{F}^*(\cdot)$, which is characterized by the following properties:

(1) in each interval $[z_{(t)}, z_{(t+1)}]$, where $t = 1, 2, \ldots, T_0 - 1$, the function $\widetilde{F}^*(\cdot)$ is a polynomial of the third degree;
(2) the function $\widetilde{F}^*(\cdot)$, as well as its 1st and 2nd order derivatives, is continuous at interval boundary points $\{z_t\}$; this condition isn't imposed on higher order derivatives;
(3) the 2nd order derivative $\widetilde{F}^{*\prime\prime}(\cdot)$ is equal to zero at boundary points $z_{(1)}$ and $z_{(T_0)}$.

Finding the optimal value of the smoothing parameter $\lambda^*_{T_0}$ is described in [6].

5.6.3 Regressograms and Median Smoothing

Similarly to the definition of a histogram, the range A of a regressor z ($z \in A \subseteq \mathbb{R}^M$) can be split into L bins:

$$A = \bigcup_{i=1}^{L} A_i, \quad A_i \cap A_j = \emptyset, \ i \neq j.$$

Definition 5.5. A regressogram (nonparametric regressor estimator) is defined as the following statistic (a piecewise constant function) [13]:

$$\widehat{F}(z) = \sum_{i=1}^{L} \nu_i \mathbf{1}_{A_i}(z), \quad \nu_i = \frac{\sum_{t=1}^{T_0} x_t \mathbf{1}_{A_i}(z_t)}{\sum_{t=1}^{T_0} \mathbf{1}_{A_i}(z_t)}.$$

Finally, let us mention nonparametric regression function estimators obtained by median smoothing [6]:

$$\widehat{F}(z) = \mathrm{Med}\left\{x_t : t \in J_k(z)\right\},$$

where $\mathrm{Med}\{a_1, \ldots, a_s\}$ is the sample median, and k, $J_k(z)$ are defined as in (5.73).

References

1. Anderson, T.: An Introduction to Multivariate Statistical Analysis. Wiley, Hoboken (2003)
2. Borovkov, A.: Mathematical Statistics. Gordon & Breach, Amsterdam (1998)
3. Borovkov, A.: Probability Theory. Gordon & Breach, Amsterdam (1998)
4. Gasser, T., Müller, H.: Estimating regression functions and their derivatives by the kernel method. Scand. J. Stat. **11**, 171–185 (1984)
5. Greblicki, W., Krzyzak, A.: Asymptotic properties of kernel estimates of a regression function. J. Stat. Plann. Infer. **4**, 81–90 (1980)
6. Haerdle, W.: Applied Nonparametric Regression. Cambridge University Press, Cambridge (1993)
7. Koroljuk, V.: Handbook on Probability Theory and Mathematical Statistics (in Russian). Nauka, Moscow (1985)
8. Pampel, F.: Logistic Regression: A Primer. Thousand Oaks, Sage (2000)
9. Parzen, E.: On the estimation of a probability density function and the mode. Ann. Math. Stat. **40**, 1063–1076 (1962)
10. Priestley, M.: Spectral Analysis and Time Series. Academic, New York (1999)
11. Seber, G., Lee, A.: Linear Regression Analysis. Wiley-Interscience, Hoboken (2003)
12. Shiryaev, A.: Probability. Springer, New York (1996)
13. Tukey, J.: Curves as parameters and touch estimation. In: Proceedings of the 4th Berkeley Symposium, pp. 681–694 (1961)

Chapter 6
Robustness of Time Series Forecasting Based on Regression Models

Abstract This chapter presents a robustness analysis of the forecasting statistics introduced in the previous chapter under the following distortion types: four functional distortion varieties of the regression function, additive outliers, and correlation between random errors. A quantitative characterization of forecasting robustness is obtained by using the robustness indicators introduced in Chap. 4, namely the forecast risk instability coefficient and the δ-admissible distortion level. Robust forecasting statistics are constructed by using Huber estimators and a specially chosen type of M-estimators for the regression function parameters. A local-median forecasting algorithm is proposed to mitigate the influence of outliers under regression models, and its robustness is evaluated.

6.1 Robustness of Least Squares Forecasting Under Functional Distortions of Multiple Linear Regression Models

6.1.1 *Formulation of the Problem*

As discussed in Sect. 5.2, construction of parametric regression models is one of the most widely used approaches to forecasting dynamic behavior of stochastic systems in engineering, economics, medicine, environmental studies, and other disciplines. This approach can be summarized as follows:

(1) a certain hypothetical parametric model of the regression function (which is usually linear in its parameters) is postulated, establishing the stochastic dependence between the dependent (endogenous) variables and the nonrandom independent (exogenous) variables (regressors or predictors); the true value of the parameter vector $\theta^0 \in \mathbb{R}^m$ of this hypothetical regression model is assumed to be unknown;

Y. Kharin, *Robustness in Statistical Forecasting*, DOI 10.1007/978-3-319-00840-0_6,

(2) based on observations $x_1, \ldots, x_T$ made until the time point $T > m$, a least squares estimator $\hat{\theta} \in \mathbb{R}^m$ for the regression function parameters is constructed;
(3) the obtained estimator $\hat{\theta}$ is substituted into the regression function, and for given values of the regressors at a future time point $T + \tau$ $(\tau \geq 1)$, a forecast of the dependent variables is computed.

This approach yields accurate results only in a setting where the hypothetical regression model of observations conforms exactly to the real-world data. Unfortunately, in applications the hypothetical model assumptions are usually distorted [1,7,9,11,13,14,22,33]:

(A_1) the hypothetical regression function is "slightly" distorted;
(A_2) observations are contaminated by outliers;
(A_3) the probability distribution of random observation errors is non-normal;
(A_4) random observation errors are dependent (or correlated);
(A_5) model parameters change with time (parameter drift).

Distortion types A_2, A_3 have been discussed in [1, 7, 11, 13, 33] and Sect. 6.2 of this book; distortion type A_4 has been investigated w.r.t. regression model identification in [9, 10, 23], as well as in Sect. 6.3; distortion type A_5 has been analyzed in [23,27] and Chap. 9. This section is devoted to distortion type A_1; some basic results related to this distortion type can be found in [14].

We are going to investigate the effect of distortion type A_1 (functional distortion, FD) on the forecast risk of the mean squares multiple linear regression forecasting and evaluate its robustness characteristics. Forecasting algorithms that are robust under FD will be constructed and analyzed in Sect. 6.5.

6.1.2 The Hypothetical Regression Model and Its Functional Distortions

Assume that in the studied stochastic system, the observations $x_t \in \mathbb{R}$ satisfy the stochastic multiple linear regression equation:

$$x_t = \sum_{i=1}^{m} \theta_i^0 \psi_i(z_t) + \lambda(z_t) + u_t, \tag{6.1}$$

where $t \in \mathbb{N} = \{1, 2, \ldots\}$ is a discrete time point, $z_t \in U \subseteq \mathbb{R}^M$ is a nonrandom observed input influence (the vector of factors or independent variables) at time t, U is the set of regressor values at time points representing the design of the experiment, $\{\psi_i(\cdot) : \mathbb{R}^M \to \mathbb{R}\}$ is a given collection of m linearly independent functions, $u_t \in \mathbb{R}$ is the random error (perturbation) in the observation made at time t, $\theta^0 = (\theta_i^0) \in \mathbb{R}^m$ is the vector of m unknown true values of the model parameters, $\lambda(\cdot) : \mathbb{R}^M \to \mathbb{R}$ is an unknown deterministic function describing the FD.

The random errors $\{u_t\}$ are assumed to be jointly independent identically distributed random variables such that

$$\mathbb{E}\{u_t\} = 0, \quad \mathbb{D}\{u_t\} = \sigma^2 < +\infty. \tag{6.2}$$

By (6.1), (6.2), for $z_t = z$ the regression function can be written as

$$f(z) = f_0(z) + \lambda(z), \quad z \in U, \quad f_0(z) = \sum_{i=1}^{m} \theta_i^0 \psi_i(z), \tag{6.3}$$

where $f_0(z)$ is the hypothetical (assumed) regression function, and the error present in its definition due to lack of prior information, complexity of the system, or specific properties of the observation process is denoted as $\lambda(z) = f(z) - f_0(z)$. If $\lambda(z) \equiv 0$, then (6.1)–(6.3) define the hypothetical (undistorted) model.

Models of the form (6.1)–(6.3) are classified as so-called semiparametric regression models [10]. Note that taking $z_t \equiv t$ yields the trend model, which is the most commonly used forecasting model [4, 6, 8, 24].

Regression forecasting is construction of a forecast $\widehat{x}_{T+\tau} \in \mathbb{R}$ for the unknown value $x_{T+\tau} \in \mathbb{R}$ at a future time point $t = T + \tau$, where $\tau \geq 1$ is the given "forecasting horizon," based on T observations $X_T = (x_1, x_2, \dots, x_T)' \in \mathbb{R}^T$ and predictor values $z_1, \dots, z_T, z_{T+\tau} \in U$.

Let us define four types of FD $\lambda(\cdot)$ in (6.1), (6.3).

FD-1. *Interval distortion:*

$$\varepsilon_-(z) \leq \lambda(z) \leq \varepsilon_+(z), \quad z \in U,$$

where $\varepsilon_\pm(z)$ are known boundary functions; in particular, for $\varepsilon_\pm(z) = \pm\varepsilon$ we have:

$$-\varepsilon \leq \lambda(z) \leq +\varepsilon, \quad z \in U, \tag{6.4}$$

where $\varepsilon \geq 0$ is the distortion level in the C-metric.

FD-2. *Relative distortion:*

$$\frac{|\lambda(z)|}{|f_0(z)|} \leq \varepsilon, \quad z \in U,$$

where $\varepsilon \geq 0$ is the level of relative distortion in the C-metric. Thus, the relative error in the definition of the regression function doesn't exceed ε.

FD-3. *Distortion in the l_p-metric:*

$$\left(\sum_{t=1}^{T} |\lambda(z_t)|^p + |\lambda(z_{T+\tau})|^p \right)^{\frac{1}{p}} \leq \varepsilon,$$

where $p \geq 1$ is a parameter, and $\varepsilon \geq 0$ is the distortion level in the l_p-metric. This definition limits the l_p norm of the accumulated over time FD by ε.

FD-4. *Distortion represented as an orthogonal expansion*:

$$\lambda(z) = \sum_{j \in J} S_j \eta_j(z), \quad \left(\sum_{j \in J} |S_j|^p \right)^{\frac{1}{p}} \leq \varepsilon, \quad z \in U,$$

where $J \subset \mathbb{N}$ is a given index subset of size $M' = |J| < +\infty$, $\varepsilon \geq 0$ is the distortion level in the l_p-metric, $\{\eta_j(\cdot) : U \to \mathbb{R}\}$ is an orthonormal function system over U. In particular, distortion type FD-4 occurs when the hypothetical regression function is a truncation of some functional series, $f_0(z) = \sum_{j=1}^{m} \theta_j^0 \eta_j(z)$, and $\lambda(z) = \sum_{j=m+1}^{w} S_j \eta_j(z)$ is a non-negligible segment of the residual part of the series, $w \in \mathbb{N}$, $w > m$.

For each of the given distortion types, a special case of piecewise distortion can be defined by assuming prior knowledge that no distortion is present in a subset $U_0 \subset U$:

$$U = U_0 \cup U_1, \quad U_0 \cap U_1 = \emptyset, \quad \lambda(z) = 0, \; z \in U_0. \tag{6.5}$$

6.1.3 Robustness Characteristics of Forecasting Algorithms

Consider an algorithm of forecasting $x_{T+\tau} \in \mathbb{R}$ based on T observations $X_T \in \mathbb{R}^T$ and predictor values $z_1, \ldots, z_T, z_{T+\tau} \in U$, which is defined by a statistic $F(\cdot)$:

$$\widehat{x}_{T+\tau} = F(X_T; z_1, \ldots, z_T, z_{T+\tau}) \in \mathbb{R}. \tag{6.6}$$

Risk of a τ-step-ahead statistical forecast (6.6) based on T observations is defined as the mean square forecast error $r(T,\tau) = \mathbb{E}\{(\hat{x}_{T+\tau} - x_{T+\tau})^2\} \geq 0$. By (6.1), (6.3), (6.6), the risk functional has the following form:

$$r(T,\tau) = \sigma^2 + \mathbb{E}\left\{(f_0(z_{T+\tau}) + \lambda(z_{T+\tau}) - \widehat{x}_{T+\tau})^2\right\}, \tag{6.7}$$

and it depends on the distortion function $\lambda(\cdot)$. The guaranteed forecast risk will be defined as the exact upper bound of the forecast risk,

$$r_+(T,\tau) = \sup_{\lambda(z_1),\ldots,\lambda(z_T),\lambda(z_{T+\tau})} r(T,\tau), \tag{6.8}$$

where the range of the function $\lambda(\cdot)$ is defined by the FD type.

Let $r_0(T, \tau) > 0$ be the risk of the statistical forecast (6.6) in the absence of FD: $\lambda(\cdot) \equiv 0$ in (6.7). We are going to assume that in the absence of distortion, the algorithm (6.6) is consistent, and thus its risk converges from the right as observation length tends to infinity: $r_0(T, \tau) \to r_0 + 0$ for $T \to \infty$, where $r_0 = \sigma^2$ is the lowest possible risk for an a priori given regression function in the absence of distortion: $\widehat{x}_{T+\tau} = f_0(z_{T+\tau})$. Motivated by this argument, let us define the instability coefficient of the forecasting algorithm (6.6) as the relative increment of the guaranteed risk (see Sect. 4.4):

$$\kappa(T, \tau) = (r_+(T, \tau) - r_0)/r_0 \geq 0. \tag{6.9}$$

For an arbitrary $\delta > 0$ and distortion types FD-1 (defined by (6.4)) to FD-4, the δ-admissible distortion level $\varepsilon_+(\delta)$ will be defined as the highest distortion level ε such that the instability coefficient (6.9) doesn't exceed the admissible level δ:

$$\varepsilon_+(\delta) = \sup\{\varepsilon : \kappa(T, \tau) \leq \delta\}. \tag{6.10}$$

Lower values of $r_+(T, \tau)$, $\kappa(T, \tau)$ and larger values of $\varepsilon_+(\delta)$ for a fixed δ correspond to higher robustness of the forecasting algorithm (6.6) under the considered distortion type.

6.1.4 Robustness Analysis of Least Squares Forecasting

As noted earlier, in parametric forecasting the most commonly used algorithm is based on the method of least squares:

$$\widehat{x}_{T+\tau} = \hat{\theta}'\psi(z_{T+\tau}), \qquad \hat{\theta} = (\Psi_T'\Psi_T)^{-1}\Psi_T'X_T, \qquad |\Psi_T'\Psi_T| \neq 0, \tag{6.11}$$

where $\psi(z) = (\psi_i(z)) \in \mathbb{R}^m$, $\Psi_T = (\psi(z_1) \vdots \ldots \vdots \psi(z_T))'$ is a $(T \times m)$-matrix such that the matrix product $\Psi_T'\Psi_T$ is nonsingular, and $\hat{\theta} \in \mathbb{R}^m$ is the least squares estimator of the parameter vector θ^0. As usual, the prime symbol $'$ denotes matrix transpositions, and $|\cdot|$ is the matrix determinant. It is known [30] that the Eicker condition on asymptotic behavior (as $T \to \infty$) of the smallest eigenvalue of the matrix $\Psi_T'\Psi_T$ is a sufficient condition for consistency of the estimator $\hat{\theta}$:

$$\lambda_{\min}(\Psi_T'\Psi_T) \to \infty. \tag{6.12}$$

Let us introduce the following notation:

$$\Lambda_T = (\lambda(z_t)) \in \mathbb{R}^T, \qquad U_T = (u_t) \in \mathbb{R}^T,$$

$$C_T = (\Psi_T'\Psi_T)^{-1}\Psi_T', \qquad g(T, \tau) = (g_t(T, \tau)) = C_T'\psi(z_{T+\tau}) \in \mathbb{R}^T,$$

$$\mathbf{I}_A(z) = \{1,\ z \in A;\ 0,\ z \notin A\}, \qquad \mathbf{I}(z) = \mathbf{I}_{(0,+\infty)}(z), \quad (z)_+ = \max(z, 0).$$

Theorem 6.1. *Forecast risk* (6.11) *under FD* $\lambda(\cdot)$, *assuming observation model* (6.1), *can be written as*

$$r(T,\tau) = \sigma^2 \left(1 + \|g(T,\tau)\|^2\right) + \left(\lambda(z_{T+\tau}) - g'(T,\tau)\Lambda_T\right)^2. \tag{6.13}$$

Proof. Taking into account (6.11), we obtain:

$$\hat{\theta} - \theta^0 = C_T \Lambda_T + C_T U_T.$$

Since $\mathbb{E}\{U_T\} = 0$, we have $\mathbb{E}\{\hat{\theta} - \theta^0\} = C_T \Lambda_T$. Applying $C_T C_T' = (\Psi_T' \Psi_T)^{-1}$ yields

$$\mathbb{E}\left\{(\hat{\theta} - \theta^0)(\hat{\theta} - \theta^0)'\right\} = C_T \Lambda_T \Lambda_T' C_T' + \sigma^2 (\Psi_T' \Psi_T)^{-1}.$$

By (6.1), (6.7), we can write

$$\begin{aligned} r(T,\tau) &= \mathbb{E}\left\{(\hat{\theta}'\psi(z_{T+\tau}) - \theta^{0\prime}\psi(z_{T+\tau}) - \lambda(z_{T+\tau}) - \xi_{T+\tau})^2\right\} \\ &= \sigma^2 \left(1 + g'(T,\tau)g(T,\tau)\right) + \left(\lambda(z_{T+\tau}) - g'(T+\tau)\Lambda_T\right)^2, \end{aligned}$$

which coincides with (6.13). □

Corollary 6.1. *In the absence of FD,* $\lambda(\cdot) \equiv 0$, *we have*

$$r_0(T,\tau) = \sigma^2 \left(1 + \|g(T,\tau)\|^2\right) \geq r_0 = \sigma^2,$$

and if the condition (6.12) *is satisfied, then* (6.11) *is a mean squares consistent forecasting algorithm.*

From (6.13) we can see that under distortion $\lambda(\cdot)$ of the model (6.1), the forecast risk is a sum of three nonnegative components:

$$r(T,\tau) = r_0 + r_1 + r_2,$$

where $r_0 = \sigma^2$ is the forecast risk under complete prior information and no distortion, $r_1 = \sigma^2 \|g(T,\tau)\|^2$ is the risk increment due to finiteness of the observation time T, and $r_2 = (\lambda(z_{T+\tau}) - g'(T,\tau)\Lambda_T)^2$ is the risk increment due to simultaneous influence of distortion and finiteness of T.

Theorem 6.2. *Under FD-1 distortions, the guaranteed risk of the forecasting algorithm* (6.11) *equals*

$$\begin{aligned} r_+(T,\tau) &= r_0(T,\tau) + \\ &+ \max\left(\sum_{t=1}^{T}\left((g_t(T,\tau))_+ \varepsilon_\pm(z_t) - (-g_t(T,\tau))_+ \varepsilon_\mp(z_t)\right) - \varepsilon_\mp(z_{T+\tau})\right)^2. \end{aligned}$$

Proof. From (6.8), continuity (in λ) of the risk function (6.13), and compactness of the range of functions $\{\lambda(z_t)\}$ we have

$$r_+(T,\tau) = r_0(T,\tau) + Q(\lambda) =$$
$$= r_0(T,\tau) + \max_{\varepsilon_-(z_t)\leq\lambda(z_t)\leq\varepsilon_+(z_t)} \left(\lambda(z_{T+\tau}) - \sum_{t=1}^{T} g_t(T,\tau)\lambda(z_t)\right)^2 .$$

The maximum value of $Q(\lambda)$ is attained if the expression

$$\sum_{t=1}^{T} g_t(T,\tau)\lambda(z_t) - \lambda(z_{T+\tau})$$

is maximized or minimized. By varying $\lambda(z_t) \in [\varepsilon_-(z_t), \varepsilon_+(z_t)]$ to obtain the maximum, we have

$$Q_{1max} = \left(\sum_{t=1}^{T} g_t(T,\tau)\left(\mathbf{I}(g_t(T,\tau))\varepsilon_+(z_t) + \mathbf{I}(-g_t(T,\tau))\varepsilon_-(z_t)\right) - \varepsilon_-(z_{T+\tau})\right)^2 .$$

Similarly, minimization yields

$$Q_{2max} = \left(\sum_{t=1}^{T} g_t(T,\tau)\left(\mathbf{I}(g_t(T,\tau))\varepsilon_-(z_t) + \mathbf{I}(-g_t(T,\tau))\varepsilon_+(z_t)\right) - \varepsilon_+(z_{T+\tau})\right)^2 .$$

Thus, $Q_{max} = \max(Q_{1max}, Q_{2max})$. □

Corollary 6.2. *Under FD-1 distortion* (6.4), *the instability coefficient* (6.9) *and the δ-admissible distortion level* (6.10) *can be written as*

$$\kappa(T,\tau) = \|g(T,\tau)\|^2 + (\varepsilon/\sigma)^2 \left(1 + \sum_{t=1}^{T} |g_t(T,\tau)|\right)^2 ,$$
$$\varepsilon_+(\delta) = \sigma\left((\delta - \|g(T,\tau)\|^2)_+\right)^{\frac{1}{2}} \Big/ \left(1 + \sum_{t=1}^{T} |g_t(T,\tau)|\right).$$

Corollary 6.3. *Under piecewise distortion* (6.5), *we have:*

$$r_+(T,\tau) = r_0(T,\tau) + \max\left(\sum_{t:z_t\in U_1} \left((g_t(T,\tau))_+ \varepsilon_\pm(z_t) - \right.\right.$$
$$\left.\left. - (-g_t(T,\tau))_+ \, \varepsilon_\mp(z_t)\right) - \varepsilon_\mp(z_{T+\tau})\mathbf{I}_{U_1}(z_{T+\tau})\right)^2 .$$

Theorem 6.3. *Under FD-2 distortion, the guaranteed risk* (6.5) *of the forecasting algorithm* (6.11) *equals*

$$r_+(T,\tau) = r_0(T,\tau) + \varepsilon^2 \left(\left|\psi'(z_{T+\tau})\theta^0\right| + \sum_{t=1}^{T} \left|g_t(T,\tau)\theta^{0\prime}\psi(z_t)\right| \right)^2 .$$

Proof. By an argument similar to the proof of Theorem 6.2, we have:

$$r_+(T,\tau) = \max\left\{ r(T,\tau) : |\lambda(z_t)| \le \varepsilon \left|\theta^{0\prime}\psi(z_t)\right| \right\}.$$

Writing the forecast risk in the form (6.13), only the second summand of the resulting expression depends on $\{\lambda(z_t)\}$. Maximizing this summand yields

$$\left(\lambda(z_{T+\tau}) - g_t(T,\tau)\lambda(z_t)\right)^2 \le$$
$$\le \varepsilon^2 \left(\left|\psi(z_{T+\tau})\prime\theta^0\right| + \sum_{t=1}^{T} \left|\psi(z_{T+\tau})'(\Psi_T'\Psi_T)^{-1}\psi(z_t)\psi(z_t)'\theta^0\right| \right)^2 .$$

□

Corollary 6.4. *Under FD-2 distortion, the instability coefficient* (6.9) *and the* δ*-admissible distortion level* (6.10) *can be written as*

$$\varepsilon_+(\delta) = \sigma \left((\delta - \|g(T,\tau)\|^2)_+ \right)^{\frac{1}{2}} / \left(\left|\psi'(z_{T+\tau})\theta^0\right| + \sum_{t=1}^{T} \left|g_t(T,\tau)\theta^{0\prime}\psi(z_t)\right| \right),$$

$$\kappa(T,\tau) = \|g(T,\tau)\|^2 + (\varepsilon/\sigma)^2 \left(\left|\psi'(z_{T+\tau})\theta^0\right| + \sum_{t=1}^{T} \left|g_t(T,\tau)\theta^{0\prime}\psi(z_t)\right| \right)^2 .$$

Corollary 6.5. *Assuming FD-2 piecewise distortion* (6.5), *we have*

$$r_+(T,\tau) = r_0(T,\tau) + \varepsilon^2 \left(\left|\psi'(z_{T+\tau})\theta^0\right| \mathbf{I}_{U_1}(z_{T+\tau}) + \sum_{t:z_t\in U_1} \left|g_t(T,\tau)\theta^{0\prime}\psi(z_t)\right| \right)^2 .$$

Let us continue by investigating the case of FD-3 distortion in the l_p-metric. Let us prove the following auxiliary statement.

Lemma 6.1. *If* $x, g \in \mathbb{R}^n$, $p > 1$, $\alpha \ge 0$, *then*

$$\max\left\{ (x'g)^2 : \sum_{i=1}^{n} |x_i|^p \le \alpha \right\} = \alpha^{\frac{2}{p}} \left(\sum_{i=1}^{n} |g_i|^{\frac{p}{p-1}} \right)^{\frac{2(p-1)}{p}} .$$

Proof. From Hölder's inequality, we have

$$(x'g)^2 \le \left(\sum_{i=1}^n |x_i|\,|g_i|\right)^2 \le \left(\sum_{i=1}^n |x_i|^p\right)^{\frac{2}{p}} \left(\sum_{i=1}^n |g_i|^{\frac{p}{p-1}}\right)^{\frac{2(p-1)}{p}}$$

$$\le \alpha^{\frac{2}{p}} \left(\sum_{i=1}^n |g_i|^{\frac{p}{p-1}}\right)^{\frac{2(p-1)}{p}}.$$

The upper bound is attained for

$$x_i = \alpha^{\frac{1}{p}} \left(\sum_{j=1}^n |g_j|^{\frac{p}{p-1}}\right)^{-\frac{1}{p}} |g_i|^{\frac{1}{p-1}} \operatorname{sign}(g_i), \quad i = 1,\ldots,n,$$

and thus

$$(x'g)^2 = \alpha^{\frac{2}{p}} \left(\sum_{i=1}^n |g_i|^{\frac{p}{p-1}}\right)^{\frac{2(p-1)}{p}}.$$

□

Theorem 6.4. *Under FD-3 distortion, the guaranteed risk of the forecasting algorithm* (6.11) *equals*

$$r_+(T,\tau) = \begin{cases} r_0(T,\tau) + \varepsilon^2 \big(\max\{1, |g_t(T,\tau)| : t = 1,\ldots,T\}\big)^2, & p = 1, \\ r_0(T,\tau) + \varepsilon^2 \left(\sum\limits_{t=1}^T |g_t(T,\tau)|^{\frac{p}{p-1}} + 1\right)^{\frac{2(p-1)}{p}}, & p > 1. \end{cases}$$

Proof. By an argument similar to the proof of Theorem 6.2, we obtain:

$$r_+(T,\tau) = \max\left\{ r(T,\tau) : \sum_{t=1}^T |\lambda(z_t)|^p + |\lambda(z_{T+\tau})|^p \le \varepsilon^p \right\}.$$

Let us maximize the summand depending on $\{\lambda(z_t)\}$ in the representation (6.9) of the forecast risk $r(T,\tau)$:

$$Q(\lambda) = \left(\lambda(z_{T+\tau}) - \sum_{t=1}^T g_t(T,\tau)\lambda(z_t)\right)^2 \le \left(|\lambda(z_{T+\tau})| + \sum_{t=1}^T |g_t(T,\tau)|\,|\lambda(z_t)|\right)^2.$$

Depending on the value of p, the following two cases will be considered separately.

(I) $p = 1$. Let us maximize $Q(\lambda)$ over the possible values of $\{\lambda\}$,

$$\sum_{t=1}^{T} |\lambda(z_t)| + |\lambda(z_{T+\tau})| \leq \varepsilon.$$

Consider the following two cases:

a. There exists an index j_0 such that $\left|g_{j_o}(T+\tau)\right| = \max_j \left|g_j(T+\tau)\right| \geq 1$. Since the coefficient in front of $\lambda(z_{j_0})$ is the largest, the maximum value of $Q(\lambda)$ is attained when $\lambda(z_{j_0})$ assumes one of its boundary values, i.e., for $\lambda(z_{j_0}) = \pm\varepsilon$ and $\lambda(z_j) = 0,\ j \neq j_0$.
b. Assume that $\max_j \left|g_j(T,\tau)\right| < 1$. In that case, a similar argument can be constructed for $\lambda(z_{T+\tau})$. Thus, we obtain

$$Q_{max} = \varepsilon^2 \left(\max\left\{1, \left|g_1(T+\tau)\right|, \ldots, \left|g_T(T+\tau)\right|\right\}\right)^2.$$

(II) $p > 1$. By Lemma 6.1, we have

$$Q_{max} = \varepsilon^p \tfrac{2}{p} \left(\sum_{t=1}^{T} |g_t(T,\tau)|^{\frac{p}{p-1}} + 1\right)^{\frac{2(p-1)}{p}}.$$

□

Corollary 6.6. *The instability coefficient* (6.9) *and the* δ*-admissible distortion level* (6.10) *can be written as*

$$\kappa(T,\tau) = \begin{cases} \|g(T,\tau)\|^2 + (\varepsilon/\sigma)^2\left(\max\{1, |g_t(T,\tau)| : t = 1,\ldots,T\}\right)^2, & p = 1, \\ \|g(T,\tau)\|^2 + (\varepsilon/\sigma)^2 \left(\sum\limits_{t=1}^{T} |g_t(T,\tau)|^{\frac{p}{p-1}} + 1\right)^{\frac{2(p-1)}{p}}, & p > 1, \end{cases}$$

$$\varepsilon_+(\delta) = \begin{cases} \sigma\left((\delta - \|g(T,\tau)\|^2)_+\right)^{\frac{1}{2}} / \max\{1, |g_t(T,\tau)| : t = 1,\ldots,T\}, & p = 1, \\ \sigma\left((\delta - \|g(T,\tau)\|^2)_+\right)^{\frac{1}{2}} / \left(\sum\limits_{t=1}^{T} |g_t(T,\tau)|^{\frac{p}{p-1}} + 1\right)^{\frac{(p-1)}{p}}, & p > 1. \end{cases}$$

Finally, let us investigate the case of distortion in the L_p-metric.

Theorem 6.5. *Under FD-4 distortion, the guaranteed risk of the algorithm* (6.11) *can be expressed as*

$$r_+(T,\tau) = \begin{cases} r_0(T,\tau) + \varepsilon^2 \max\limits_{j\in J} \left(\sum\limits_{t=1}^{T} g_t(T,\tau)\eta_j(z_t) - \eta_j(z_{T+\tau})\right)^2, & p = 1, \\ r_0(T,\tau) + \varepsilon^2 \left(\sum\limits_{j\in J} \left|\sum\limits_{t=1}^{T} g_t(T,\tau)\eta_j(z_t) - \eta_j(z_{T+\tau})\right|^{\frac{p}{p-1}}\right)^{\frac{2(p-1)}{p}}, & p > 1. \end{cases}$$

Proof. As in the proof of Theorem 6.2, we have

$$r_{+}(T,\tau) = \max_{\lambda(z_t)=\sum\limits_{j\in J} S_j\eta_j(z_t),\ \sum\limits_{j\in J}|S_j|^p\le\varepsilon^p} r(T,\tau).$$

Let us maximize the summand depending on $\{\lambda(z_t)\}$ in the expression (6.13) for $r(T,\tau)$:

$$Q(\lambda) = \left(\lambda(z_{T+\tau}) - \sum_{t=1}^{T} g_t(T,\tau)\lambda(z_t)\right)^2 =$$

$$= \left(\sum_{j\in J} S_j\left(\sum_{t=1}^{T} g_t(T,\tau)\eta_j(z_t) - \eta_j(z_{T+\tau})\right)\right)^2.$$

(I) $p = 1$. Let

$$j_0 = \arg\max_{j\in J}\left|\sum_{t=1}^{T} g_t(T,\tau)\eta_j(z_t) - \eta_j(z_{T+\tau})\right|,$$

then, since the coefficient in front of S_{j_0} is maximal, the maximum value $Q(\lambda)$ is obtained for $|S_{j_0}| = \varepsilon$, $S_j = 0$, $j \ne j_0$. This leads to the expression

$$Q_{max} = \varepsilon^2 \max_{j\in J}\left(\sum_{t=1}^{T} g_t(T,\tau)\eta_j(z_t) - \eta_j(z_{T+\tau})\right)^2.$$

(II) $p > 1$. Then we have

$$S'\gamma\gamma' S \to \max,\quad \lambda(z_t) = \sum_{j\in J} S_j\eta_j(z_t),\quad \sum_{j\in J}|S_j|^p \le \varepsilon^p,$$

where $S = (S_j)$, $\gamma = (\gamma_j)$, $j \in J$, and $\gamma_j = \sum\limits_{t=1}^{T} g_t(T,\tau)\eta_j(z_t) - \eta_j(z_{T+\tau})$. Now, by Lemma 6.1, we have

$$Q_{max} = \varepsilon^{p\frac{2}{p}}\left(\sum_{j\in J}|\gamma_j|^{\frac{p}{p-1}}\right)^{\frac{2(p-1)}{p}}.$$

□

Corollary 6.7. *Under FD-4 distortion, the instability coefficient* (6.9) *and the δ-admissible distortion level* (6.10) *can be written as*

$$\kappa(T,\tau) =$$

$$= \begin{cases} \|g(T,\tau)\|^2 + \left(\frac{\varepsilon}{\sigma}\right)^2 \max\limits_{j\in J}\left(\sum\limits_{t=1}^{T} g_t(T,\tau)\eta_j(z_t) - \eta_j(z_{T+\tau})\right)^2, & p = 1,\\ \|g(T,\tau)\|^2 + \left(\frac{\varepsilon}{\sigma}\right)^2 \left(\sum\limits_{j\in J}\left|\sum\limits_{t=1}^{T} g_t(T,\tau)\eta_j(z_t) - \eta_j(z_{T+\tau})\right|^{\frac{p}{p-1}}\right)^{\frac{2(p-1)}{p}}, & p > 1; \end{cases}$$

$$\varepsilon_+(\delta) =$$

$$= \begin{cases} \sigma\sqrt{(\delta - \|g(T,\tau)\|^2)_+}\left(\max\limits_{j\in J}\left|\sum\limits_{t=1}^{T} g_t(T,\tau)\eta_j(z_t) - \eta_j(z_{T+\tau})\right|\right)^{-1}, & p = 1,\\ \sigma\sqrt{(\delta - \|g(T,\tau)\|^2)_+}\left(\sum\limits_{j\in J}\left|\sum\limits_{t=1}^{T} g_t(T,\tau)\eta_j(z_t) - \eta_j(z_{T+\tau})\right|^{\frac{p}{p-1}}\right)^{-\frac{(p-1)}{p}}, & p > 1. \end{cases}$$

6.2 Robustness of Least Squares Forecasting Under Functional Distortions of Multivariate Linear Regression Models

6.2.1 *Mathematical Description of Model Distortions*

Consider a multivariate linear regression model under FD, which is a generalization of the model (6.1), (6.2) studied in the previous section.

Let the observed d-variate time series $x_t \in \mathbb{R}^d$ be defined by the following stochastic equation [17, 20]:

$$x_t = \sum_{i=1}^{m} \theta^0_{(i)}\psi_i(z_t) + \lambda(z_t) + u_t = \theta^0\psi(z_t) + \lambda(z_t) + u_t, \quad t = 1,\dots,T,\dots, \tag{6.14}$$

where $u_t \in \mathbb{R}^d$ is the random error vector at time t; the vector

$$z_t \in U \subseteq \mathbb{R}^M$$

is composed of M a priori known values of the independent variables (regressors) at time t; the set U contains the possible regressor values; $\{\psi_i(z) : \mathbb{R}^M \to \mathbb{R}^1\}$ is a given system of m linearly independent functions such that

$$\left|\sum_{t=1}^{T} \psi(z_t)\psi'(z_t)\right| \neq 0, \quad \psi(z) = \left(\psi_i(z)\right) \in \mathbb{R}^m;$$

$\theta^0 = (\theta^0_{ij}) = (\theta^0_{(1)} \vdots \ldots \vdots \theta^0_{(m)}) \in \mathbb{R}^{d\times m}$ is the $(d \times m)$-matrix of unknown true values of the regression coefficients, $\theta^0_{(i)} \in \mathbb{R}^d$ denotes the ith column of the matrix θ^0; $\lambda(\cdot) : \mathbb{R}^M \to \mathbb{R}^d$ is an unknown deterministic vector function describing the multivariate functional distortion (MFD) of the hypothetical multivariate linear regression. Random errors $\{u_t\}$ are assumed to be jointly independent and identically distributed with

$$\mathbb{E}\{u_t\} = \mathbf{0}_d, \quad \mathbf{Cov}\{u_t, u_t\} = \mathbb{E}\{u_t u_t'\} = \Sigma, \quad |\Sigma| \neq 0. \tag{6.15}$$

Taking $\lambda(z) \equiv 0$ in (6.14), the expression (6.15) becomes the hypothetical model M_o—a multivariate (or multiple for $d = 1$) linear regression model of time series considered in Sect. 5.1:

$$f_0(z) = \left(f_{0i}(z)\right) = \theta^0 \psi(z).$$

Let us define the commonly encountered MFD types.

MFD-1. *Interval distortion:*

$$\varepsilon_{i-}(z) \le \lambda_i(z) \le \varepsilon_{i+}(z), \quad z \in U, \ i = 1, \ldots, d,$$

where $\varepsilon_{\pm}(z) = (\varepsilon_{i\pm}(z)) \in \mathbb{R}^d$ are some boundaries defined as vector functions.

MFD-2. *Relative distortion:*

$$\frac{|\lambda_i(z)|}{|f_{0i}(z)|} \le \varepsilon_i, \quad \varepsilon_i \ge 0, \ z \in U, \ i = 1, \ldots, d,$$

where a given vector $\varepsilon = (\varepsilon_i)$ defines the relative distortion levels for the components of the multivariate linear regression function; the relative specification error of the ith component is at most ε_i.

MFD-3. *Distortion in the l_p-metric:*

$$\left(\sum_{t=1}^{T}\sum_{i=1}^{d} |\lambda_i(z_t)|^p + \sum_{i=1}^{d} |\lambda_i(z_{T+\tau})|^p\right)^{1/p} \le \varepsilon,$$

where $p \in \mathbb{N}$ is a given positive integer, $\varepsilon \ge 0$ is the given distortion level; and the l_p norm of the FD accumulated over time doesn't exceed ε.

Distortion type MFD-4—distortion due to discarded remainders of functional expansions—has been studied in [18], but won't be discussed here due to space considerations.

6.2.2 Robustness Evaluation of Least Squares Forecasting

In this subsection, we are going to study the sensitivity of the mean squares forecast risk

$$r(T,\tau) = \mathbb{E}\left\{(\widehat{x}_{T+\tau} - x_{T+\tau})'(\widehat{x}_{T+\tau} - x_{T+\tau})\right\} \geq 0 \tag{6.16}$$

to MFD-1, MFD-2, and MFD-3 distortions. We will be evaluating the respective instability coefficient (6.9) and the δ-admissible distortion level (6.10).

Let us introduce the following matrix notation:

$$A = \sum_{t=1}^{T} \psi(z_t)\psi'(z_t) \in \mathbb{R}^{m\times m}, \qquad C = \sum_{t=1}^{T} x_t\psi'(z_t) \in \mathbb{R}^{d\times m},$$

$$L = \sum_{t=1}^{T} \lambda(z_t)\psi'(z_t) \in \mathbb{R}^{d\times m}, \qquad G_{T+\tau} = LA^{-1}\psi(z_{T+\tau}) \in \mathbb{R}^{d},$$

$$\alpha_t = \psi'(z_t)A^{-1}\psi(z_{T+\tau}),\ t = 1,\dots,T, \qquad K_{T\tau} = \psi'(z_{T+\tau})A^{-1}\psi(z_{T+\tau}) > 0. \tag{6.17}$$

First of all, note that in the absence of distortion ($\lambda(z_t) \equiv 0$) and assuming prior knowledge of θ^0, the minimal risk $r_0 = \mathrm{tr}\,(\Sigma)$ is attained for the optimal forecasting statistic (see Sect. 5.1):

$$\widehat{x}^0_{T+\tau} = \theta^0\psi(z_{T+\tau}).$$

As a rule, in applied problems where θ^0 is unknown, a nonsingular design of the experiment $|A| \neq 0$ is assumed, and a least squares forecasting algorithm based on the plug-in approach is constructed (see Sect. 5.2):

$$\widehat{x}_{T+\tau} = \widehat{\theta}\psi(z_{T+\tau}), \quad \widehat{\theta} = CA^{-1}, \tag{6.18}$$

where $\widehat{\theta} = (\widehat{\theta}_{(1)} \vdots \dots \vdots \widehat{\theta}_{(m)}) \in \mathbb{R}^{d\times m}$ is the least squares estimator for the matrix θ^0.

Lemma 6.2. *In a multivariate linear regression model under FD* (6.14), (6.15), *the risk functional* (6.16) *of a least squares forecast* (6.18) *can be written as*

$$r(T,\tau) = r_0(T,\tau) + \|G_{T+\tau} - \lambda(z_{T+\tau})\|^2, \quad r_0(T,\tau) = (1 + K_{T\tau})\mathrm{tr}\,(\Sigma), \tag{6.19}$$

where $r_0(T,\tau)$ is the forecast risk in the absence of distortion.

Proof. Substitute (6.18) in (6.16) and use the notation (6.17). □

As in Sect. 6.1, if the design of the experiment $\{z_1, \dots, z_T\} \subset U$ is chosen so that for the minimum eigenvalue we have $\lambda_{min}(A) \to +\infty$ as $T \to +\infty$ (i.e., the Eicker condition is satisfied), then the forecasting statistic (6.18) is asymptotically optimal:

$$K_{T\tau} \to 0, \quad r(T, \tau) \to r_0.$$

Theorem 6.6. *Under MFD-1 distortion, the risk instability coefficient for the least squares forecasting statistic* (6.18) *can be written as*

$$\kappa(T, \tau) = K_{T\tau} + \\ + (\mathrm{tr}\, \Sigma)^{-1} \sum_{k=1}^{d} \max \left(\sum_{t=1}^{T} \left((\alpha_t)_+ \varepsilon_{k\pm}(z_t) - (-\alpha_t)_+ \varepsilon_{k\mp}(z_t) \right) - \varepsilon_{k\mp}(z_{T+\tau}) \right)^2. \tag{6.20}$$

Proof. The proof is based on Lemma 6.2; it can be constructed from (6.16) to (6.19) similarly to the proofs of Theorems 6.1, 6.2. □

The expression (6.20) for the risk instability coefficient contains two summands. Under the Eicker condition, the first summand satisfies $K_{T\tau} \to 0$ as $T \to \infty$; it isn't affected by the level of MFD-1 distortion and reflects the stochastic error in the consistent estimator for the matrix θ^0 of regression coefficients based on a finite undistorted sample of size T. The second summand is due to MFD-1 distortion; it grows as the contribution of systematic errors $\{\varepsilon_{k\pm}(z)\}$ increases in relation to the total variance tr (Σ) of the random error u_t.

Theorem 6.7. *Under MFD-2 distortion, the risk instability coefficient for the least squares forecasting algorithm* (6.18) *equals*

$$\kappa(T, \tau) = K_{T\tau} + \widetilde{\varepsilon}' \chi / \mathrm{tr}\, (\Sigma),$$

where $\widetilde{\varepsilon} = (\varepsilon_1^2, \dots, \varepsilon_d^2)'$, $\chi = (\chi_1, \dots, \chi_d)'$,

$$\chi_k = \left(\sum_{t=1}^{T} \left| (\theta^0 \psi(z_t))_k \right| |\alpha_t| + \left| (\theta^0 \psi(z_{T+\tau}))_k \right| \right)^2, \quad k = 1, \dots, d.$$

Proof. The proof is similar to the proof of Theorem 6.3. □

Theorem 6.8. *Under d-variate distortion MFD-3, the risk instability coefficient of the least squares forecasting algorithm* (6.18) *can, depending on the order p of the l_p-metric, be expressed by one of the following formulas:*

$$p = 1: \quad \kappa(T,\tau) = K_{T\tau} + \varepsilon^2 \big(\max\{1, |\alpha_1|, \ldots, |\alpha_T|\}\big)^2 (\mathrm{tr}\,\Sigma)^{-1};$$

$$1 < p < 2: \quad \kappa(T,\tau) \le K_{T\tau} + \varepsilon^2 \left(\sum_{t=1}^{T} |\alpha_t|^{\frac{p}{p-1}} + 1\right)^{\frac{2(p-1)}{p}} (\mathrm{tr}\,\Sigma)^{-1};$$

$$p \ge 2: \quad \kappa(T,\tau) \le K_{T\tau} + \varepsilon^2 d^{\frac{2(p-1)}{p}} \left(\sum_{t=1}^{T} |\alpha_t|^{\frac{p}{p-1}} + 1\right)^{\frac{2(p-1)}{p}} (\mathrm{tr}\,\Sigma)^{-1}.$$

Proof. The proof is similar to the proof of Theorem 6.4. □

The expressions for the risk instability coefficient obtained in Theorems 6.6–6.8 can be used to estimate the δ-admissible distortion level (6.10). For instance, under MFD-3 distortion in the l_1-metric ($p = 1$) we have

$$\varepsilon_+(\delta) = \frac{\sqrt{(\mathrm{tr}\,\Sigma)(\delta - K_{T\tau})_+}}{\max\{1, |\alpha_1|, \ldots, |\alpha_T|\}}.$$

6.3 Robustness of Least Squares Forecasting Under Outliers

In this section, we are going to consider time series under outliers (distortions of type A_2 described in Sect. 6.1.1).

Let the observed time series be one-dimensional and assume that it includes additive distortion in the form of outliers [13, 16]:

$$x_t = (\theta_0)' \psi(z_t) + u_t + \xi_t v_t, \quad t \in \mathbb{N}. \tag{6.21}$$

Here, as in Sect. 6.1, $z_t \in U \subseteq \mathbb{R}^M$ is a nonrandom regressor; $\psi(\cdot) = (\psi_i(\cdot))$ is a column vector of m linearly independent functions; $\{u_t\}$ is a sequence of independent random observation errors such that $\mathbb{E}\{u_t\} = 0$, $\mathbb{D}\{u_t\} = \sigma^2$; the column vector $\theta^0 = (\theta_i^0) \in \mathbb{R}^m$ consists of the m unknown multiple linear regression coefficients; $\{\xi_t\}$ is a sequence of independent Bernoulli random variables,

$$\mathbb{P}\{\xi_t = 1\} = 1 - \mathbb{P}\{\xi_t = 0\} = \varepsilon; \tag{6.22}$$

$\varepsilon \in [0, \varepsilon_+]$ is the assumed outlier probability, $0 \le \varepsilon_+ < 1/2$ is a given upper bound on the outlier probability; $\{v_t\}$ are i.i.d. random variables,

$$\mathbb{E}\{v_t\} = a, \quad \mathbb{D}\{v_t\} = K\sigma^2, \quad K \ge 0; \tag{6.23}$$

the random variables $\{\xi_t\}$, $\{u_t\}$, $\{v_t\}$ are jointly independent. Under this model, if $\xi_t = 1$ for some time point t, then an outlier is present, and an additional random summand $v_t \in \mathbb{R}^1$ appears in (6.21); the case $\xi_t = 0$ corresponds to an undistorted observation at time t. We are going to distinguish two special cases of distortion (6.21): distortion in variance ($a = 0$, $K > 1$) and in the mean value ($K = 0$, $a \neq 0$).

In the notation of Sect. 6.1, let us evaluate the robustness of the traditional least squares forecasting algorithm (6.11).

Theorem 6.9. *If the observed time series x_t satisfies the model* (6.21)–(6.23) *under outliers, then for $T > m$, $|\Psi_T'\Psi_T| \neq 0$, the forecast risk of the least squares forecasting algorithm* (6.11) *equals*

$$r_\varepsilon(T,\tau) = r_0(T,\tau) + \varepsilon\left(K\sigma^2||g(T,\tau)||^2 + a^2\left((1-\varepsilon)||g(T,\tau)||^2 + \varepsilon(g'(T,\tau)\mathbf{1}_T)^2\right)\right), \tag{6.24}$$

and the instability coefficient can be written as

$$\kappa(T,\tau) = ||g(T,\tau)||^2 + \\ + \varepsilon_+\left(K||g(T,\tau)||^2 + \frac{a^2}{\sigma^2}\left((1-\varepsilon_+)||g(T,\tau)||^2 + \varepsilon_+(g'(T,\tau)\mathbf{1}_T)^2\right)\right). \tag{6.25}$$

Proof. Substituting (6.11) into the expression for the risk functional and applying (6.21)–(6.23), we obtain

$$r_\varepsilon(T,\tau) = \mathbb{E}\left\{\left((\widehat{\theta}-\theta^0)'\psi(z_{T+\tau}) - u_{T+\tau}\right)^2\right\} = \sigma^2 + \psi'(z_{T+\tau})V_{(\varepsilon)}\psi(z_{T+\tau}), \tag{6.26}$$

where

$$V_{(\varepsilon)} = \mathbb{E}\left\{(\widehat{\theta}-\theta^0)(\widehat{\theta}-\theta^0)'\right\} \tag{6.27}$$

is the covariance matrix of the least squares estimator $\widehat{\theta}$ defined by (6.11) under level ε outliers. From (6.21) we have:

$$X_T = \Psi_T\theta^0 + U_T + \Xi_T V_T,$$

where

$$X_t = (x_t) \in \mathbb{R}^{T\times 1}, \qquad U_T = (u_t) \in \mathbb{R}^{T\times 1},$$

$$\Xi_T = \operatorname{diag}\{\xi_1,\ldots,\xi_T\} \in \mathbb{R}^{T\times T}, \qquad V_T = (v_t) \in \mathbb{R}^{T\times 1}, \quad t = 1,\ldots,T.$$

By (6.11), we can write

$$\widehat{\theta} - \theta^0 = C_T U_T + C_T \Xi_T V_T, \quad C_T = (\Psi_T'\Psi_T)^{-1}\Psi_T'.$$

Now from (6.22), (6.23), and (6.27) we have

$$V_{(\varepsilon)} = \left(\sigma^2 + \varepsilon(K\sigma^2 + (1-\varepsilon)a^2)\right) C_T C_T' + \varepsilon^2 a^2 C_T \mathbf{1}(C_T \mathbf{1}_T)'. \qquad (6.28)$$

Substituting (6.28) into (6.26), performing the obvious matrix transformations, and applying the notation leads to (6.24).

By computing a partial derivative of (6.24) in $\varepsilon \in [0, 1/2)$, we obtain

$$\frac{\partial r_\varepsilon(T,\tau)}{\partial \varepsilon} = K\sigma^2 \|g(T,\tau)\|^2 + a^2 \left((1-2\varepsilon)||g(T,\tau)||^2 + 2\varepsilon(g'(T,\tau)\mathbf{1}_T)^2\right) > 0.$$

Thus, the risk $r_\varepsilon(T,\tau)$ is monotonous increasing as ε increases. This, together with the definition of the instability coefficient (6.9), leads to (6.25). □

From (6.25) we can see that the risk instability coefficient can be split into three additive components:

$$\kappa(T,\tau) = \kappa_1(T,\tau) + \kappa_2(T,\tau) + \kappa_3(T,\tau),$$

where the component

$$\kappa_1(T,\tau) = \|g(T,\tau)\|^2 > 0$$

is due to finiteness of the observation length T when estimating θ; the size of the component

$$\kappa_2(T,\tau) = \varepsilon_+ K \|g(T,\tau)\|^2 \geq 0$$

is determined by the variance of the outliers w.r.t. the random error variance u_t, $\mathbb{D}\{\upsilon_t\}/\mathbb{D}\{u_t\} = K$; and the size of

$$\kappa_3(T,\tau) = \varepsilon_+ \frac{a^2}{\sigma^2} \left((1-\varepsilon_+)||g(T,\tau)||^2 + \varepsilon_+(g'(T,\tau)\mathbf{1}_T)^2\right) \geq 0$$

is influenced by the relation between the squared expectation of the outliers and the variance of random errors u_t: $(\mathbb{E}\{\upsilon_t\})^2/\mathbb{D}\{u_t\} = a^2/\sigma^2$.

Corollary 6.8. *Under outliers in the variance ($a = 0$, $K > 1$), the risk instability coefficient equals $\kappa(T,\tau) = \kappa_1(T,\tau) + \kappa_2(T,\tau)$, and the δ-admissible distortion level ($\delta > 0$) can be written as*

$$\varepsilon_+(\delta) = \min\left(\frac{1}{2}, \frac{1}{K}\left(\frac{\delta}{\|g(T,\tau)\|^2} - 1\right)_+\right).$$

Proof. It is sufficient to assume $a = 0$ in (6.25) and apply (6.10). □

Corollary 6.9. *Under outliers in the mean value* ($K = 0$, $a \neq 0$)*, we have* $\kappa(T,\tau) = \kappa_1(T,\tau) + \kappa_3(T,\tau)$*, and the* δ*-admissible distortion level* $\varepsilon_+(\delta)$ *is the smallest positive root of the quadratic equation*

$$\|g(T,\tau)\|^2 + \varepsilon_+ \frac{a^2}{\sigma^2}\left((1-\varepsilon_+)\|g(T,\tau)\|^2 + \varepsilon_+ g'(T,\tau)\mathbf{1}_T)^2\right) = \delta.$$

Corollary 6.10. *As* $T \to +\infty$*, assuming* $a = 0$ *and Eicker asymptotics for the minimal eigenvalue of the matrix* $\Psi_T'\Psi_T$*,*

$$\lambda_{\min}(\Psi_T'\Psi_T) \to +\infty, \tag{6.29}$$

the least squares forecasting algorithm (6.11) *is asymptotically robust:*

$$\kappa(T,\tau) \to 0, \quad \tau \geq 1. \tag{6.30}$$

Proof. Applying well-known properties of matrix eigenvalues [25] to (6.29) yields

$$\lambda_{\max}\left((\Psi_T'\Psi_T)^{-1}\right) \to 0.$$

Thus, $\forall z \in \mathbb{R}^m$ such that $|z| \leq C < +\infty$, for the quadratic form defined by $\Psi_T'\Psi_T$ we have

$$z'(\Psi_T'\Psi_T)^{-1}z \leq C^2\lambda_{\max}\left((\Psi_T'\Psi_T)^{-1}\right) \to 0.$$

This implies that $\|g(T,\tau)\| \to 0$, and therefore, (6.25) leads to (6.30). □

Note that for $a \neq 0$, the asymptotic robustness condition (6.30) for the least squares forecasting algorithm requires a stricter asymptotic condition on the design of the experiment $\{z_t\}$:

$$\frac{1}{\sqrt{T}}\lambda_{\min}(\Psi_T'\Psi_T) \to +\infty.$$

6.4 Impact of Correlation Between Observation Errors on Forecast Risk

In all of the time series regression models considered in Sects. 6.1–6.3, it is assumed that the random errors $\{u_t\}$ are uncorrelated:

$$\mathbf{Cov}\{u_t, u_{t'}\} = 0, \quad t \neq t'. \tag{6.31}$$

In applications, the condition (6.31) is often not satisfied, and $\{u_t\}$ are some weakly correlated random variables (i.e., the model is subject to distortions of type A_4

defined in Sect. 6.1.1). Let us evaluate the impact of such distortions on the risk of the traditional least squares forecasting algorithm (6.11).

In the notation of Sect. 6.3, let us define the mathematical model of the observed time series $x_t \in \mathbb{R}^1$ under distortions of type A_4:

$$x_t = \theta^{0'} \psi(z_t) + u_t, \quad t \in \mathbb{N}, \tag{6.32}$$

where $\{u_t\}$ are identically distributed weakly correlated random variables which form a strictly stationary random sequence [5]:

$$\mathbb{E}\{u_t\} = 0, \quad \mathbb{D}\{u_t\} = \sigma^2, \quad \mathbf{Cov}\{u_t, u_{t+\tau}\} = \mathbb{E}\{u_t u_{t+\tau}\} = \sigma^2 \varrho(\tau). \tag{6.33}$$

It is convenient to introduce the following two classes of weak correlation by imposing the following conditions on the correlation function $\varrho(\cdot)$:

- uniform ε-boundedness of the correlation function:

$$0 \leq |\varrho(\tau)| \leq \varepsilon, \quad \tau \neq 0; \tag{6.34}$$

- 1-dependence of the random sequence $\{u_t\}$:

$$|\varrho(\tau)| = \begin{cases} 1, & \text{if } \tau = 0, \\ \varepsilon, & \text{if } \tau \in \{-1, +1\}, \\ 0, & \text{if } |\tau| \geq 2. \end{cases} \tag{6.35}$$

In the relations (6.34) and (6.35), $\varepsilon \in [0, 1]$ is the maximum allowed distortion level. If $\varepsilon = 0$, then $\{u_t\}$ are uncorrelated, and no distortion of type A_4 is present.

Let us introduce the following notation for $i, j = 1, \ldots, T$:

$$\begin{aligned} U_T &= (u_t) \in \mathbb{R}^{T\times 1}, & C_T &= (\Psi_T' \Psi_T)^{-1} \Psi_T' \in \mathbb{R}^{m\times T}, \\ \Sigma_T &= (\sigma_{Tij}) \in \mathbb{R}^{T\times T}, & \sigma_{Tij} &= \sigma^2 \varrho(|i-j|), \\ G_{T\tau} &= (g_{T\tau i}) \in \mathbb{R}^{T\times 1}, & g_{T\tau i} &= \varrho(|T+\tau-i|), \\ g(T,\tau) &= C_T' \psi(z_{T+\tau}) \in \mathbb{R}^{T\times 1}. \end{aligned} \tag{6.36}$$

Theorem 6.10. *If the observed time series x_t satisfies the model* (6.32), (6.33) *under distortion in the form of correlated random observation errors $\{u_t\}$, and*

$$T > m, \quad |\Psi_T' \Psi_T| \neq 0,$$

then the risk of the least squares forecasting algorithm (6.11) *equals*

$$r_\varepsilon(T,\tau) = \sigma^2 + (g(T,\tau))' \Sigma_T \, g(T,\tau) - 2 G_{T\tau}' g(T,\tau). \tag{6.37}$$

Proof. Similarly to (6.26), we have:

$$\begin{aligned}\widehat{x}_{T+\tau} - x_{T+\tau} &= (\widehat{\theta} - \theta^0)'\psi(z_{T+\tau}) - u_{T+\tau},\\ r_\varepsilon(T,\tau) &= \mathbb{E}\left\{(\widehat{x}_{T+\tau} - x_{T+\tau})^2\right\} =\\ &= \sigma^2 + \psi'(z_{T+\tau})V_{(\varepsilon)}\psi(z_{T+\tau}) - 2\mathbb{E}\left\{u_{T+\tau}(\widehat{\theta} - \theta^0)'\right\}\psi(z_{T+\tau}).\end{aligned} \tag{6.38}$$

Similarly to (6.28), the expressions (6.33), (6.36) allow us to write

$$\widehat{\theta} - \theta^0 = C_T U_T, \quad V_{(\varepsilon)} = C_T \mathbb{E}\{U_T U_T'\}C_T' = C_T \Sigma_T C_T',$$
$$\mathbb{E}\left\{u_{T+\tau}(\widehat{\theta} - \theta^0)\right\} = C_T \mathbb{E}\{u_{T+\tau}U_T\} = C_T G_{T\tau}.$$

Substituting these equalities into (6.38) leads to (6.37). □

Corollary 6.11. *For a uniformly ε-bounded correlation function* (6.34)*, the forecast risk lies in the following range:*

$$r_-(T,\tau) \le r_\varepsilon(T,\tau) \le r_+(T,\tau), \tag{6.39}$$

where the interval boundaries are

$$r_\pm(T,\tau) = r_0(T,\tau) \pm 2\varepsilon\sigma^2\left(\sum_{i=1}^{T}|g_i(T,\tau)| + \sum_{i<j}|g_i(T,\tau)||g_j(T,\tau)|\right).$$

Here $r_0(T,\tau)$ denotes the forecast risk for $\varepsilon = 0$—i.e., under the noncorrelatedness assumption (6.31)*—which can be written as $r_0(T,\tau) = \sigma^2(1 + \|g(T,\tau)\|^2)$.*

Proof. By (6.37), (6.34) we have

$$\begin{aligned} r(T,\tau) &= \sigma^2 + \sum_{i,j=1}^{T} g_i(T,\tau)g_j(T,\tau)\varrho(|i-j|) - 2\sum_{i=1}^{T} g_i(T,\tau)\varrho(i+\tau-1) \le\\ &\le \sigma^2 + \sum_{i=1}^{T}(g_i(T,\tau))^2\sigma^2 + \sigma^2\varepsilon\sum_{i\ne j}|g_i(T,\tau)g_j(T,\tau)| + 2\sigma^2\varepsilon\sum_{i=1}^{T}|g_i(T,\tau)|,\end{aligned}$$

leading to the second inequality in (6.39). The first inequality of (6.39) is proved similarly. □

Corollary 6.12. *For a 1-dependent random observation sequence* (6.35)*, we have*

$$r_-(T,\tau) \le r_\varepsilon(T,\tau) \le r_+(T,\tau),$$

$$r_\pm(T,\tau) = r_0(T,\tau) \pm 2\sigma^2\varepsilon \left| \sum_{i=1}^{T-1} g_i(T,\tau) g_{i+1}(T,\tau) - \delta_{\tau,1} g_1(T,\tau) \right|. \tag{6.40}$$

From (6.37), (6.39), and (6.12), we can see that, unlike distortions of types A_1 and A_2 discussed earlier, distortions of type A_4 do not necessarily lead to an increased forecast risk compared to the absence of distortion ($\varepsilon = 0$). Under certain conditions, distortions of type A_4 can even yield a reduction of the forecast risk.

6.5 Robust Forecasting Based on M-Estimators Under Functional Distortion

6.5.1 Construction of a Robust Forecasting Algorithm

To construct a robust forecasting algorithm under the regression model (6.1), let us consider a family of forecasting algorithms based on M-estimators [7, 11, 13, 22, 28, 33]:

$$\hat{x}_{T+\tau} = \hat{\theta}'\psi(z_{T+\tau}), \quad \hat{\theta} = \arg\min_\theta \sum_{t=1}^{T} \varrho\left(x_t - \theta'\psi(z_t)\right), \tag{6.41}$$

where $\varrho(z)$ is a loss function which is assumed to be convex, even, and doubly differentiable almost everywhere on $\mathbb{R}$. Note that the least squares forecasting algorithm (6.11) can be viewed as a special case of the M-estimator method with $\varrho(z) = z^2$.

Let us introduce the following notation: $\mathbf{1}_T$ is a T-vector of ones; $\mathbf{0}_m$ is an m-vector of zeroes;

$$\mu(z) = \frac{d\varrho(z)}{dz}, \quad \nu(z) = \frac{d^2\varrho(z)}{dz^2},$$

$$\boldsymbol{\nu}(\theta) = (\nu_{tt'}(\theta)) = \mathrm{diag}\left(\nu(x_t - \theta'\psi(u_t))\right) \in \mathbb{R}^{T\times T}, \quad t, t' = 1, \dots, T, \tag{6.42}$$

$$\boldsymbol{\mu}(\theta) = (\mu_t(\theta)) = \left(\mu(x_t - \theta'\psi(u_t))\right) \in \mathbb{R}^T, \quad t = 1, \dots, T,$$

$$D(\theta) = \Psi'_T \boldsymbol{\nu}(\theta)\Psi_T, \quad M(\theta) = \Psi'_T \boldsymbol{\mu}(\theta). \tag{6.43}$$

Computing $\hat{\theta}$ in (6.41) requires solving a convex optimization problem. We are going to apply the standard approach of constructing an iterative solver [11, 13].

The necessary optimality condition, which is also sufficient due to convexity, has the form $M(\hat{\theta}) = \mathbf{0}_m$. Let us apply Taylor's linear formula for $\boldsymbol{\mu}(\hat{\theta})$ in a neighborhood of $\theta = \theta_{(n)}$ (an approximation for $\hat{\theta}$ in the nth step, $n \in \mathbb{N}$):

$$\boldsymbol{\mu}(\hat{\theta}) = \boldsymbol{\mu}(\theta_{(n)}) - \boldsymbol{\nu}(\theta_{(n)})\Psi_T(\hat{\theta} - \theta_{(n)}) + o\left(\left\|\hat{\theta} - \theta_{(n)}\right\|\right)\mathbf{1}_T,$$

$$M(\theta_{(n)}) - D(\theta_{(n)})(\hat{\theta} - \theta_{(n)}) + o\left(\left\|\hat{\theta} - \theta_{(n)}\right\|\right)\mathbf{1}_T = 0.$$

An existence condition for the above iteration steps is defined in the following theorem.

Theorem 6.11. *If we have* $\operatorname{rank}(\boldsymbol{\nu}(\theta_{(n)})) = k$, $m \le k \le T$, *then the approximation* $\theta_{(n+1)}$ *for* $\hat{\theta}$ *in the* $(n+1)$*th step exists and is given by*

$$\theta_{(n+1)} = \theta_{(n)} + (D(\theta_{(n)}))^{-1}M(\theta_{(n)}). \tag{6.44}$$

Proof. Denote $B = \boldsymbol{\nu}^{\frac{1}{2}}\Psi_T$. Since the diagonal matrix $\boldsymbol{\nu}^{\frac{1}{2}}$ has exactly $T-k$ zero rows, the matrix B also has $T-k$ zero rows, and the remaining k rows are nonzero and differ from the corresponding rows of the matrix Ψ_T by a positive multiplier. Thus, by the definition of rank,

$$\operatorname{rank}(B) = \min(m, k) = m.$$

From matrix property 2.17.2 in [25], we have

$$\operatorname{rank}(D(\theta_{(n)})) = \operatorname{rank}(B'B) = \operatorname{rank}(B) = m.$$

Then $\left|D(\theta_{(n)})\right| \ne 0$, and the system of equations $M(\theta_{(n)}) - D(\theta_{(n)})(\hat{\theta} - \theta_{(n)}) = 0$ has a solution (6.44). □

To accommodate for functional distortions FD-1 defined by (6.4), let us choose a loss function $\varrho(\cdot)$ in (6.41) to have the following special form:

$$\varrho(z) = 0.5\,\mathbf{1}(|z| - \delta_\varepsilon)(z - \delta_\varepsilon\ \operatorname{sign}(z))^2, \tag{6.45}$$

where $\delta_\varepsilon \ge 0$ is a parameter of the algorithm (not yet defined); taking $\delta_\varepsilon = 0$ yields the least squares forecasting algorithm (6.11). In the notation (6.42), (6.43), let us define the following functions for $z \ne \pm\delta_\varepsilon$:

$$\mu(z) = \mathbf{1}(|z| - \delta_\varepsilon)(z - \delta_\varepsilon\ \operatorname{sign}(z)), \quad \nu(z) = \mathbf{1}(|z| - \delta_\varepsilon). \tag{6.46}$$

Note that the conditions of Theorem 6.11 require the parameter δ_ε in (6.45) to be chosen in a special way.

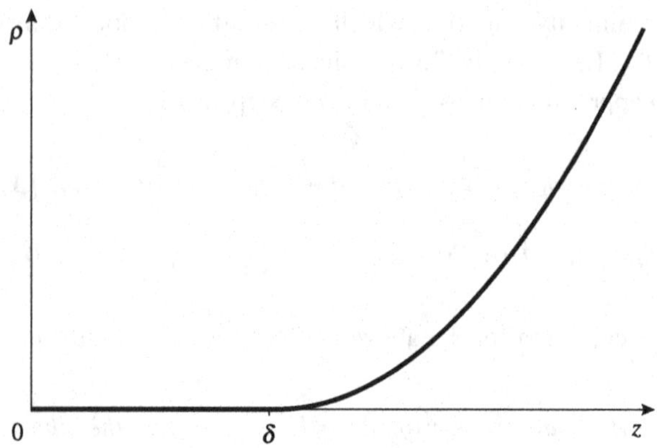

Fig. 6.1 Loss function (6.45)

Let us also consider a loss function of the following form ($z \geq 0$):

$$\varrho(z) = f_1(z)\mathbf{I}_{[0,\delta_\varepsilon-d]}(z) + g(z)\mathbf{I}_{(\delta_\varepsilon-d,\delta_\varepsilon)}(z) + f_2(z)\mathbf{I}_{[\delta_\varepsilon,+\infty)}(z),$$

where $f_1(z) = az^2$ is a parabola, $g(z)$ is a connecting function, $f_2(z) = b(z-\alpha)^2+\beta$ is another parabola, δ is the transition point between the loss function intervals, d is the length of the connecting interval. By varying the coefficients a and b, we can attain low, but nonzero, sensitivity to small distortions. The connecting function $g(z)$ is chosen as a cubic parabola satisfying the condition of double differentiability of $\varrho(z)$ on $\mathbb{R}$ ($z \geq 0$):

$$\varrho(z) = \begin{cases} az^2, \quad 0 \leq z \leq \delta_\varepsilon - d, \\ \dfrac{b-a}{3d}z^3 + \dfrac{bd + \delta_\varepsilon(a-b)}{d}z^2 + \dfrac{(b-a)(\delta_\varepsilon - d)^2}{d}z + \\ \qquad\qquad + \dfrac{(a-b)(\delta_\varepsilon - d)^3}{3d}, \quad \delta_\varepsilon - d < z < \delta_\varepsilon, \\ b\left(z - \dfrac{2(b-a)\delta_\varepsilon + d(a-b)}{2b}\right)^2 + \\ \qquad + \dfrac{12\delta_\varepsilon(\delta_\varepsilon - d)a(b-a) + d^2(b+3a)(b-a)}{12b}, \quad z \geq \delta_\varepsilon. \end{cases} \tag{6.47}$$

Figures 6.1 and 6.2 present graphs of loss functions (6.47) and (6.45) respectively. In applications, it is easier to use the loss function (6.47) compared to the function (6.45) since the former satisfies the conditions of Theorem 6.11, and thus all iterations of the algorithm (6.44) exist for any value of the parameter δ_ε.

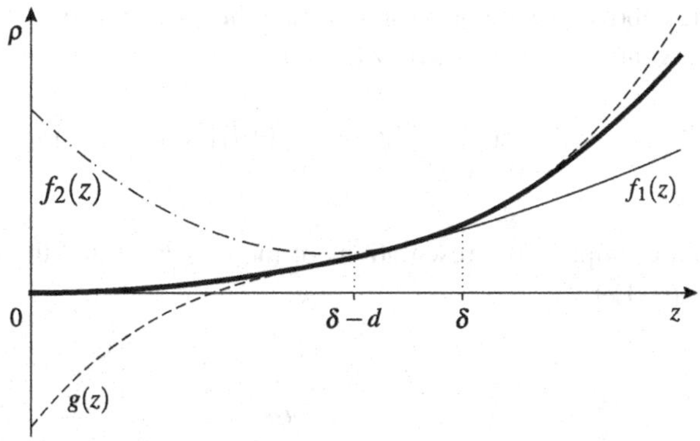

Fig. 6.2 Loss function (6.47)

6.5.2 *Evaluation of the Constructed Robust Forecasting Algorithm*

Let us study the recurrence procedure (6.44)–(6.46):

$$\theta_{(n+1)} = G(\theta_{(n)}) ::= \theta_{(n)} + \left(\Psi_T' \nu(\theta_{(n)}) \Psi_T\right)^{-1} \Psi_T' \mu(\theta_{(n)}), \quad n = 0, 1, 2, \dots .$$

We are going to investigate a single step of this procedure in a neighborhood of the true value θ^0, assuming $\theta_{(0)} ::= \theta^0$. The random variation of the "one-step" estimator $\hat{\theta} = \theta_{(1)} = G(\theta_{(0)})$ can be written as $\hat{\theta} - \theta^0 = A_T^{-1} B_T$, where by (6.1), (6.42), (6.43), (6.46) we have:

$$A_T = T^{-1} \sum_{t=1}^{T} \nu(u_t + \lambda(z_t)) \psi(z_t) \psi'(z_t),$$

$$B_T = T^{-1} \sum_{t=1}^{T} \mu(u_t + \lambda(z_t)) \psi(z_t) \in \mathbb{R}^T. \tag{6.48}$$

This technique of considering only the local properties of estimators in some neighborhood of the true value is commonly used in asymptotic statistical analysis [2].

Using (6.7), we obtain the following representation of the risk:

$$r = r(T, \tau) = \sigma^2 + \lambda^2(z_{T+\tau}) - 2\lambda(z_{T+\tau}) \psi'(z_{T+\tau}) b\{\hat{\theta}\} + \psi'(z_{T+\tau}) V\{\hat{\theta}\} \psi(z_{T+\tau}),$$

where $b\{\hat{\theta}\} = \mathbb{E}_{\theta^0}\{\hat{\theta} - \theta^0\}$ is the bias vector and $V\{\hat{\theta}\} = \mathbb{E}_{\theta^0}\{(\hat{\theta} - \theta^0)(\hat{\theta} - \theta^0)'\}$ is the mean square error matrix. The parameter δ_ε of the forecasting algorithm will be chosen to minimize the guaranteed risk:

$$r_+(T, \tau) \to \min_{\delta_\varepsilon}.$$

As mentioned above, we are going to consider the case (6.4) of FD-1 distortion. Let us find a maximum of the risk over $\lambda(z_{T+\tau})$:

$$r_+(T,\tau) = \sigma^2 + \varepsilon^2 + 2\varepsilon \sup_{\lambda(z_1),\dots,\lambda(z_T)} \left(\left| \psi'(z_{T+\tau}) b\{\hat{\theta}\} \right| + \psi'(z_{T+\tau}) V\{\hat{\theta}\} \psi(z_{T+\tau}) \right). \tag{6.49}$$

Let us obtain asymptotic representations of the bias $b\{\hat{\theta}\}$ and the mean square error matrix $V\{\hat{\theta}\}$. Let

$$\varphi(z \mid 0, \sigma^2) = \frac{1}{\sqrt{2\pi}\sigma} e^{-\frac{z^2}{2\sigma^2}}$$

be the $\mathcal{N}(0, \sigma^2)$ normal probability density, and let

$$\varphi(z) ::= \varphi(z \mid 0, 1), \quad \Phi(z) = \frac{1}{\sqrt{2\pi}} \int_{-\infty}^{z} e^{-0.5x^2} dx$$

be, respectively, the standard normal probability density and the standard normal distribution function. Let us introduce the notation:

$$\tilde{\delta}_\varepsilon = \delta_\varepsilon / \sigma, \quad \tilde{\lambda}_t = \lambda(z_t)/\sigma, \quad \mu_t = \mathbb{E}\{\mu(\lambda(z_t) + u_t)\},$$
$$\nu_t = \mathbb{E}\{\nu(\lambda(z_t) + u_t)\} = \mathbb{P}\{|u_t + \lambda(z_t)| > \delta_\varepsilon\},$$
$$H_T = T^{-1} \sum_{t=1}^{T} \psi(z_t)\psi(z_t)', \quad \overline{\psi} = T^{-1} \sum_{t=1}^{T} \psi(z_t) \in \mathbb{R}^m, \quad \tilde{\varepsilon} = \varepsilon/\sigma,$$
$$K_\varepsilon = 2\sigma^2 \Big((\tilde{\delta}_\varepsilon^2 + \tilde{\varepsilon}^2 + 1)\Phi(-\tilde{\delta}_\varepsilon) -$$
$$-2\tilde{\varepsilon}^2 \Phi^2(-\tilde{\delta}_\varepsilon) - \tilde{\delta}_\varepsilon \varphi(\tilde{\delta}_\varepsilon) \Big) \left(2\Phi(-\tilde{\delta}_\varepsilon) + \tilde{\varepsilon}^2 \tilde{\delta}_\varepsilon \varphi(\tilde{\delta}_\varepsilon) \right)^{-2}.$$

We are going to investigate asymptotic behavior of the sequences A_T, B_T defined in (6.48) as $T \to \infty$.

Lemma 6.3. *If the functions* $\{\psi_i(u)\}$ *in the regression model* (6.1) *are bounded,*

$$\sup_{u \in U} |\psi_i(u)| \le c_i < +\infty, \tag{6.50}$$

and u_t *has the normal probability density* $\varphi(z \mid 0, \sigma^2)$*, then the sequences of matrices* A_T *and vectors* B_T *defined by* (6.48) *satisfy the strong law of large numbers:*

$$A_T - T^{-1} \sum_{t=1}^{T} \nu_t \psi(z_t)\psi'(z_t) \xrightarrow{\text{a.s.}} \mathbf{0}_{m \times m}, \; B_T - T^{-1} \sum_{t=1}^{T} \mu_t \psi(z_t) \xrightarrow{\text{a.s.}} \mathbf{0}_m, \quad T \to \infty.$$

Proof. Taking the (i, j)th element of the matrix A_T and the ith element of the vector B_T, $i, j = 1, \ldots, m$, let us prove that the sequences A_T and B_T satisfy the strong law of large numbers by verifying Kolmogorov's sufficient condition [3], taking into account (6.46) and (6.48). Since $\nu(\cdot) \in \{0, 1\}$ is a Bernoulli random variable, the condition (6.50) implies that

$$K_A = \sum_{t=1}^{\infty} \frac{\mathbb{D}\{\nu(\lambda(z_t) + u_t)\psi_i(z_t)\psi_j(z_t)\}}{t^2} \leq c_i^2 c_j^2 \sum_{t=1}^{\infty} \frac{1}{4} t^{-2} < +\infty.$$

Since $\left|\tilde{\lambda}_t\right| \leq \tilde{\varepsilon}$, we have

$$\mathbb{D}\{\mu(\lambda(z_t) + u_t)\} < 2\sigma^2 \left((\tilde{\delta}_\varepsilon + \tilde{\varepsilon})^2 + 1 + \tilde{\delta}_\varepsilon + \tilde{\varepsilon}\right).$$

Then by (6.50), taking into account (6.46) and (6.48), we obtain:

$$K_B = \sum_{t=1}^{\infty} \mathbb{D}\{\mu(\lambda(z_t) + u_t)\psi_i(z_t)\} t^{-2} < 2\sigma^2 \left((\tilde{\delta}_\varepsilon + \tilde{\varepsilon})^2 + 1 + \tilde{\delta}_\varepsilon + \tilde{\varepsilon}\right) c_i^2 \sum_{t=1}^{\infty} t^{-2} < +\infty.$$

Convergence of this series proves the lemma. □

In numerical simulations [22] it was observed (and we can also see that from Theorem 6.2) that the forecast risk grows as the level of distortion increases. That motivates studying the case of the maximum possible distortion in the class (6.4): $\lambda(z_t) = \pm\varepsilon$, $t = 1, \ldots, T$. We are going to need the following two auxiliary lemmas.

Lemma 6.4. *If u_t has the normal probability density $\varphi(z \mid 0, \sigma^2)$, then:*

$$\begin{aligned}
\nu_t &= \Phi(-\tilde{\delta}_\varepsilon - \tilde{\lambda}_t) + \Phi(-\tilde{\delta}_\varepsilon + \tilde{\lambda}_t), \\
\mu_t &= \sigma\big((\tilde{\delta}_\varepsilon + \tilde{\lambda}_t)\Phi(-\tilde{\delta}_\varepsilon - \tilde{\lambda}_t) + \\
&\quad + (-\tilde{\delta}_\varepsilon + \tilde{\lambda}_t)\Phi(-\tilde{\delta}_\varepsilon + \tilde{\lambda}_t) + \varphi(\tilde{\delta}_\varepsilon - \tilde{\lambda}_t) - \varphi(\tilde{\delta}_\varepsilon + \tilde{\lambda}_t)\big), \\
\mathbb{E}\{\mu^2(\lambda(z_t) + u_t)\} &= \sigma^2\Big(\big((\tilde{\delta}_\varepsilon + \tilde{\lambda}_t)^2 + 1\big)\Phi(-\tilde{\delta}_\varepsilon - \tilde{\lambda}_t) - (\tilde{\delta}_\varepsilon + \tilde{\lambda}_t)\varphi(\tilde{\delta}_\varepsilon + \tilde{\lambda}_t) + \\
&\quad + \big((-\tilde{\delta}_\varepsilon + \tilde{\lambda}_t)^2 + 1\big)\Phi(-\tilde{\delta}_\varepsilon + \tilde{\lambda}_t) + (-\tilde{\delta}_\varepsilon + \tilde{\lambda}_t)\varphi(-\tilde{\delta}_\varepsilon + \tilde{\lambda}_t)\Big);
\end{aligned}$$

and asymptotically as $\tilde{\varepsilon} \to 0$ we have:

$$\mu_t = 2\sigma\tilde{\lambda}_t\Phi(-\tilde{\delta}_\varepsilon) + o(\tilde{\varepsilon}^2), \quad \nu_t = 2\Phi(-\tilde{\delta}_\varepsilon) + \tilde{\delta}_\varepsilon\varphi(\tilde{\delta}_\varepsilon)\tilde{\lambda}_t^2 + o(\tilde{\varepsilon}^2),$$

$$\mathbb{E}\{\mu^2(\lambda(z_t) + u_t)\} = 2\sigma^2\Big((\tilde{\delta}_\varepsilon^2 + 1)\Phi(-\tilde{\delta}_\varepsilon) - \tilde{\delta}_\varepsilon\varphi(\tilde{\delta}_\varepsilon) + \tilde{\lambda}_t^2\Phi(-\tilde{\delta}_\varepsilon)\Big) + o(\tilde{\varepsilon}^2).$$

Proof. The proof is based on the following simple chain of equalities:

$$\mu_t = \int_{-\infty}^{-\lambda(z_t)-\delta_\varepsilon} (\lambda(z_t) + \delta_\varepsilon + x)\varphi(x \mid 0, \sigma^2)dx +$$

$$+ \int_{\delta_\varepsilon-\lambda(z_t)}^{+\infty} (\lambda(z_t) - \delta_\varepsilon + x)\varphi(x \mid 0, \sigma^2)dx =$$

$$= \sigma\Big((\tilde{\delta}_\varepsilon + \tilde{\lambda}_t)\Phi(-\tilde{\delta}_\varepsilon - \tilde{\lambda}_t) + \varphi(\tilde{\delta}_\varepsilon - \tilde{\lambda}_t) - \varphi(\tilde{\delta}_\varepsilon + \tilde{\lambda}_t)\Big);$$

$$\nu_t = \int_{-\infty}^{-\lambda(z_t)-\delta_\varepsilon} \varphi(x \mid 0, \sigma^2)dx + \int_{\delta_\varepsilon-\lambda(z_t)}^{+\infty} \varphi(x \mid 0, \sigma^2)dx =$$

$$= \Phi(-\tilde{\delta}_\varepsilon - \tilde{\lambda}_t) + \Phi(-\tilde{\delta}_\varepsilon + \tilde{\lambda}_t);$$

$$\mathbb{E}\{\mu^2(\lambda(z_t) + u_t)\} = \int_{-\infty}^{-\lambda(z_t)-\delta_\varepsilon} (\lambda(z_t) + \delta_\varepsilon + x)^2\varphi(x \mid 0, \sigma^2)dx +$$

$$+ \int_{\delta_\varepsilon-\lambda(z_t)}^{+\infty} (\lambda(z_t) - \delta_\varepsilon + x)^2\varphi(x \mid 0, \sigma^2)dx =$$

$$= \sigma^2\Big(((\tilde{\delta}_\varepsilon + \tilde{\lambda}_t)^2 + 1)\Phi(-\tilde{\delta}_\varepsilon - \tilde{\lambda}_t) - (\tilde{\delta}_\varepsilon + \tilde{\lambda}_t)\varphi(\tilde{\delta}_\varepsilon + \tilde{\lambda}_t) +$$

$$+ ((\tilde{\delta}_\varepsilon - \tilde{\lambda}_t)^2 + 1)\Phi(-\tilde{\delta}_\varepsilon + \tilde{\lambda}_t) + (\tilde{\delta}_\varepsilon - \tilde{\lambda}_t)\varphi(\tilde{\delta}_\varepsilon - \tilde{\lambda}_t)\Big);$$

$$\mu_t = \sigma\Big((\tilde{\delta}_\varepsilon + \tilde{\lambda}_t)\Phi(-\tilde{\delta}_\varepsilon - \tilde{\lambda}_t) + \varphi(\tilde{\delta}_\varepsilon - \tilde{\lambda}_t) - \varphi(\tilde{\delta}_\varepsilon + \tilde{\lambda}_t)\Big) =$$

$$= 2\sigma\tilde{\lambda}_t\Phi(-\tilde{\delta}_\varepsilon) + o(\tilde{\lambda}_t^2);$$

$$\nu_t = \Phi(-\tilde{\delta}_\varepsilon - \tilde{\lambda}_t) + \Phi(-\tilde{\delta}_\varepsilon + \tilde{\lambda}_t) = 2\Phi(-\tilde{\delta}_\varepsilon) + \tilde{\delta}_\varepsilon\varphi(\tilde{\delta}_\varepsilon)\tilde{\lambda}_t^2 + o(\tilde{\lambda}_t^2).$$

Now we can write

$$\mathbb{E}\{\mu^2(\lambda(z_t) + u_t)\} = \sigma^2\Big(((\tilde{\delta}_\varepsilon + \tilde{\lambda}_t)^2 + 1)\Phi(-\tilde{\delta}_\varepsilon - \tilde{\lambda}_t) - (\tilde{\delta}_\varepsilon + \tilde{\lambda}_t)\varphi(\tilde{\delta}_\varepsilon + \tilde{\lambda}_t) +$$

$$+ ((\tilde{\delta}_\varepsilon - \tilde{\lambda}_t)^2 + 1)\Phi(-\tilde{\delta}_\varepsilon + \tilde{\lambda}_t) + (\tilde{\delta}_\varepsilon - \tilde{\lambda}_t)\varphi(\tilde{\delta}_\varepsilon - \tilde{\lambda}_t)\Big) =$$

$$= \sigma^2\Big((2\tilde{\delta}_\varepsilon^2 + 1)\Phi(-\tilde{\delta}_\varepsilon) - 2\tilde{\delta}_\varepsilon\varphi(\tilde{\delta}_\varepsilon) + 2\tilde{\lambda}_t^2\Phi(-\tilde{\delta}_\varepsilon)\Big) + o(\tilde{\lambda}_t^2).$$

Applying the convergence $\left|\tilde{\lambda}_t\right| \leq \tilde{\varepsilon} \to 0$ concludes the proof. □

Lemma 6.5. *Let $\lambda_t = \pm\varepsilon$, $t = 1,\ldots,T$, $\tilde{\varepsilon} \to 0$ and assume that there exists a matrix limit $H_0 = \lim_{T\to\infty} T^{-1}\sum_{t=1}^{T}\psi(z_t)\psi(z_t)'$, $|H_0| \neq 0$. Let u_t have the normal probability density $\varphi(z \mid 0,\sigma^2)$, then the following asymptotic expansions are satisfied:*

$$\mathbf{A}_\infty :: = \lim_{T\to\infty} T^{-1}\sum_{t=1}^{T}\nu_t\psi(z_t)\psi'(z_t) =$$

$$= \left(2\Phi(-\tilde{\delta}_\varepsilon) + \tilde{\varepsilon}^2\tilde{\delta}_\varepsilon\varphi(\tilde{\delta}_\varepsilon)\right)H_0 + o(\tilde{\varepsilon}^2)\mathbf{1}_{m\times m},$$

$$\mathbb{E}\{B_T\} = \pm 2\sigma\tilde{\varepsilon}\Phi(-\tilde{\delta}_\varepsilon)\overline{\psi} + o(\tilde{\varepsilon}^2)\mathbf{1}_m,$$

$$\mathbb{E}\{B_T B_T'\} = 4\sigma^2\tilde{\varepsilon}^2\Phi^2(\tilde{\delta}_\varepsilon)\overline{\psi\psi}' + 2\sigma^2/T\left((\tilde{\delta}_\varepsilon^2 + \tilde{\varepsilon}^2 + 1)\Phi(-\tilde{\delta}_\varepsilon) -\right.$$

$$\left.-2\tilde{\varepsilon}^2\Phi^2(-\tilde{\delta}_\varepsilon) - \tilde{\delta}_\varepsilon\varphi(\tilde{\delta}_\varepsilon)\right)H_0 + o(\tilde{\varepsilon}^2)\mathbf{1}_{m\times m}.$$

Proof. By (6.48) and Lemma 6.4, for $\lambda_t = \pm\varepsilon$ we have the expression for $\mathbf{A}_\infty$, and

$$\mathbb{E}\{B_T\} = T^{-1}\sum_{t=1}^{T}2\sigma(\pm\tilde{\varepsilon})\Phi(-\tilde{\delta}_\varepsilon)\psi(z_t) + o(\tilde{\varepsilon}^2)1_m = \pm 2\sigma\tilde{\varepsilon}\Phi(-\tilde{\delta}_\varepsilon)\overline{\psi} + o(\tilde{\varepsilon}^2)\mathbf{1}_m.$$

Taking into account the convergence $T^{-1}\sum_{t=1}^{T}\psi(z_t)\psi'(z_t) \underset{T\to\infty}{\to} H_0$, we obtain:

$$\mathbb{E}\{B_T B_T'\} = \frac{1}{T^2}\left(\sum_{t=1}^{T}\left(\mathbb{E}\{u_t + \mu^2(\lambda(z_t))\} - \mu_t^2\right)\psi(z_t)\psi'(z_t) + \right.$$

$$\left. + \sum_{t=1}^{T}\mu_t\psi(z_t)\left(\sum_{t=1}^{T}\mu_t\psi(z_t)\right)'\right) = \mathbb{E}\{B_T\}\mathbb{E}\{B_T\}' +$$

$$+ \frac{1}{T^2}\sum_{t=1}^{T}\left(2\sigma^2\left((\tilde{\delta}_\varepsilon^2 + 1)\Phi(-\tilde{\delta}_\varepsilon) - \tilde{\delta}_\varepsilon\varphi(\tilde{\delta}_\varepsilon) + \tilde{\varepsilon}^2\Phi(-\tilde{\delta}_\varepsilon)\right) + \right.$$

$$\left. + o(\tilde{\varepsilon}^2) - 4\sigma^2\tilde{\varepsilon}^2\Phi^2(-\tilde{\delta}_\varepsilon)\right)\psi(z_t)\psi'(z_t) = 4\sigma^2\tilde{\varepsilon}^2\Phi^2(-\tilde{\delta}_\varepsilon)\overline{\psi\psi}' +$$

$$+ \frac{2\sigma^2}{T}\left((\tilde{\delta}_\varepsilon^2 + \tilde{\varepsilon}^2 + 1)\Phi(-\tilde{\delta}_\varepsilon) - 2\tilde{\varepsilon}^2\Phi^2(-\tilde{\delta}_\varepsilon) - \tilde{\delta}_\varepsilon\varphi(\tilde{\delta}_\varepsilon)\right)H_0 +$$

$$+ o(\tilde{\varepsilon}^2)\mathbf{1}_{m\times m}.$$

□

By Lemmas 6.3–6.5, the following asymptotic relation holds for the forecast risk of the algorithm (6.44)–(6.46) under the condition $\tilde{\varepsilon} = \varepsilon/\sigma \to 0$ (i.e., if the FDs $\lambda(z_t)$ are much smaller than the observation errors u_t) [22]:

$$r_+ = \sigma^2 \left(1 + \tilde{\varepsilon}^2 \left(1 + |\psi'(z_{T+\tau}) H_0^{-1} \overline{\psi}|\right)^2\right) + o(\tilde{\varepsilon}^2) \quad \text{as} \quad T \to \infty.$$

Now consider the case of "large" values of $\tilde{\varepsilon} = \varepsilon/\sigma$. Then, by (6.48), choosing $\lambda(z_t) = \pm\varepsilon$ yields:

$$\mathbb{E}\{A_T\} = \left(\Phi(-\tilde{\delta}_\varepsilon - \tilde{\varepsilon}) + \Phi(-\tilde{\delta}_\varepsilon + \tilde{\varepsilon})\right) H_T,$$

$$\mathbb{E}\{B_T\} = \pm\sigma\left((\tilde{\delta}_\varepsilon + \tilde{\varepsilon})\Phi(-\tilde{\delta}_\varepsilon - \tilde{\varepsilon}) + (-\tilde{\delta}_\varepsilon + \tilde{\varepsilon})\Phi(-\tilde{\delta}_\varepsilon + \tilde{\varepsilon}) + \right.$$
$$\left. + \varphi(\tilde{\delta}_\varepsilon - \tilde{\varepsilon}) - \varphi(\tilde{\delta}_\varepsilon + \tilde{\varepsilon})\right)\overline{\psi}.$$

Let us define a function of two variables:

$$G(x, y) = \frac{(x + y)\Phi(-x - y) + (-x + y)\Phi(-x + y) + \varphi(x - y) - \varphi(x + y)}{\Phi(-x - y) + \Phi(-x + y)}.$$

Note that the function $G(x, y) \geq 0$ decreases in x for $y \geq 0$. By (6.49), the guaranteed mean square forecast risk can be expressed asymptotically as $T \to \infty$ in the following form [22]:

$$r_+(T, \tau) = \sigma^2 + \varepsilon^2 + 2\varepsilon\sigma G(\tilde{\delta}_\varepsilon, \tilde{\varepsilon}) \left|\psi'(z_{T+\tau}) H_0^{-1} \overline{\psi}\right| +$$
$$+ \sigma^2 G(\tilde{\delta}_\varepsilon, \tilde{\varepsilon})^2 \left(\psi'(z_{T+\tau}) H_0^{-1} \overline{\psi}\right)^2.$$

Let us choose $\tilde{\delta}_\varepsilon$ so that the asymptotic guaranteed risk is minimal. By the properties of the function $G(x, y)$, such $\tilde{\delta}_\varepsilon$ can be found by solving the following minimization problem:

$$G(\tilde{\delta}_\varepsilon, \tilde{\varepsilon}) \to \min_{\tilde{\delta}_\varepsilon \geq 0}.$$

Since the function $G(\tilde{\delta}_\varepsilon, \tilde{\varepsilon})$ is decreasing in $\tilde{\delta}_\varepsilon$ for $\tilde{\varepsilon} \geq 0$, minimization of this function implies taking the maximum value of $\tilde{\delta}_\varepsilon$. However, as seen from (6.46)–(6.48), increasing $\tilde{\delta}_\varepsilon$ leads to a decrease in the rank of the matrix A_T: the tth observation gives a nonzero contribution (is included) in the matrix A_T with probability $p_t = \nu_t = \Phi(-\tilde{\delta}_\varepsilon - \tilde{\varepsilon}) + \Phi(-\tilde{\delta}_\varepsilon + \tilde{\varepsilon})$. The inclusion process is a sequence of random Bernoulli trials, and the random number of included observations

$$\zeta = \sum_{t=1}^{T} \mathbf{1}\left(|\lambda(z_t) + u_t| - \delta_\varepsilon\right)$$

has the following expectation:

$$L(\tilde{\delta}_\varepsilon) = \mathbb{E}\{\zeta\} = \sum_{t=1}^{T} p_t = T(\Phi(-\tilde{\delta}_\varepsilon - \tilde{\varepsilon}) + \Phi(-\tilde{\delta}_\varepsilon + \tilde{\varepsilon})).$$

This $L(\tilde{\delta}_\varepsilon)$ decreases in $\tilde{\delta}_\varepsilon$ since $L'_{\tilde{\delta}_\varepsilon}(\tilde{\delta}_\varepsilon) = -T(\varphi(\tilde{\delta}_\varepsilon + \tilde{\varepsilon}) + \varphi(-\tilde{\delta}_\varepsilon + \tilde{\varepsilon})) < 0$. Thus, let us introduce the following restriction:

$$L(\tilde{\delta}_\varepsilon) \geq sT,$$

where s characterizes the "safety margin" of the algorithm w.r.t. inclusion of the observations, $0.5 < s < 1$, $s \geq mT^{-1}$. Then we have the following optimization problem:

$$G(\tilde{\delta}_\varepsilon, \tilde{\varepsilon}) \to \min_{\tilde{\delta}_\varepsilon \geq 0, L(\tilde{\delta}_\varepsilon) \geq sT}.$$

The functions $G(\tilde{\delta}_\varepsilon, \tilde{\varepsilon})$, $L(\tilde{\delta}_\varepsilon)$ decrease in $\tilde{\delta}_\varepsilon$, therefore, the solution of this problem, denoted as $\tilde{\delta}^*_\varepsilon$, can be found from the following equation

$$\Phi(-\tilde{\delta}_\varepsilon - \tilde{\varepsilon}) + \Phi(-\tilde{\delta}_\varepsilon + \tilde{\varepsilon}) = s. \tag{6.51}$$

Since the function $\Phi(-x - y) + \Phi(-x + y)$ decreases in x and increases in y for $x > 0$, this equation implies that an increase in the distortion level $\tilde{\varepsilon}$ leads to an increase of the optimal value $\tilde{\delta}^*_\varepsilon$. Note that if the distortion $\tilde{\varepsilon}$ is sufficiently small, then

$$\tilde{\delta}^*_\varepsilon \approx \Phi^{-1}(1 - 0.5s).$$

Observe that the function $\tilde{L}(\tilde{\delta}_\varepsilon) = \Phi(-\tilde{\delta}_\varepsilon - \tilde{\varepsilon}) + \Phi(-\tilde{\delta}_\varepsilon + \tilde{\varepsilon})$ is decreasing with a point of inflexion at $\tilde{\delta}_\varepsilon = \tilde{\varepsilon}$, and it tends to a step function $\tilde{L}_\infty(\tilde{\delta}_\varepsilon) = \mathbf{1}(\tilde{\varepsilon} - \tilde{\delta}_\varepsilon)$ as $\tilde{\varepsilon} \to +\infty$. In the latter case, the solution of (6.51) can be given as

$$\tilde{\delta}^*_\varepsilon \approx \tilde{\varepsilon} - \Phi^{-1}(s).$$

To conclude the subsection, let us note that the above approach to robust forecasting under FDs of multiple linear regression models (6.1) can be easily adapted to multivariate regression models (6.14). Vector forecasts can be made similarly to (6.41):

$$\hat{x}_{T+\tau} = \hat{\theta}\psi(z_{T+\tau}), \quad \hat{\theta} = \arg\min_{\theta} \sum_{t=1}^{T} \varrho(\|x_t - \theta\psi(z_t)\|),$$

where the loss function $\varrho(\cdot)$ is defined by (6.47).

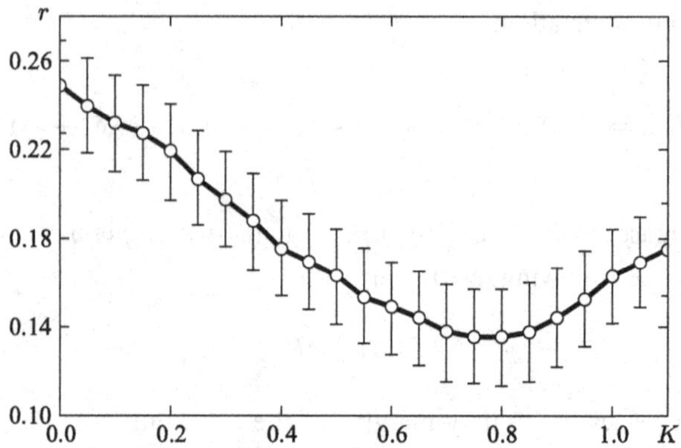

Fig. 6.3 Dependence between the risk and K for the loss function (6.45)

6.5.3 Numerical Examples

Robustness of the proposed algorithm (6.44) was evaluated in comparison with the traditional least squares forecasting algorithm by performing Monte-Carlo simulation experiments, where loss functions were chosen as (6.45) and (6.47).

Example 1. The experimental setting used the following model parameters:

$$T = 12; \quad \sigma^2 = 0.01; \quad \varepsilon = 0.5; \quad f_0(t) = 3 - 0.5t + 0.05t^2; \quad \tau = 1;$$
$$m = 3; \quad \lambda(t) = \varepsilon \cos(t); \quad \delta_\varepsilon = K\varepsilon, \quad K = 0, 0.05, 0.10, \ldots, 1.10;$$

the number of realizations was 1,000; the loss function was chosen as (6.45).

For each value of K, point and interval estimators (at confidence level 0.9) of the forecast risk $r(T, \tau)$ were computed (Fig. 6.3). Applying the algorithm (6.44) for $K = 0.8$ resulted in a 44 % reduction of the forecast risk (from 0.24 to 0.13) compared to least squares forecasting ($K = 0$). This difference is explained by discarding observations falling in the area of probable distortion.

Example 2. The experimental setting used the following model parameters:

$$T = 20; \quad \sigma^2 = 0.01; \quad \varepsilon = 0.5; \quad f_0(t) = 3 - 0.5t + 0.05t^2; \quad \tau = 1;$$
$$m = 3; \quad \lambda(t) = \varepsilon \cos t; \quad \delta_\varepsilon = K\varepsilon, \quad K = 0, 0.05, 0.10, \ldots, 1.50;$$

the number of realizations was 1,000; the loss function was chosen as (6.47) with $a = 0.1, b = 1$.

Interval estimators of the forecast risk were computed at 95 % confidence. Figure 6.4 shows that the forecast risk of the algorithm (6.44) for $K = 0.8$ was 24 %

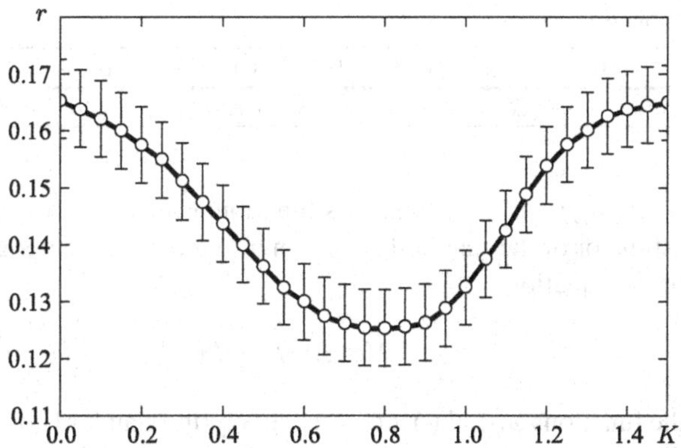

Fig. 6.4 Dependence between the risk and K for the loss function (6.47)

lower compared to the risk of the traditional least squares forecasting algorithm (0.125 vs. 0.165) due to reduced responsiveness to observations falling in the area of probable distortion.

6.6 Robust Regression Forecasting Under Outliers Based on the Huber Estimator

Let us consider the problem of constructing and evaluating a robust forecasting statistic in a multiple linear regression model distorted by outliers (6.21). In Sect. 6.3, we have studied the effect of outliers on the least squares forecasting statistic. To construct a robust forecasting statistic, let us use the M-estimator approach described in Sect. 6.5. Let us choose the Huber function as the loss function $\varrho(\cdot)$, following the approach of [13]:

$$\tilde{x}_{T+\tau} = \tilde{\theta}'\psi(z_{T+\tau}), \quad \tilde{\theta} = \arg\min_{\theta} \sum_{t=1}^{T} \varrho_H(r_t/\tilde{\sigma}), \tag{6.52}$$

$$r_t = r_t(\theta) = x_t - \theta'\psi(z_t),$$

$$\varrho_H(z) = \begin{cases} -Lz, & z \le -L, \\ z^2/2, & -L < z < L, \\ Lz, & z \ge L, \end{cases}$$

$$q(z) = \varrho_H'(z) = \max\{-L, \min\{L, z\}\} = \begin{cases} -L, & z \le -L, \\ z, & -L < z < L, \\ L, & z \ge L, \end{cases}$$

Table 6.1 Values $L = L(\varepsilon)$

ε	0	0.001	0.005	0.01	0.02	0.05	0.10	0.15	0.20	0.25	0.30	0.40
$L(\varepsilon)$	$+\infty$	2.63	2.16	1.95	1.72	1.40	1.14	0.98	0.86	0.77	0.69	0.55

where $\varrho_H(\cdot)$ is the Huber loss function; $\tilde{\theta}$ is the Huber estimator; $\tilde{\sigma}$ is some robust estimator of the standard deviation σ (e.g., the estimator given in [13]); L is the root of the equation

$$2(\varphi(L)/L - \Phi(-L)) = \varepsilon(1-\varepsilon); \tag{6.53}$$

the functions $\varphi(\cdot)$, $\Phi(\cdot)$ are, respectively, the standard normal probability density and distribution function; and $\varepsilon \in (0, 1/2)$ is a given distortion level, i.e., the probability of an outlier. Table 6.1 presents values of $L = L(\varepsilon)$ for some distortion levels ε.

We are going to use the notation of Sect. 6.1 together with some additional symbols:

$$H = (h_{ij}) = \Psi_T(\Psi_T'\Psi_T)^{-1}\Psi_T' = \Psi_T C_T \tag{6.54}$$

is a symmetric $(T \times T)$-matrix,

$$h = \max_{1 \le i \le T} h_{ii} > 0$$

is the maximum diagonal element of the matrix H. As shown in [13], evaluation of the asymptotic behavior of the estimator $\tilde{\theta}$ (and hence, the forecast risk of the algorithm (6.52)) under outliers may be restricted to a simplified case where the standard deviation σ is a priori known: $\tilde{\sigma} = \sigma$.

Theorem 6.12. *Under the regression model* (6.21) *with outliers in the variance* ($a_t \equiv 0$, $K \ge 2L^2$), $T > m$, $|\Psi_T'\Psi_T| > 0$, *let the random variables* u_t, v_t *have a normal distribution, and assume that the algorithm* (6.52) *with a known value of* σ *is being used for forecasting. Under the asymptotics*

$$\varepsilon \to 0, \quad T \to +\infty, \quad m \to +\infty, \quad h \to 0, \quad hm^2 \to 0, \tag{6.55}$$

the mean square risk of forecasting $x_{T+\tau}$ *satisfies the following expansion:*

$$r_\varepsilon(T,\tau) = r_0(T,\tau) + \varepsilon\sigma^2(K - 2L^2)\|g(T,\tau)\|^2 + O(\varepsilon^2) + o(h), \tag{6.56}$$

where $r_0(T,\tau) = \sigma^2(1 + \|g(T,\tau)\|^2) > 0$ *is the least squares forecast risk in the absence of outliers* ($\varepsilon = 0$) *defined in Sect. 6.1.*

Proof. Given the asymptotics (6.55), the mean square error matrix of the Huber estimator $\tilde{\theta}$ in (6.52) satisfies the following asymptotic expansion [13]:

$$V_{(\varepsilon)} = \mathbb{E}\{(\tilde{\theta}-\theta)(\tilde{\theta}-\theta)'\} = \frac{\mathbb{E}\{q^2(r_t(\theta))\}}{(\mathbb{E}\{q'(r_t(\theta))\})^2}(\Psi_T\Psi_T')^{-1} + o(h)\,\mathbf{1}_{m\times m}. \tag{6.57}$$

Applying (6.21), (6.52)–(6.55) yields

$$\frac{\mathbb{E}\{q^2(r_t(\theta))\}}{\left(\mathbb{E}\{q'(r_t(\theta))\}\right)^2} = \sigma^2\left(1 + (K - 2L^2)\varepsilon\right) + O(\varepsilon^2).$$

Substituting this expansion into (6.57) and using the general expression (6.26) for the risk of a plug-in forecasting algorithm leads to (6.56). □

Corollary 6.13. *The risk instability coefficient* (6.9) *of the forecasting algorithm* (6.52) *based on the Huber estimator satisfies the following asymptotic expansion:*

$$\kappa_H(T,\tau) = \frac{r_+(T,\tau) - r_0}{r_0} = \|g(T,\tau)\|^2 + \varepsilon_+(K-2L^2)\|g(T,\tau)\|^2 + O(\varepsilon_+^2) + o(h). \tag{6.58}$$

Proof. Maximizing (6.56) in $\varepsilon \in [0, \varepsilon_+]$ and applying the equality $r_0 = \sigma^2$ yields the expansion (6.58). □

As established in Sect. 6.3, the risk instability coefficient of the least squares forecasting algorithm under outliers in the variance can be written as

$$\kappa_{LS}(T,\tau) = \|g(T,\tau)\|^2 + \varepsilon_+ K\|g(T,\tau)\|^2. \tag{6.59}$$

A comparison of (6.58) and (6.59) shows that replacing the classical least squares estimator $\hat{\theta}$ defined by (6.11) with the Huber estimator $\tilde{\theta}$ defined by (6.52) in the plug-in forecasting algorithm leads to a decreased risk instability coefficient:

$$\kappa_{LS}(T,\tau) - \kappa_H(T,\tau) = \varepsilon_+ 2L^2\|g(T,\tau)\|^2 + O(\varepsilon_+^2) + o(h).$$

A numerical robustness comparison of forecasting algorithms (6.11) and (6.52) was performed via two series of Monte-Carlo simulations [15]. Ten independent experiments were performed for each combination of independent parameters.

Series 1. The time series x_t was simulated according to (6.21) with the following conditions: $T = 15$, $\tau \in \{1, 2, \ldots, 5\}$, $z_t = t$ (trend model), $m = 3$,

$$\psi'(z_t) = \psi'(t) = (1, t, t^2), \quad \theta^{0\prime} = (1, 0.1, 0.01), \quad \varepsilon \in \{0, 0.1, 0.2, 0.25, 0.3\},$$
$$\mathcal{L}\{u_t\} = N(0, \sigma^2), \quad \mathcal{L}\{v_t\} = N(0, K\sigma^2), \quad \sigma^2 = 0.09, \quad K = 50.$$

Table 6.2 presents sample estimators of the risks $\hat{r}_{LS}$, $\hat{r}_H$ for the least squares forecasting algorithm and the algorithm based on the robust Huber estimator, as well as the ratio of these two risk values denoted as

$$\Delta = \hat{r}_{LS}/\hat{r}_H,$$

for different distortion levels ε and forecasting depths τ.

Table 6.2 Robustness characteristics, Series 1

τ	Robustness characteristics	ε 0.00	0.10	0.20	0.25	0.30
1	r_{LS}	0.161	0.518	0.875	1.054	1.233
	r_H	0.162	0.201	0.247	0.274	0.307
	Δ	0.994	2.577	3.543	3.847	4.016
2	r_{LS}	0.207	0.791	1.375	1.667	1.959
	r_H	0.208	0.272	0.346	0.392	0.445
	Δ	0.995	2.908	3.974	4.253	4.402
3	r_{LS}	0.272	1.185	2.097	2.553	3.009
	r_H	0.274	0.374	0.490	0.561	0.644
	Δ	0.993	3.168	4.280	4.551	4.672
4	r_{LS}	0.363	1.730	3.097	3.780	4.464
	r_H	0.366	0.515	0.690	0.796	0.920
	Δ	0.918	3.359	4.488	4.749	4.852
5	r_{LS}	0.485	2.460	4.435	5.422	6.410
	r_H	0.489	0.704	0.956	1.110	1.290
	Δ	0.992	3.494	4.639	4.885	4.969

Table 6.3 Robustness characteristics, Series 2

τ	Robustness characteristics	ε 0.00	0.10	0.20	0.25	0.30
1	r_{LS}	0.116	0.248	0.380	0.446	0.512
	r_H	0.117	0.131	0.148	0.158	0.170
	Δ	0.991	1.893	2.568	2.823	3.012
2	r_{LS}	0.105	0.181	0.258	0.296	0.334
	r_H	0.105	0.114	0.123	0.129	0.136
	Δ	1.000	1.158	2.098	2.295	2.456
3	r_{LS}	0.102	0.161	0.221	0.250	0.280
	r_H	0.102	0.108	0.116	0.121	0.126
	Δ	1.000	1.491	1.905	2.066	2.222
4	r_{LS}	0.123	0.287	0.450	0.532	0.614
	r_H	0.123	0.141	0.162	0.175	0.190
	Δ	1.000	2.035	2.778	3.040	3.232
5	r_{LS}	0.103	0.168	0.233	0.265	0.297
	r_H	0.103	0.110	0.118	0.123	0.129
	Δ	1.000	1.527	1.975	2.154	2.302

Series 2. In the experimental setting of Series 1, the polynomial trend of the time series x_t is replaced by the following harmonic trend:

$$\psi'(t) = \big(1, \cos(t), \cos(2t)\big), \quad \theta^{0\prime} = (1, 0.5, 0.6).$$

Numerical results of this series of experiments are shown in Table 6.3.

Tables 6.2 and 6.3 show that using the Huber estimator in plug-in forecasting algorithms leads to a significant decrease of the risk. However, calculation of the Huber estimator $\tilde{\theta}$ requires exact knowledge of the outlier probability ε.

6.7 Local-Median (LM) Forecasting and Its Properties

6.7.1 *Description of the Method*

In Sects. 6.5 and 6.6, we have used the plug-in forecasting statistic

$$\hat{x}_{T+\tau} = f_0(x_1, \ldots, x_T; \tilde{\theta})$$

to construct robust algorithms of forecasting $x_{T+\tau}$. Here $f_0(x_1, \ldots, x_T; \theta)$ is the mean square optimal forecasting statistic under prior knowledge of the model parameter θ and without model distortion; $\tilde{\theta}$ is some robust estimator of the parameter θ (in Sects. 6.5 and 6.6, we have considered the family of M-estimators defined by the loss function $\varrho(\cdot)$). Let us note some disadvantages of this approach:

(1) computational difficulty of finding the estimator $\tilde{\theta}$ (in particular, iteration-based computation of $\tilde{\theta}$ is sensitive to the initial approximation);
(2) knowledge of the distortion level ε and other unknown characteristics is often essential.

This section presents a different approach to robust forecasting, initially proposed for robust trend forecasting [21]. Let us illustrate this approach and evaluate its robustness in the setting of regression forecasting under outliers discussed in Sect. 6.3.

Assume that observations $\{x_t\}$ of the investigated stochastic dynamic system are modeled by a regression equation under outliers:

$$x_t = \theta^{0\prime}\psi(z_t) + u_t + \xi_t v_t, \tag{6.60}$$

where $t \in Z$ is a discrete time point; $z_t \in \mathbb{R}^M$ is a nonrandom regressor (predictor) vector observed at time t; $\psi(z) = (\psi_i(z)) : \mathbb{R}^M \to \mathbb{R}^m$ is a vector of m linearly independent functions; $\theta^0 = (\theta_i^0) \in \mathbb{R}^m$ is the m-vector of unknown true model parameter values (regression coefficients); $u_t \in \mathbb{R}$ is a random error at time t and $v_t \in \mathbb{R}$ is an outlier at time t; $\{\xi_t\}$ are i.i.d. Bernoulli random variables that determine presence or absence of outliers, $\mathbb{P}\{\xi_t = 1\} = \varepsilon$, $\mathbb{P}\{\xi_t = 0\} = 1 - \varepsilon$, where $\varepsilon \in [0; 0.5)$ is the outlier probability (the average proportion of outliers in the observed sample).

It is assumed that random errors $\{u_t\}$ are i.i.d. random variables, $\mathbb{E}\{u_t\} = 0$, $\mathbb{D}\{u_t\} = \sigma^2 < +\infty$; the outliers $\{v_t\}$ are independent random variables such that

$\mathbb{E}\{\upsilon_t\} = a_t$, $\mathbb{D}\{\upsilon_t\} = K\sigma^2 < +\infty$, $K \geq 0$; the random variables $\{u_t\}$, $\{\upsilon_t\}$, $\{\xi_t\}$ are jointly independent.

Let us introduce the following notation: T is the observation length;

$$\Upsilon^{(l)} = \{t_1^{(l)}, \dots, t_n^{(l)}\} \subset N_T ::= \{1, 2, \dots, T\}, \quad l = 1, \dots, L,$$

are subsets of n ($m \leq n \leq T$) observed time points, where L is the number of all different subsets of time points ($m \leq L \leq L_+ = C_T^n$);

$$\Psi = (\psi_j(z_t)), \quad j = 1, \dots, m, \; t = 1, \dots, T;$$
$$\Psi_n^{(l)} = (\psi_j(z_{t_i^{(l)}})), \quad i = 1, \dots, n, \; j = 1, \dots, m,$$

is the $(n \times m)$-matrix obtained from the $(T \times m)$-matrix Ψ,

$$\left|\Psi_n^{(l)\prime}\Psi_n^{(l)}\right| \neq 0;$$

$X = (x_1, x_2, \dots, x_T)' \in \mathbb{R}^T$ is the observed sample;

$$X_n^{(l)} = \left(x_{t_1^{(l)}}, x_{t_2^{(l)}}, \dots, x_{t_n^{(l)}}\right)' \in \mathbb{R}^n$$

is a subsample of size n of the sample X; $a^{(l)} = (a_i^{(l)}) = (a_{t_1^{(l)}}, a_{t_2^{(l)}}, \dots, a_{t_n^{(l)}})' \in \mathbb{R}^n$.

As in (6.11), let us define the lth local least squares estimator for θ^0 based on the lth subsample $X_n^{(l)}$:

$$\hat{\theta}^{(l)} = (\Psi_n^{(l)\prime}\Psi_n^{(l)})^{-1}\Psi_n^{(l)\prime}X_n^{(l)}, \quad l = 1, \dots, L, \tag{6.61}$$

and use these local least squares estimators to construct a sample of L local forecasts of the future state $x_{T+\tau}$ for $\tau \geq 1$:

$$\hat{x}_{T+\tau}^{(l)} = \hat{\theta}^{(l)\prime}\psi(z_{T+\tau}), \quad l = 1, \dots, L. \tag{6.62}$$

The LM forecasting statistic was proposed in [21] as the sample median of the L local forecasts (6.62):

$$\hat{x}_{T+\tau} = S(X) = \text{Med}\left\{\hat{x}_{T+\tau}^{(1)}, \dots, \hat{x}_{T+\tau}^{(L)}\right\}. \tag{6.63}$$

The subsample size n and the number of subsamples L are the parameters of the LM method. If $n = T$, $L = 1$, then the LM forecast coincides with the traditional least squares forecast (6.11). If $L = L_+$, then the LM forecast (6.61)–(6.63) is based on all of the possible subsamples of size n from the initial sample of size T.

6.7.2 The Breakdown Point

Let us evaluate robustness of LM forecasting (6.63) by estimating the breakdown point in the Hampel sense. As in Sect. 4.4, the breakdown point is defined as the maximal proportion ε^* of "arbitrarily large outliers" in the sample X such that the forecasting statistic $S(X)$ cannot be made arbitrarily big by varying the values of the outliers [11]:

$$\varepsilon^* = \max\left\{\varepsilon \in [0,1] : \forall X_{(\varepsilon)} \ |S(X_{(\varepsilon)})| \le C < +\infty\right\}, \tag{6.64}$$

where $X_{(\varepsilon)} = \{x_t : \xi_t = 1, 1 \le t \le T\}$ is the set of observations (6.60) distorted by outliers.

Theorem 6.13. *If* $L = L_+ = C_T^n$*, then the breakdown point* (6.64) *of the LM forecast* (6.63) *under the distorted model* (6.60) *can be found as the unique root* $\varepsilon \in [0, 1 - nT^{-1}]$ *of the following algebraic equation of order* n*:*

$$\prod_{t=0}^{n-1}(1-\varepsilon-tT^{-1}) = (1-\alpha)\prod_{t=0}^{n-1}(1-tT^{-1}), \tag{6.65}$$

where

$$\alpha = \lfloor (L-1)/2 \rfloor / L = 1/2 + O\left(1/C_T^n\right). \tag{6.66}$$

Proof. First, let us show that (6.65) only has a unique root, and that this root ε_r lies in the interval $[0, 1 - nT^{-1}]$, i.e., it is sufficient to search for the breakdown point in this interval. Let us introduce the function

$$f(\varepsilon) = \prod_{t=0}^{n-1}(1-\varepsilon-t/T) - (1-\alpha)\prod_{t=0}^{n-1}(1-t/T)$$

to define an equivalent form of (6.65): $f(\varepsilon) = 0$. Since the derivative can be written as

$$f'(\varepsilon) = -\sum_{p=0}^{n-1}\ \prod_{t=0,t\neq p}^{n-1}(1-\varepsilon-t/T) < 0$$

for $\varepsilon \le 1 - n/T$, the function $f(\varepsilon)$ is strictly monotonous decreasing in ε in the interval $[0, 1 - nT^{-1}]$.

Consider the following cases:

(a) $n = T$; in this case, $\varepsilon_r = 0$, and thus ε_r lies in the interval $[0, 1 - nT^{-1}]$;
(b) $n = 1$, $T > 1$; in that case, $\varepsilon_r = \alpha \in [0, 1 - T^{-1}]$ since $\alpha \le 1/2$;
(c) $2 \le n \le T - 1$. Then we have

$$f(0) = \alpha \prod_{t=0}^{n-1} (1 - t/T) > 0, \quad f(1 - nT^{-1}) = n!\,T^{-n}\left(1 - (1-\alpha)C_T^n\right) < 0,$$

since $T > 2$, $n \le T - 1$, $\alpha \le 1/2$. The root ε_r of (6.65) exists and is unique due to strict monotonicity of the function $f(\varepsilon)$.

Let us proceed by obtaining an explicit form of (6.65). In [11], it was shown that the breakdown point α of the median calculated for a sample of local forecasts (6.62) equals

$$\alpha = \lfloor (L-1)/2 \rfloor / L = \begin{cases} 1/2 - L^{-1}, & \text{for } L \text{ even}, \\ 1/2 - (2L)^{-1}, & \text{for } L \text{ odd}. \end{cases}$$

Denote the number of distorted observations as $\beta = \varepsilon T \in N$. If $\varepsilon > 1 - nT^{-1}$, then $\beta > T - n$, and all local forecasts based on subsamples of size n are distorted. Thus, the LM forecast (6.63) is also distorted. Therefore, in (6.64) it is sufficient to consider $\varepsilon \in [0, 1 - nT^{-1}]$.

The total number of subsamples which only contain undistorted observations equals $C_{T-\beta}^n$. Assuming that the local-median breakdown point is equal to α implies that the number of local forecasts based on distorted subsamples should not exceed αL [11]. This leads to a condition on the maximum number of undistorted observations in the sample X: $C_{T-\beta}^n \ge (1-\alpha)C_T^n$. Applying (6.64) and performing equivalent transformations leads to (6.65), proving the theorem. □

Let us compute the breakdown point ε^* in some special cases.

Corollary 6.14. *If $n \le \alpha T$, then $\varepsilon^* \ge T^{-1} > 0$.*

Corollary 6.15. *If $n = m = 1$, then $\varepsilon^* = \alpha$; if $n = 2$, then we have*

$$\varepsilon^* = 1 - (2T)^{-1} - \left(\left(1 - (2T)^{-1}\right)^2 - \alpha\left(1 - T^{-1}\right)\right)^{1/2}.$$

Corollary 6.16. *If n is fixed, and $T \to \infty$, then $\varepsilon^* \to 1 - 2^{-1/n}$, and the subsample size which is optimal w.r.t. maximization of the breakdown point* (6.64) *is equal to the number of unknown parameters in the model* (6.60)*: $n^* = m$.*

6.7.3 Probability Distribution of the LM Forecast

Let us introduce the following notation:

$$\begin{aligned} g^{(l)} &= (g_i^{(l)}) = \Psi_n^{(l)}(\Psi_n^{(l)\prime}\Psi_n^{(l)})^{-1}\psi(z_{T+\tau}) \in \mathbb{R}^n, \quad l = 1, \ldots, L; \\ x_{T+\tau}^0 &= \theta^{0\prime}\psi(z_{T+\tau}); \\ \varphi(z \mid \mu; \sigma^2) &= (2\pi\sigma^2)^{-1/2} \exp\left(-(z-\mu)^2/(2\sigma^2)\right), \quad z \in \mathbb{R}, \end{aligned} \tag{6.67}$$

where $\varphi(\cdot)$ is the normal probability density function (PDF); $\Phi(\cdot)$ is the corresponding probability distribution function; $N_n = \{1, 2, \ldots, n\}$;

$$\Pi_r^n = \left\{A_j^{(r,n)} : j = 1, \ldots, C_n^r\right\}$$

is the set of C_n^r different ordered combinations

$$A_j^{(r,n)} = \left\{i_1^{(j)}, \ldots, i_r^{(j)} : \; i_1^{(j)} < i_2^{(j)} < \cdots < i_r^{(j)}, \quad i_k^{(j)} \in N_n, \; k = 1, \ldots, r\right\}$$

of r elements, and $\bar{A}_j^{(r,n)} ::= N_n \backslash A_j^{(r,n)}$;

$$G^{(n,r)(l)} = \sum_{k=r}^{n} (g_k^{(l)})^2, \quad r = 1, \ldots, n, \qquad G^{(n,r)(l)} = 0, \quad r > n,$$

and $G^{(l)} = G^{(n,1)(l)}$ is the squared length of the vector $g^{(l)}$, $l = 1, \ldots, L$.

Theorem 6.14. *Under the distorted regression model* (6.60)*, if the random variables* $\{u_t\}$*,* $\{v_t\}$ *are normally distributed, then the probability density of the* l*th local forecast* (6.62) *is a mixture of* 2^n *normal probability densities:*

$$\begin{aligned} p_{\hat{x}_{T+\tau}^{(l)}}(z) &= \sum_{r=0}^{n} (1-\varepsilon)^r \varepsilon^{n-r} \sum_{(k_1,\ldots,k_r)\in\Pi_r^n} p_{k_1,\ldots,k_r}(z), \quad z \in \mathbb{R}, \; l = 1, \ldots, L, \\ p_{k_1,\ldots,k_r}(z) &= \varphi\left(z \mid x_{T+\tau}^0 + \sum_{i=r+1}^{n} g_{k_i}^{(l)} a_{k_i}^{(l)} \; ; \; \sigma^2\left(G^{(l)} + KG^{(n,r+1)(l)}\right)\right), \end{aligned} \tag{6.68}$$

where $\{k_{r+1}, \ldots, k_n\} = N_n \backslash \{k_1, \ldots, k_r\}$ *is the subset of the elements in* N_n *which aren't included in the collection* $(k_1, \ldots, k_r)$.

Proof. Let us use the characteristic function approach. By model assumptions, the characteristic functions of u_t, v_t, $\xi_t v_t$, and $\eta_t = u_t + \xi_t v_t$ are written as

$$f_{u_t}(\lambda) = \exp\left(-\frac{1}{2}\sigma^2\lambda^2\right), \quad f_{v_t}(\lambda) = \exp\left(ia_t\lambda - \frac{K}{2}\sigma^2\lambda^2\right),$$

$$f_{\xi_t v_t}(\lambda) = (1-\varepsilon) + \varepsilon \exp\left(ia_t\lambda - \frac{K}{2}\sigma^2\lambda^2\right),$$

$$f_{\eta_t}(\lambda) = f_{u_t}(\lambda) f_{\xi_t v_t}(\lambda) = (1-\varepsilon)\exp\left(-\frac{1}{2}\sigma^2\lambda^2\right) +$$

$$+\varepsilon \exp\left(ia_t\lambda - \frac{K+1}{2}\sigma^2\lambda^2\right), \quad \lambda \in \mathbb{R}, \; t = 1, \ldots, T. \tag{6.69}$$

From (6.60), (6.62), (6.67), we have

$$\hat{x}^{(l)}_{T+\tau} = x^0_{T+\tau} + g^{(l)\prime}\eta^{(l)}, \tag{6.70}$$

where $\eta^{(l)} = (\eta_{t_1^{(l)}}, \ldots, \eta_{t_n^{(l)}})'$. Equations (6.69), (6.70), the independence of $\{\eta_t\}$, and the properties of the characteristic functions lead to

$$f_{\hat{x}^{(l)}_{T+\tau}}(\lambda) = e^{ix^0_{T+\tau}\lambda} \prod_{k=1}^{n} \left((1-\varepsilon)e^{-\frac{1}{2}(g_k^{(l)})^2\sigma^2\lambda^2} + \varepsilon e^{ia_k^{(l)} g_k^{(l)}\lambda - \frac{K+1}{2}\sigma^2(g_k^{(l)})^2\lambda^2}\right). \tag{6.71}$$

Using the inverse transformation formula [3] yields the probability density

$$\hat{x}^{(l)}_{T+\tau}: \quad p_{\hat{x}^{(l)}_{T+\tau}}(z) = \frac{1}{2\pi}\int_{-\infty}^{+\infty} e^{-iz\lambda} f_{\hat{x}^{(l)}_{T+\tau}}(\lambda)d\lambda, \quad z \in \mathbb{R}.$$

Now substituting (6.71) and performing equivalent transformations leads to (6.68).
□

Corollary 6.17. *For $l \in \{1, \ldots, L\}$, the probability density function of the lth local forecast satisfies the following asymptotic expansion as $\varepsilon \to 0$:*

$$p_{\hat{x}^{(l)}_{T+\tau}}(z) = \varphi\left(z \mid x^0_{T+\tau}; \sigma^2 G^{(l)}\right) +$$

$$+\varepsilon\left(\sum_{i=1}^{n} \varphi\left(z \mid x^0_{T+\tau} + g_i^{(l)}a_i^{(l)}; \sigma^2 G^{(l)} + \sigma^2 K (g_i^{(l)})^2\right) - n\varphi\left(z \mid x^0_{T+\tau}; \sigma^2 G^{(l)}\right)\right) +$$

$$+\varepsilon^2 \sum_{\substack{i,j=1,\\ i<j}}^{n} \varphi\left(z \mid x^0_{T+\tau} + g_i^{(l)}a_i^{(l)} + g_j^{(l)}a_j^{(l)}; \sigma^2 G^{(l)} + \sigma^2 K\left((g_i^{(l)})^2 + (g_j^{(l)})^2\right)\right) -$$

$$-\varepsilon^2(n-1)\sum_{i=1}^{n} \varphi\left(z \mid x^0_{T+\tau} + g_i^{(l)}a_i^{(l)}; \sigma^2 G^{(l)} + \sigma^2 K (g_i^{(l)})^2\right) +$$

$$+\frac{1}{2}\varepsilon^2 n(n-1)\varphi\left(z \mid x^0_{T+\tau}; \sigma^2 G^{(l)}\right) + o(\varepsilon^2). \tag{6.72}$$

Corollary 6.18. *For $l \in \{1, \ldots, L\}$, the probability distribution function of the lth local forecast satisfies the following asymptotic expansion as $\varepsilon \to 0$:*

$$F_{\hat{x}^{(l)}_{T+\tau}}(z) = \Phi(z \mid x^0_{T+\tau}\,; \sigma^2 G^{(l)}) +$$

$$+\,\varepsilon\left(\sum_{i=1}^{n} \Phi\left(z \mid x^0_{T+\tau} + g_i^{(l)} a_i^{(l)}\,; \sigma^2 G^{(l)} + \sigma^2 K (g_i^{(l)})^2\right) - n\Phi\left(z \mid x^0_{T+\tau}\,; \sigma^2 G^{(l)}\right)\right) +$$

$$+\,\varepsilon^2 \sum_{\substack{i,j=1,\\ i<j}}^{n} \Phi\left(z \mid x^0_{T+\tau} + g_i^{(l)} a_i^{(l)} + g_j^{(l)} a_j^{(l)}\,; \sigma^2 G^{(l)} + \sigma^2 K\left((g_i^{(l)})^2 + (g_j^{(l)})^2\right)\right) -$$

$$-\,\varepsilon^2 (n-1) \sum_{i=1}^{n} \Phi\left(z \mid x^0_{T+\tau} + g_i^{(l)} a_i^{(l)}\,; \sigma^2 G^{(l)} + \sigma^2 K (g_i^{(l)})^2\right) +$$

$$+\,\frac{1}{2}\varepsilon^2 n(n-1) \Phi\left(z \mid x^0_{T+\tau}\,; \sigma^2 G^{(l)}\right) + o(\varepsilon^2). \tag{6.73}$$

By (6.63), the LM forecast is an order statistic of a sample that is made up of L local forecasts. To find its probability distribution, assume that the local forecasts (6.62) are constructed so that they are jointly independent. The independence condition is satisfied, for instance, if $\bigcap_{l=1}^{L} \Upsilon^{(l)} = \emptyset$.

Let $\mathbf{1}(z) = \{1,\ z > 0;\ 0,\ z \le 0\}$, and let $M(z) = \sum_{i=1}^{L} \mathbf{1}(z - \hat{x}^{(i)}_{T+\tau})$ be the number of local forecasts below the level z; also denote

$$F^{(0)}_{\hat{x}^{(l)}_{T+\tau}}(z) = \Phi\left(z \mid x^0_{T+\tau}\,; \sigma^2 G^{(l)}\right); \tag{6.74}$$

$$U_{A,B}(z) = \left(\prod_{p \in A} F^{(0)}_{\hat{x}^{(p)}_{T+\tau}}(z)\right)\left(\prod_{q \in B}\left(1 - F^{(0)}_{\hat{x}^{(q)}_{T+\tau}}(z)\right)\right); \tag{6.75}$$

$$F^{(1)}_{\hat{x}^{(l)}_{T+\tau}}(z) = \sum_{i=1}^{n} \Phi\left(z \mid x^0_{T+\tau} + g_i^{(l)} a_i^{(l)}; \sigma^2 G^{(l)} + \sigma^2 K (g_i^{(l)})^2\right) - \tag{6.76}$$

$$-\,n\Phi\left(z \mid x^0_{T+\tau}; \sigma^2 G^{(l)}\right); \tag{6.77}$$

$$F^{(2)}_{\hat{x}^{(l)}_{T+\tau}}(z) = \tag{6.78}$$

$$= \sum_{\substack{i,j=1,\\ i<j}}^{n} \Phi\left(z \mid x^0_{T+\tau} + g_i^{(l)} a_i^{(l)} + g_j^{(l)} a_j^{(l)}; \sigma^2 G^{(l)} + \sigma^2 K\left((g_i^{(l)})^2 + (g_j^{(l)})^2\right)\right) -$$

$$-\,(n-1)\sum_{i=1}^{n} \Phi\left(z \mid x^0_{T+\tau} + g_i^{(l)} a_i^{(l)}; \sigma^2 G^{(l)} + \sigma^2 K (g_i^{(l)})^2\right) +$$

$$+\,(1/2) n(n-1) \Phi\left(z \mid x^0_{T+\tau}; \sigma^2 G^{(l)}\right).$$

Theorem 6.15. *If L local forecasts* (6.62) *are jointly independent, and L is odd, then, under the conditions of Theorem 6.14, the probability distribution function of the local-median forecast* (6.63) *satisfies the following asymptotic expansion as* $\varepsilon \to 0$*:*

$$
\begin{aligned}
F_{\hat{x}_{T+\tau}}(z) = &\sum_{k=\frac{L+1}{2}}^{L} \sum_{j=1}^{C_L^k} U_{A_j^{(k,L)}, \bar{A}_j^{(k,L)}}(z) + \\
&+ \varepsilon \sum_{k=\frac{L+1}{2}}^{L} \sum_{j=1}^{C_L^k} \Bigg(\sum_{\alpha \in A_j^{(k,L)}} \left(U_{A_j^{(k,L)} \setminus \{\alpha\}, \bar{A}_j^{(k,L)}}(z) F^{(1)}_{\hat{x}^{(\alpha)}_{T+\tau}}(z) \right) - \\
&- \sum_{\alpha \in \bar{A}_j^{(k,L)}} \left(U_{A_j^{(k,L)}, \bar{A}_j^{(k,L)} \setminus \{\alpha\}}(z) F^{(1)}_{\hat{x}^{(\alpha)}_{T+\tau}}(z) \right) \Bigg) + \\
&+ \varepsilon^2 \sum_{k=\frac{L+1}{2}}^{L} \sum_{j=1}^{C_L^k} \Bigg(\sum_{\alpha \in A_j^{(k,L)}} \left(U_{A_j^{(k,L)} \setminus \{\alpha\}, \bar{A}_j^{(k,L)}}(z) F^{(2)}_{\hat{x}^{(\alpha)}_{T+\tau}}(z) \right) - \\
&- \sum_{\alpha \in \bar{A}_j^{(k,L)}} \left(U_{A_j^{(k,L)}, \bar{A}_j^{(k,L)} \setminus \{\alpha\}}(z) F^{(2)}_{\hat{x}^{(\alpha)}_{T+\tau}}(z) \right) + \\
&+ \sum_{\alpha \in A_j^{(k,L)}} \sum_{\beta \in A_j^{(k,L)} \setminus \{\alpha\}} \left(U_{A_j^{(k,L)} \setminus \{\alpha,\beta\}, \bar{A}_j^{(k,L)}}(z) F^{(1)}_{\hat{x}^{(\alpha)}_{T+\tau}}(z) F^{(1)}_{\hat{x}^{(\beta)}_{T+\tau}}(z) \right) - \\
&- \sum_{\alpha \in A_j^{(k,L)}} \sum_{\beta \in \bar{A}_j^{(k,L)}} \left(U_{A_j^{(k,L)} \setminus \{\alpha\}, \bar{A}_j^{(k,L)} \setminus \{\beta\}}(z) F^{(1)}_{\hat{x}^{(\alpha)}_{T+\tau}}(z) F^{(1)}_{\hat{x}^{(\beta)}_{T+\tau}}(z) \right) + \\
&+ \sum_{\alpha \in \bar{A}_j^{(k,L)}} \sum_{\beta \in \bar{A}_j^{(k,L)} \setminus \{\alpha\}} \left(U_{A_j^{(k,L)}, \bar{A}_j^{(k,L)} \setminus \{\alpha,\beta\}}(z) F^{(1)}_{\hat{x}^{(\alpha)}_{T+\tau}}(z) F^{(1)}_{\hat{x}^{(\beta)}_{T+\tau}}(z) \right) \Bigg) + o(\varepsilon^2).
\end{aligned}
\tag{6.79}
$$

Proof. Consider the jth order statistic $x_{(j)}$ computed for a sample of L local forecasts (6.62). Since the events $\{z > \hat{x}^{(j)}_{T+\tau}\}$ and $\{M(z) \geq j\}$ coincide, the probability distribution function of the jth order statistic equals

$$
F_{x_{(j)}}(z) = \mathbb{P}\{z > x_{(j)}\} = \sum_{k=j}^{L} \mathbb{P}\{M(z) = k\}.
$$

By (6.73), the random variable $\chi_i = \mathbf{1}(z - \hat{x}^{(i)}_{T+\tau})$ follows a Bernoulli probability distribution defined as

$$p_i = \mathbb{P}\{\chi_i = 1\} = 1 - \mathbb{P}\{\chi_i = 0\} = F_{\hat{x}^{(i)}_{T+\tau}}(z).$$

The quantity $M(z)$ is a sum of not identically distributed Bernoulli random variables. Since the local forecasts (6.62) are jointly independent, we obtain the characteristic function for $M(z)$:

$$f_M(\lambda) = \prod_{j=1}^{L} f_{\chi_j}(\lambda) = \prod_{j=1}^{L} \left(1 + p_j(e^{i\lambda} - 1)\right) =$$

$$= (1-p_1)\dots(1-p_L) + \sum_{k=1}^{L}(1-p_1)\dots(1-p_L)e^{i\lambda}p_k(1-p_k)^{-1} +$$

$$+ \sum_{\substack{k,l=1,\\ k<l}}^{L} (1-p_1)\dots(1-p_L)e^{2i\lambda}p_k p_l(1-p_k)^{-1}(1-p_l)^{-1} + \dots + p_1\dots p_L e^{Li\lambda}.$$

Hence, we have

$$\mathbb{P}\{M(z) = k\} = \left(\sum_{A_j^{(k,L)},\ j=1,\dots,C_L^k} \frac{p_{i_1}\dots p_{i_k}}{(1-p_{i_1})\dots(1-p_{i_k})} \right) \prod_{i=1}^{L}(1-p_i),$$

and thus the probability distribution function is equal to

$$F_{x_{(j)}}(z) = \prod_{i=1}^{L}(1-p_i) \sum_{k=j}^{L} \left(\sum_{A_j^{(k,L)},\ j=1,\dots,C_L^k} \frac{p_{i_1}\dots p_{i_k}}{(1-p_{i_1})\dots(1-p_{i_k})} \right),$$

leading to

$$F_{x_{(j)}}(z) = \prod_{i=1}^{L}\left(1-F_{\hat{x}^{(i)}_{T+\tau}}(z)\right) \sum_{k=j}^{L} \left(\sum_{\substack{A_j^{(k,L)}\\ j=1,\dots,C_L^k}} \frac{F_{\hat{x}^{(i_1)}_{T+\tau}}(z)\dots F_{\hat{x}^{(i_k)}_{T+\tau}}(z)}{\left(1-F_{\hat{x}^{(i_1)}_{T+\tau}}(z)\right)\dots\left(1-F_{\hat{x}^{(i_k)}_{T+\tau}}(z)\right)} \right).$$

Since the sample median is an order statistic, the probability distribution function of the LM forecast (6.63) has the following form:

$$F_{\hat{x}_{T+\tau}}(z) = \prod_{i=1}^{L}\left(1 - F_{\hat{x}^{(i)}_{T+\tau}}(z)\right) \sum_{k=\frac{L+1}{2}}^{L} \left(\sum_{\substack{A_j^{(k,L)} \\ j=1,\dots,C_L^k}} \frac{F_{\hat{x}^{(i_1)}_{T+\tau}}(z) \dots F_{\hat{x}^{(i_k)}_{T+\tau}}(z)}{\left(1 - F_{\hat{x}^{(i_1)}_{T+\tau}}(z)\right) \dots \left(1 - F_{\hat{x}^{(i_k)}_{T+\tau}}(z)\right)} \right). \tag{6.80}$$

Applying the expansion (6.73) with the notation (6.74)–(6.78), and then gathering the expansion terms of orders $O(1)$, $O(\varepsilon)$, and $O(\varepsilon^2)$ in (6.80) proves (6.79).

□

6.7.4 Risk of the LM Forecast

The risk of a τ-step-ahead statistical LM forecast (6.63) based on T observations is equal to the mean square forecast error:

$$r(T, \tau) = \mathbb{E}\left\{(\hat{x}_{T+\tau} - x_{T+\tau})^2\right\} \geq 0. \tag{6.81}$$

Let us consider the following special case (outliers in variance):

$$\mathbb{E}\{v_t\} = a_t = 0, \quad t = 1, \dots, T. \tag{6.82}$$

Let the positive-definite symmetric matrix $(\frac{1}{n}\Psi_n^{(l)\prime}\Psi_n^{(l)})^{-1} = Q_n$ be independent of the subsample number l. Then we have

$$G^{(n,r)(l)} = G^{(n,r)}, \quad l = 1, \dots, L, \ r = 1, 2, \dots \ . \tag{6.83}$$

Theorem 6.16. *Under the conditions of Theorem 6.14, assuming* (6.82), (6.83), *and independence of the local forecasts* (6.62), *the forecast risk* (6.81) *of the LM method* (6.63) *satisfies the following asymptotic relation as $T \to \infty$, $L \to \infty$:*

$$r(T, \tau) - \left(\sigma^2 + \varepsilon K \sigma^2 + \frac{\pi \sigma^2}{2L} \left(\sum_{r=0}^{n} \frac{(1-\varepsilon)^r \varepsilon^{n-r} \binom{n}{r}}{\sqrt{G^{(n,1)} + K G^{(n,r+1)}}} \right)^{-2} \right) \to 0. \tag{6.84}$$

Proof. By (6.60), (6.67), the forecast risk functional has the following form:

$$r(T, \tau) = \mathbb{E}\left\{(\hat{x}_{T+\tau} - x^0_{T+\tau} - u_{T+\tau} - \xi_{T+\tau} v_{T+\tau})^2\right\}.$$

Since the random variables $u_{T+\tau}$, $\xi_{T+\tau}$, $v_{T+\tau}$, and the statistic $\hat{x}_{T+\tau}$ are jointly independent, $a_t = 0$, $t = 1, \dots, T$, we have

$$r(T, \tau) = \mathbb{E}\left\{(\hat{x}_{T+\tau} - x^0_{T+\tau})^2\right\} + \mathbb{E}\left\{(u_{T+\tau} + \xi_{T+\tau} v_{T+\tau})^2\right\}.$$

It can be shown that $\mathbb{E}\{(u_{T+\tau} + \xi_{T+\tau} v_{T+\tau})^2\} = (\varepsilon K + 1)\sigma^2$.

By applying Theorem 6.14 and the conditions of this theorem, we obtain that the local forecasts (6.62) are identically distributed and have the probability density function

$$\tilde{p}(z) = \sum_{r=0}^{n}(1-\varepsilon)^r \varepsilon^{n-r} \sum_{(k_1,\dots,k_r)\in\Pi_r^n} \varphi\left(z \mid x^0_{T+\tau}; \sigma^2\left(G^{(n,1)} + KG^{(n,r+1)}\right)\right), \quad z \in \mathbb{R}. \tag{6.85}$$

Since the LM forecast (6.63) is a median of local forecasts with the PDF (6.85), its probability distribution is asymptotically normal as $L \to \infty$ [2] with the asymptotic expectation equal to $x^0_{T+\tau}$ and the asymptotic variance given by

$$\mathbb{D}\{\hat{x}^2_{T+\tau}\} = \left(4L\tilde{p}^2(\zeta_{1/2})\right)^{-1},$$

where $\zeta_{1/2} = x^0_{T+\tau}$ is the median of the distribution (6.85). Straightforward transformations lead to the following expression for the asymptotic variance of the LM forecast:

$$\mathbb{D}\{\hat{x}^2_{T+\tau}\} = \frac{\pi\sigma^2}{2L}\left(\sum_{r=0}^{n} \frac{(1-\varepsilon)^r \varepsilon^{n-r}\binom{n}{r}}{\sqrt{G^{(n,1)} + KG^{(n,r+1)}}}\right)^{-2}.$$

Applying the expressions obtained above for $\mathbb{E}\{(u_{T+\tau} + \xi_{T+\tau} v_{T+\tau})^2\}$ and $\mathbb{D}\{\hat{x}^2_{T+\tau}\}$ yields (6.84). □

Corollary 6.19. *Under the conditions of Theorem 6.16, the forecast risk of the LM method satisfies the asymptotic relation*

$$\begin{aligned} r(T,\tau) = \sigma^2 + \frac{\pi\sigma^2}{2L}G^{(n,1)} + \\ + \varepsilon\left(K\sigma^2 + \frac{\pi\sigma^2 n}{L}G^{(n,1)}\left(1 - \frac{\sqrt{G^{(n,1)}}}{\sqrt{G^{(n,1)} + KG^{(n,n)}}}\right)\right) + o(\varepsilon) \end{aligned} \tag{6.86}$$

as $\varepsilon \to 0$, $T \to \infty$, $L \to \infty$.

Proof. As ε tends to zero, we have

$$\begin{aligned} \mathbb{D}\{\hat{x}^2_{T+\tau}\}\frac{2L}{\pi\sigma^2} &= \left(\sum_{r=0}^{n} \frac{(1-\varepsilon)^r \varepsilon^{n-r}\binom{n}{r}}{\sqrt{G^{(n,1)} + KG^{(n,r+1)}}}\right)^{-2} = \\ &= G^{(n,1)}\left(1 + 2n\varepsilon\left(1 - \frac{\sqrt{G^{(n,1)}}}{\sqrt{G^{(n,1)} + KG^{(n,n)}}}\right)\right) + o(\varepsilon), \end{aligned}$$

which leads to (6.86). □

6.7.5 Robustness of LM Forecasting Compared to the Traditional Least Squares Method

It follows from the results of Sect. 6.3 that the forecast risk of the least squares forecasting algorithm is equal to

$$r_{LS}(T,\tau) = \sigma^2 + G_T\sigma^2 + \varepsilon(K\sigma^2 + G_T K\sigma^2),$$

where $G_T = \psi(z_{T+\tau})'(\Psi_T'\Psi_T)^{-1}\psi(z_{T+\tau}) > 0$.

Let us assume that the design of the experiment is orthogonal:

$$(\Psi_T'\Psi_T)^{-1} = T^{-1}\mathbf{I}_m,$$

where $\mathbf{I}_m$ is the $(m \times m)$ identity matrix, $(\Psi_n^{(l)\prime}\Psi_n^{(l)})^{-1} = n^{-1}\mathbf{I}_m$,

$$G_T = T^{-1}\psi(z_{T+\tau})'\psi(z_{T+\tau}), \quad G^{(n,1)} = n^{-1}\psi(z_{T+\tau})'\psi(z_{T+\tau}).$$

Denote $\varrho = G^{(n,n)}/G^{(n,1)} \in (0,1)$. Then by (6.86) we have:

$$r_{LS}(T,\tau) = \sigma^2 + \varepsilon K\sigma^2 + \sigma^2 T^{-1}\psi(z_{T+\tau})'\psi(z_{T+\tau}) + \varepsilon K\sigma^2 T^{-1}\psi(z_{T+\tau})'\psi(z_{T+\tau}),$$

$$R_{LM}(T,\tau) = \sigma^2 + \varepsilon K\sigma^2 + \frac{\pi\sigma^2}{2Ln}\psi(z_{T+\tau})'\psi(z_{T+\tau}) +$$
$$+\varepsilon\frac{\pi\sigma^2}{L}\psi(z_{T+\tau})'\psi(z_{T+\tau})\left(1 - \frac{1}{\sqrt{1+K\varrho}}\right) + o(\varepsilon).$$

Taking $L = Tn^{-1}$ (so that the independence condition on local forecasts is satisfied) yields

$$R_{LM}(T,\tau) = \sigma^2 + \varepsilon K\sigma^2 + \frac{\pi}{2}\sigma^2 T^{-1}\psi(z_{T+\tau})'\psi(z_{T+\tau}) +$$
$$+\varepsilon\pi\sigma^2 T^{-1} n\psi(z_{T+\tau})'\psi(z_{T+\tau})\left(1 - \frac{1}{\sqrt{1+K\varrho}}\right) + o(\varepsilon).$$

Discarding the remainder term and solving the inequality $r_{LM}(T,\tau) < r_{LS}(T,\tau)$ in ε leads to a criterion for preferring LM forecasting to least squares forecasting:

$$\varepsilon > \varepsilon^*(K) = \frac{\pi/2 - 1}{K - \pi n\left(1 - \frac{1}{\sqrt{1+K\varrho}}\right)} \quad \text{for} \quad K - \pi n\left(1 - \frac{1}{\sqrt{1+K\varrho}}\right) > 0.$$

Figure 6.5 shows the preference threshold curves for the LM method in the special case $n^* = m$, $\varrho = m^{-1}$, which are defined as

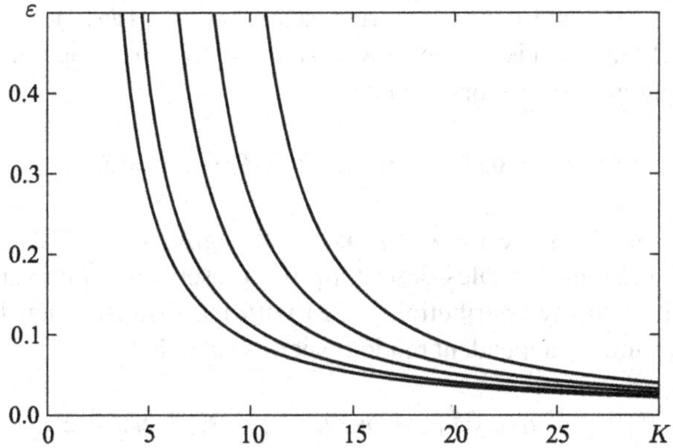

Fig. 6.5 Preference thresholds for the LM forecasting method, $m = 2, 3, 5, 7, 10$ (*left to right*)

$$\varepsilon = \varepsilon^*(K) = \frac{\pi/2 - 1}{K - \pi m \left(1 - (1 + K/m)^{-\frac{1}{2}}\right)}$$

under the assumption

$$K > m\left(2\pi - 1 - \sqrt{1 + 4\pi}\right)/2 \approx 0.8m$$

for $m = 2, 3, 5, 7, 10$.

Note that the function

$$d(\varepsilon, K, n) = 1 + \varepsilon K - \frac{\pi}{2} - \varepsilon \pi n \left(1 - \frac{1}{\sqrt{1 + K\varrho}}\right)$$

is monotonous increasing in ε and K. Also note that this function is monotonous decreasing in n, affirming the optimal subsample size $n^* = m$ obtained in Corollary 6.16.

6.7.6 A Generalization of the LM Method for Multivariate Regression

Let us consider a multivariate regression model under outliers, which is a generalization of the model (6.21):

$$x_t = \theta^0 \psi(z_t) + u_t + \xi_t \vartheta_t, \quad t \in \mathbb{N}, \tag{6.87}$$

where $z_t \in U \subseteq \mathbb{R}^M$ are nonrandom regressors; $\psi(\cdot) = (\psi_i(\cdot))$ is an m-column vector of linearly independent functions; $\{u_t\} \subset \mathbb{R}^d$ is a sequence of independent random observation error vectors such that

$$\mathbb{E}\{u_t\} = \mathbf{0}_d, \quad \mathbf{Cov}\{u_t, u_t\} = \mathbb{E}\{u_t u_t'\} = \Sigma;$$

$\theta^0 = (\theta^0_{ij}) \in \mathbb{R}^{d \times m}$ is the $(d \times m)$-matrix of regression coefficients; $\{\xi_t\}$ are i.i.d. Bernoulli random variables describing the presence of multivariate outliers following the probability distribution (6.22) with the distortion level equal to ε; $\{\vartheta_t\} \subset \mathbb{R}^d$ are jointly independent random vectors such that

$$\mathbb{E}\{\vartheta_t\} = a_t \in \mathbb{R}^d, \quad \mathbf{Cov}\{\vartheta_t, \vartheta_t\} = \mathbb{E}\{\vartheta_t \vartheta_t'\} = \widetilde{\Sigma};$$

and the random vectors $\{u_t\}, \{\vartheta_t\}, \{\xi_t\}$ are jointly independent.

Using the same approach as in Sect. 6.7.1, let us construct L subsets of time points $\Upsilon^{(l)} = \{t_1^{(l)}, \ldots, t_n^{(l)}\} \subseteq \{1, 2, \ldots, T\}$ and some auxiliary matrices:

$$C^{(l)} = \sum_{t \in \Upsilon^{(l)}} x_t \psi'(z_t) \in \mathbb{R}^{d \times m},$$

$$A^{(l)} = \sum_{t \in \Upsilon^{(l)}} \psi(z_t)\psi'(z_t) \in \mathbb{R}^{m \times m}, \quad l = 1, \ldots, L.$$

Assuming that $n = |\Upsilon^{(l)}| \geq m$ and the matrices $\{A^{(l)}\}$ are nonsingular, $|A^{(l)}| \neq 0$, let us construct the lth local least squares estimator for the regression coefficient matrix θ^0 based on a subsample $\{x_t : t \in \Upsilon^{(l)}\}$:

$$\hat{\theta}^{(l)} = C^{(l)}(A^{(l)})^{-1}, \quad l = 1, \ldots, L. \tag{6.88}$$

These least squares estimators will be used to construct a family of L local forecasts of the future state $x_{T+\tau}$ for a given forecasting horizon $\tau \in \mathbb{N}$:

$$\hat{x}^{(l)}_{T+\tau} = \hat{\theta}^{(l)}\psi(z_{T+\tau}) \in \mathbb{R}^d, \quad l = 1, \ldots, L. \tag{6.89}$$

Let us define the LM forecast similarly to the univariate case $d = 1$ (6.63) as the multivariate sample median of L local forecasts (6.89):

$$\hat{x}_{T+\tau} = S\left(\hat{x}^{(1)}_{T+\tau}, \ldots, \hat{x}^{(L)}_{T+\tau}\right) = \text{Med}\left\{\hat{x}^{(1)}_{T+\tau}, \ldots, \hat{x}^{(L)}_{T+\tau}\right\}. \tag{6.90}$$

Note that, unlike the univariate case, the multivariate median Med$\{\cdot\}$ in (6.90) isn't uniquely defined for $d > 1$ [26]. Let us briefly describe the five main methods of defining and computing the multivariate median: the sample L_1-median, the modified sample L_1-median, the Tukey sample median, the simplex sample median, and the Oja sample median.

Sample L_1-median

The sample L_1-median is defined as the point minimizing the sum of Euclidean distances between this point and the observations:

$$S = S(z_1, \ldots, z_T) = \arg\min_x \sum_{t=1}^{T} \|(x - z_T)\| / T. \tag{6.91}$$

For centered (at zero) distributions, the L_1-median equals zero. For $d = 1$, the median (6.91) coincides with the classical univariate median. Except for some singular cases, the minimum in (6.91) is attained at a single point. The sample L_1-median is a statistic with a limited Hampel influence function and a breakdown point at 1/2, but it isn't affine-invariant. It is difficult to compute for higher dimensions d [31].

Modified Sample L_1-Median

Several modifications of the L_1-median have been developed. These modifications are highly effective computationally, affine invariant, and can be easily computed for higher dimensions [12]. Hettmansperger's modified sample L_1-median $\widehat{\theta}_{L_1^H} \in \mathbb{R}^d$ is defined as a solution of the following system of equations:

$$S(\widehat{\theta}_{L_1^H}, \widehat{A}_\theta) = T^{-1} \sum_{t=1}^{T} \widehat{A}_\theta \left(z_t - \widehat{\theta}_{L_1^H}\right) \left\| \widehat{A}_\theta \left(z_t - \widehat{\theta}_{L_1^H}\right) \right\|^{-1} = \mathbf{0}_d, \quad \mathbf{0}_d \in \mathbb{R}^N,$$

$$T^{-1} \sum_{t=1}^{T} \widehat{A}_\theta \left(z_t - \widehat{\theta}_{L_1^H}\right) \left(z_t - \widehat{\theta}_{L_1^H}\right)' \widehat{A}_\theta' \left\| \widehat{A}_\theta \left(z_t - \widehat{\theta}_{L_1^H}\right) \right\|^{-2} = d^{-1} \mathbf{I}_d, \tag{6.92}$$

where $\widehat{A}_\theta \succ 0$ is a $(d \times d)$ upper triangle matrix with a one in the top left corner. This matrix is an M-estimator of Tayler's multivariate sample scale [32]. In a nonsingular case, the system of equations (6.92) has a unique solution [12].

The L_1-median is modified by the following scaling of the observations $z_1, \ldots, z_T$:

$$y_t = \widehat{A}_\theta z_t, \quad t = 1, \ldots, T.$$

The L_1-median of the scaled sample $\{y_t : t = 1, \ldots, T\}$ is then computed:

$$\widehat{\theta}_y = \mathrm{Med}_{L_1}\{y_1, \ldots, y_T\},$$

and the modified sample L_1-median is defined as the statistic

$$\widehat{\theta}_{L_1^H} = \widehat{A}_\theta^{-1} \widehat{\theta}_y.$$

For centered (at zero) distributions, the modified sample L_1-median equals zero; for $d = 1$ it coincides with the classical univariate median. The modified sample L_1-median is easily computable and affine invariant; its breakdown point lies in the interval $[1/(d+1), 1/d]$.

A simple iterative algorithm can be defined for computing the modified L_1-median [12]

$$d_{tj} = \|A_j(z_t - \theta_j)\|^2, \quad u_1(x) = 1/\sqrt{x}, \quad u_2(x) = d/x,$$

$$\theta_{j+1} = \left(\sum_{t=1}^{T} u_1(d_{tj})\right)^{-1} \sum_{t=1}^{T} u_1(d_{tj}) z_t,$$

$$\Sigma_j = T^{-1} \sum_{t=1}^{T} u_2(d_{tj})(z_t - \theta_{j+1})(z_t - \theta_{j+1})',$$

$$A_{j+1} : \Sigma_{j+1} = (A'_{j+1} A_{j+1})^{-1}. \tag{6.93}$$

Initial values for the algorithm (6.93) are chosen as follows: $\theta_1 = \mathbf{0}_d$, $A_1 = \mathbf{I}_d$. Several more involved algorithms, which converge more quickly, can be found in [12].

Tukey (halfspace) Sample Median

Let $U = \{u \in \mathbb{R}^d : |u| = 1\}$. For a pair of points $p \in \mathbb{R}^d$ and $u \in U$ define the following closed halfspace: $H(p, u) = \{y \in \mathbb{R}^d : u'y \geq u'p\}$. The number of observations lying outside the halfspace $H(p, u)$ equals

$$D(p, u) = \sum_{t=1}^{T} \mathbf{1}_{\mathbb{R}^d \setminus H(p,u)}(z_t).$$

The Tukey depth (or halfspace depth) of a point $p \in \mathbb{R}^d$ is the minimum number of observations that lie outside the subspaces $H(p, u)$, where $u \in U$:

$$HD(p) = \min_{u \in U} D(p, u).$$

The halfspace median (or the Tukey median) is defined as any point of $\mathbb{R}^d$ such that its halfspace depth is minimal:

$$HM = HM(z_1, \ldots, z_T) = \arg\min_{S \in \mathbb{R}^d} HD(S) \in \mathbb{R}^d.$$

The halfspace median is affine invariant, but it isn't uniquely defined since it can be equal to any point in the region $\arg\min_{T \in \mathbb{R}^d} HD(T) \subset \mathbb{R}^d$. Computational complexity of the halfspace median can be estimated as $O(T^2 \log^2 T)$ [29]. As shown in [31], the breakdown point of the halfspace median lies in the interval $[1/(d+1), 1/3]$.

Simplex Sample Median

Consider the following collections of $d+1$ indices:

$$\Upsilon^{(l)} = \{i_1^{(l)}, \ldots, i_{d+1}^{(l)} : 1 \le i_1^{(l)} < \ldots < i_{d+1}^{(l)} \le T\}, \quad l = 1, \ldots, C_T^{d+1}.$$

Based on these collections, let us construct d-dimensional simplices (or tetrahedra) $S^{(l)} = S^{(l)}\left(z_{i_1^{(l)}}, \ldots, z_{i_{d+1}^{(l)}}\right)$ with the vertices $z_{i_1^{(l)}}, \ldots, z_{i_{d+1}^{(l)}}$.

The simplex sample depth of a point $z \in \mathbb{R}^d$ is defined as the functional

$$SD(z) = \left(C_T^{d+1}\right)^{-1} \sum_{l=1}^{C_T^{d+1}} \mathbf{1}_{S^{(l)}}(z).$$

The simplex sample median is defined as the value that maximizes $SD(z)$:

$$SM = SM(z_1, \ldots, z_T) = \arg\max_{S \in \mathbb{R}^d} SD(S) \in \mathbb{R}^d.$$

In other words, the simplex median is the point lying in a maximal number of d-dimensional simplices constructed from the sample $z_1, \ldots, z_T$.

The simplex median is affine invariant. Its computational complexity is rather high [31]: $O(T^5 \log T)$ operations in the general case and $O(T^4)$ operations in two dimensions ($d = 2$).

Oja Sample Median

Consider collections of indices $\Upsilon^{(l)} = \{i_1^{(l)}, \ldots, i_d^{(l)} : 1 \le i_1^{(l)} < \ldots < i_d^{(l)} \le T\}$, where $l = 1, \ldots, C_T^d$, and the corresponding functionals

$$h_{\Upsilon^{(l)}}(z) = \det \begin{pmatrix} 1 & \ldots & 1 & 1 \\ z_{i_1^{(l)}} & \ldots & z_{i_d}^{(l)} & z \end{pmatrix}.$$

The collection $\Upsilon^{(l)}$ defines a hyperplane $H(\Upsilon^{(l)}) = \{z \in \mathbb{R}^d : h_{\Upsilon^{(l)}}(z) = 0\}$.

Denote

$$V_{\Upsilon^{(l)}}(z) = (d!)^{-1} |h_{\Upsilon^{(l)}}(z)|, \quad OD(Z) = \sum_{l=1}^{C_T^d} V_{\Upsilon^{(l)}}(z)/C_T^d.$$

The Oja sample median is defined as the statistic minimizing $OD(z)$:

$$OM = OM(z_1, \dots, z_T) = \arg\min_{S \in \mathbb{R}^d} OD(S) \in \mathbb{R}^d.$$

The Oja median is computationally efficient and affine invariant. It has a breakdown point at zero and a limited Hampel influence function; its computational complexity has been estimated in [26, 31] as $O(T^3 \log^2 T)$.

Some properties of the Oja median are:

- it lies in the intersection of the hyperplanes $\{H(\Upsilon^{(l)})\}$;
- the number of hyperplanes such that the Oja median lies in a positive halfspace w.r.t. them is the same as the number of hyperplanes satisfying the opposite (negative halfspace) condition.

In a simulation-based comparison of different multivariate medians for $d = 2$ (including the L_1-median, the Tukey median, the Oja median, and the simplex median, but not including the modified L_1-median), the standard L_1-median was found to be optimal in a series of experiments, which included measurements of the estimators' proximity to the center of the hypothetical distribution, as well as their performance and robustness [31].

6.7.7 Numerical Results

The performance of LM forecasting was evaluated by taking three examples including real-world and simulated data.

Example 1. Under a univariate regression model with outliers (6.21), a time series x_t was simulated using the following model parameters:

$$T = 15, \quad \tau \in \{1, 2, \dots, 5\}, \quad z_t = t \text{ (trend model)}, \quad m = 3,$$

$$\psi'(z_t) = \psi'(t) = (1, t, t^2), \quad \theta' = (1, 0.1, 0.01),$$

$$\mathcal{L}\{u_t\} = N(0, \sigma^2), \quad \mathcal{L}\{\vartheta_t\} = N(0, K\sigma^2), \quad \sigma^2 = 0.09, \quad K = 50.$$

Figure 6.6 presents the dependence of the sample forecast risk estimator $\hat{r}_{LM}$ computed over 10 independent realizations of a 4-step-ahead ($\tau = 4$) LM algorithm (6.63) on the size of local subsamples $n \in \{3, 4, \dots, 15\}$ for two simulated distortion levels: $\varepsilon = 0$ (no outliers) and $\varepsilon = 0.3$ (on average, 30 % of outliers are present in the time series). Dotted lines in Fig. 6.6 correspond to the risk of the least squares forecasting algorithm for $\varepsilon = 0$ and $\varepsilon = 0.3$. We can see that under outliers, the lowest LM forecast risk is attained for $n = m = 3$.

In Fig. 6.7, the sample risk $\hat{r}$ of the LM algorithm (solid lines) and the least squared algorithm (dashed lines) is plotted against the forecasting depth τ for

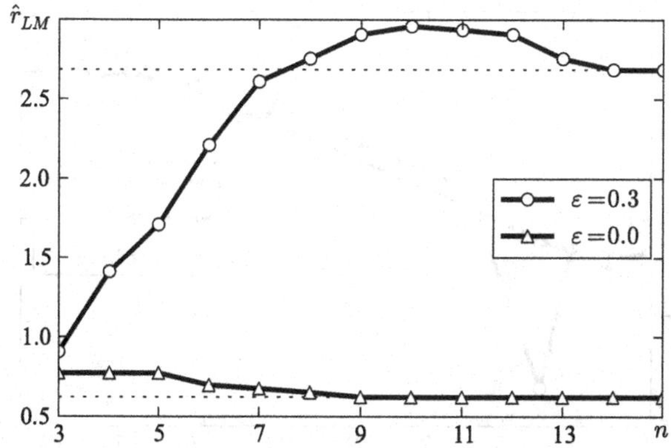

Fig. 6.6 Dependence between the forecast risk of the LM method and the local subsample size n

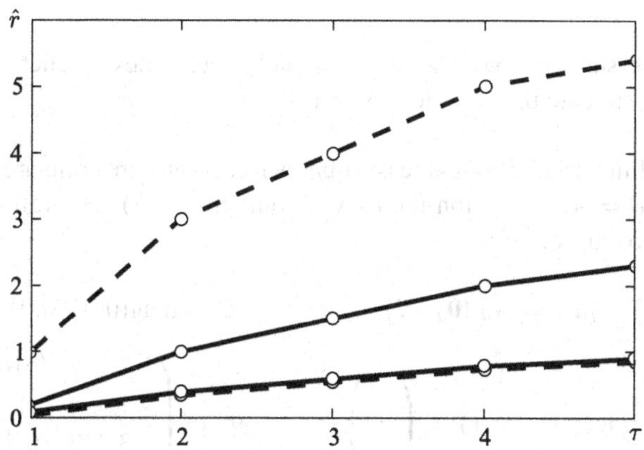

Fig. 6.7 Dependence of forecast risk on forecast depth τ

$n^* = 3$, $\varepsilon = 0$, and $\varepsilon = 0.21$. It is easy to see that for $\tau \in \{1, 2, 3, 4, 5\}$, $\varepsilon = 0.21$, the sample risk of the LM algorithm is approximately three times smaller compared to the least squares algorithm.

Example 2. In this example, the LM forecasting method was used to forecast the future revenue of "Polimir" chemical plant (a division of Belneftekhim oil concern) using a special case of the model (6.21) and real statistical data [15, 19]:

$$d = 1, \quad m = 3, \quad \psi(t) = (1, t, t^2)', \quad T = 14, \quad \tau \in \{1, 2, 3\}.$$

The time series $\{x_t\}$ was composed of logarithms of monthly revenues. A dataset collected over 14 months was used to make a forecast for the next 1–3 months. Graphs of the observed time series, the least squares forecast, and the LM forecast (for $n = m = 3$) are shown on Fig. 6.8. The absolute forecast error of the

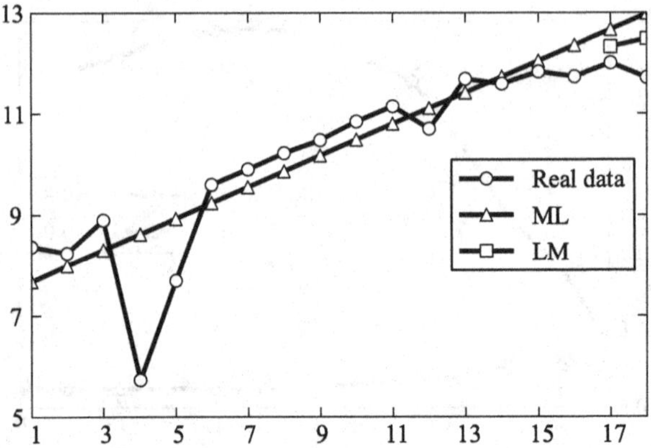

Fig. 6.8 Revenue forecasting: past and "future" values taken from a real-world dataset; the least squares and LM forecasts

LM forecasting statistic (6.63) is approximately four times smaller compared to the forecasting statistic based on least squares.

Example 3. Multivariate LM and least squares forecasts were compared under a d-variate (with $d = 4$) regression model with outliers (6.87). The following model parameters were chosen:

$$T = 25, \qquad \mathcal{L}\{u_t\} = N_d\left(\mathbf{0}_d, \Sigma\right), \qquad \Sigma = \text{diag}(0.01, 0.01, 0.01, 0.01),$$

$$m = 3, \qquad \psi(z_t) = \psi(t) = \begin{pmatrix} 1 \\ t \\ t^2 \end{pmatrix}, \qquad \theta^0 = \begin{pmatrix} 2 & -7 & 0.16 \\ -1 & -4 & 1 \\ 3 & 2 & 0.4 \\ 1 & 5 & 0.7 \end{pmatrix},$$

$$\tau = 1, \qquad n \in \{3, 4, \ldots, 25\}.$$

The fourth ($t = 4$) observation was replaced by an outlier $v_4 = (-1000, 0, 0, 0)$. For each value of n, 100 independent realizations of the time series were simulated; point and interval (at confidence level 0.95) estimators of the forecast risk $\hat{r}$ were computed for each realization. The forecasting statistic (6.63) was based on Hettmansperger's median (6.92).

The solid line in Fig. 6.9 shows the dependence of $\widehat{r}$ on the subsample size n. A larger-scale plot of the segment $n \in \{3, 4, 5, 6\}$ is presented in Fig. 6.10. Error bars indicate the confidence limits.

Figures 6.9 and 6.10 show that the minimum value of the LM forecast risk is attained for $n = 5$: $\widehat{r}_{LM} = 0.0607$. For this value of n, the empirical risk of the least squares forecast $\widehat{r}_{LS} = 170.27$ is roughly 3,000 times higher than the LM forecast risk.

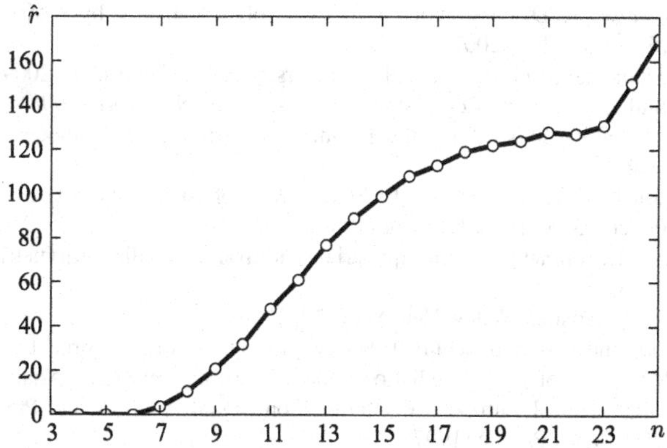

Fig. 6.9 Dependence of LM forecast risk on the subsample size

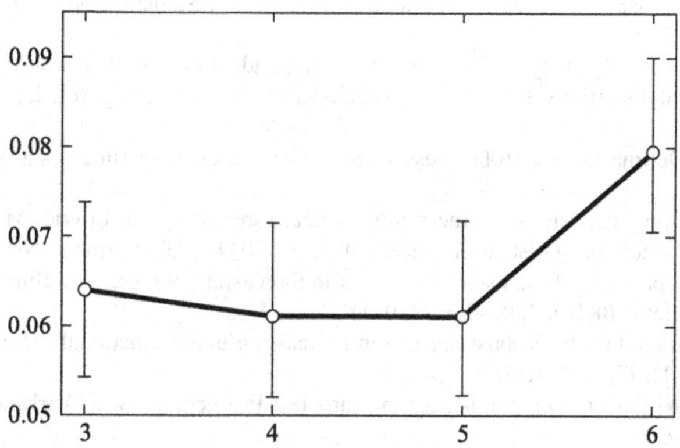

Fig. 6.10 LM forecast risk for $n \in \{3, 4, 5, 6\}$

Based on multiple simulations of the LM algorithm, we recommend to choose the subsample size n equal to the number of basis functions, $n^* = m$.

References

1. Atkinson, A., Riani, M.: Robust Diagnostic Regression Analysis. Springer, New York (2000)
2. Borovkov, A.: Mathematical Statistics. Gordon & Breach, Amsterdam (1998)
3. Borovkov, A.: Probability Theory. Gordon & Breach, Amsterdam (1998)
4. Brockwell, P., Davis, R.: Introduction to Time Series and Forecasting. Springer, New York (2002)
5. Bulinsky, A., Shyryaev, A.: Theory of Random Processes (in Russian). Fizmatlit, Moscow (2003)
6. Chatfield, C.: Time Series Forecasting. Chapman & Hall/CRC, Boca Raton (2001)

7. Croux, C., Filzmoser, P.: Discussion of "A survey of robust statistics" by S. Morgenthaler. Stat. Meth. Appl. **15**(3), 271–293 (2007)
8. Davison, A.: Statistical Models. Cambridge University Press, Cambridge (2009)
9. Draper, N., Smith, H.: Applied Regression Analysis. Wiley, New York (1998)
10. Green, P.: Penalized likelihood for general semi-parametric regression models. Int. Stat. Rev. **55**, 245–249 (1987)
11. Hampel, F., Ronchetti, E., Rousseeuw, P., Stahel, W.: Robust Statistics: The Approach Based on Influence Functions. Wiley, New York (1986)
12. Hettmansperger, T.: A practical affine equivariant multivariate median. Biometrika **89**(4), 851–860 (2002)
13. Huber, P.: Robust Statistics. Wiley, New York (1981)
14. Kharin, Y.: Robustness in Statistical Pattern Recognition. Kluwer, Dordrecht (1996)
15. Kharin, Y.: Robustness of signal prediction under distortions. In: Signal Analysis and Prediction. Proceedings of the I European Conference on Signal Analysis and Prediction, vol. 1, pp. 219–221. ICT Press, Prague (1997)
16. Kharin, Y.: Robust forecasting of parametric trends. Stud. Classif. Data Anal. **17**, 197–206 (2000)
17. Kharin, Y.: Robustness analysis in forecasting of time series. In: Dutter, R., Filzmoser, P., Gather, U., Rousseeuw, P. (eds.) Developments in Robust Statistics, pp. 180–193. Springer, Heidelberg (2003)
18. Kharin, Y.: Sensitivity analysis of risk of forecasting under multivariate linear regression model with functional distortions. In: Computer Data Analysis and Modeling, vol. 2, pp. 50–55. BSU, Minsk (2004)
19. Kharin, Y.: Optimality and Robustness in Statistical Forecasting (in Russian). BSU, Minsk (2008)
20. Kharin, Y.: Optimality and robustness in statistical forecasting. In: Lovric, M. (ed.) International Encyclopedia of Statistical Science, vol. 1, pp. 1034–1037. Springer, New York (2011)
21. Kharin, Y.: Robustness of the mean square risk in forecasting of regression time series. Comm. Stat. Theor. Meth. **40**(16), 2893–2906 (2011)
22. Kharin, Y., Maevskiy, V.: Robust regression forecasting under functional distortions. Autom. Rem. cont. (11), 125–136 (2002)
23. Launer, R., Wilkinson, G.: Robustness in Statistics: Proceedings of a Workshop. Academic, New York (1979)
24. Makridakis, S., Hyndman, R., Wheelwright, S.: Forecasting: Methods and Applications. Wiley, New York (1998)
25. Marcus, M., Minc, H.: A Survey of Matrix Theory and Matrix Inequalities. Allyn and Bacon, Boston (1964)
26. Oja, H.: Descriptive statistics for multivariate distributions. Stat. Probab. Lett. **1**, 327–332 (1983)
27. Pena, D., Tiao, G., Tsay, R.: A Course in Time Series Analysis. Wiley, New York (2001)
28. Rieder, H.: Robust Asymptotic Statistics. Springer, New York (1994)
29. Rousseeuw, P., Leroy, A.: Robust Regression and Outlier Detection. Chapman & Hall, London (1987)
30. Seber, G., Lee, A.: Linear Regression Analysis. Wiley-Interscience, Hoboken (2003)
31. Small, C.: A survey of multidimensional medians. Int. Stat. Rev. **58**, 263–277 (1990)
32. Tayler, D.: A distribution-free M-estimator of multivariate scatter. Ann. Stat. **15**, 234–251 (1987)
33. Varmuza, K., Filzmoser, P.: Introduction to Multivariate Statistical Analysis in Chemometrics. Taylor & Francis—CRC Press, Boca Raton (2009)

Chapter 7
Optimality and Robustness of ARIMA Forecasting

Abstract This chapter discusses robustness of univariate time series forecasting based on ARIMA time series models. Under complete prior knowledge, optimal forecasting statistics are constructed for the following undistorted hypothetical models: stationary time series models, AR(p), MA(q), ARMA(p, q), and ARIMA(p, d, q) models. Plug-in forecasting statistics are constructed for different types of prior uncertainty. Robustness of the obtained forecasting algorithms is evaluated under the following distortion types: parametric model specification errors, functional distortions of the innovation process in the mean value, heteroscedasticity, AO and IO outliers, bilinear autoregression distortions.

7.1 Kolmogorov's Method

Let us start this chapter by discussing two fundamental results of Andrey Kolmogorov. In his 1933 paper [17], Kolmogorov studied optimization of a k-predictor linear regression forecast based on a fixed size random sample, proving that "it is necessary to limit the number of predictors k, as well as the number of the variables that may be chosen as predictors; in this case the danger of obtaining an artificially inflated correlation coefficient between the predicted value and the predictors (and, consequently, underestimating the forecast risk) can be significantly reduced." This conclusion had a significant impact on applied forecasting methods in meteorology and geophysics, and it stays relevant to modern applied research. The paper [18], published in 1941, solves, in a general form, the problem of optimal linear forecasting of stationary random sequences if the covariance function or, equivalently, the spectral density of the predicted random sequence is a priori known.

Y. Kharin, *Robustness in Statistical Forecasting*, DOI 10.1007/978-3-319-00840-0_7,

Assume that a strictly stationary random sequence $x_t \in \mathbb{R}$, $t \in \mathbb{Z}$, with a zero expectation $\mathbb{E}\{x_t\} = 0$ (speaking without loss of generality) and an a priori known covariance function given as

$$\sigma_\tau \equiv \sigma_{-\tau} = \mathbb{E}\{x_t x_{t+\tau}\}, \quad \tau \in \mathbb{Z};\ \sigma_0 > 0,$$

is defined over a probability space $(\Omega, \mathcal{F}, \mathbb{P})$. We are going to study the problem of constructing a mean square optimal linear forecast of the future value $x_{T+\tau}$ based on $n+1$ observations at time T and earlier time points $\{x_T, x_{T-1}, \ldots, x_{T-n}\}$, where $\tau \geq 1$, $n \geq 0$ are given constants.

Let us introduce the following matrix notation:

$$X_T^{T-n} = \begin{pmatrix} x_T \\ x_{T-1} \\ \vdots \\ x_{T-n} \end{pmatrix} \in \mathbb{R}^{n+1} \quad \text{is a column vector of observations;}$$

$$\sigma_{(\tau)}^{(\tau+n)} = \begin{pmatrix} \sigma_\tau \\ \sigma_{\tau+1} \\ \vdots \\ \sigma_{\tau+n} \end{pmatrix} \in \mathbb{R}^{n+1} \quad \text{is a column vector of covariances;}$$

$$\Sigma_{(n+1)} = (\sigma_{ij}) \in \mathbb{R}^{(n+1)\times(n+1)} \quad \text{is a Toeplitz covariance matrix of order n,}$$

where $\sigma_{ij} = \sigma_{|i-j|}$, $i, j = 0, 1, \ldots, n$.

We are going to search for the optimal forecast in the family of all possible linear forecasts:

$$\hat{x}_{T+\tau} = a' X_T^{T-n}, \tag{7.1}$$

where $a = (a_i) \in \mathbb{R}^{n+1}$ is an arbitrary column vector of forecast coefficients. Also let

$$r(n+1, \tau) = \mathbb{E}\left\{(\hat{x}_{T+\tau} - x_{T+\tau})^2\right\} \tag{7.2}$$

be the mean square forecast risk of the forecasting statistic (7.1), and let

$$r_0(n+1, \tau) = \inf\{r(n+1, \tau) : a \in \mathbb{R}^{n+1}\} \tag{7.3}$$

be the exact lower bound on the forecast risk in the family of linear forecasts (7.1). Obviously, the function $r_0(n+1, \tau)$ is nonincreasing in n.

Theorem 7.1. *If* $\left|\Sigma_{(n+1)}\right| \neq 0$, *then for the stationary random sequence model defined above, the minimal mean square risk in the family of linear forecasts* (7.1) *is equal to*

$$r_0(n+1,\tau) = \sigma_0 - \left(\sigma_{(\tau)}^{(\tau+n)}\right)' \Sigma_{(n+1)}^{-1} \sigma_{(\tau)}^{(\tau+n)} \tag{7.4}$$

and is obtained for the forecasting statistic

$$\hat{x}_{T+\tau}^* = (a^*)' X_T^{T-n}, \quad a^* = \Sigma_{(n+1)}^{-1} \sigma_{(\tau)}^{(\tau+n)}. \tag{7.5}$$

Proof. By (7.1), (7.2), and the notation, we can establish a quadratic dependence between the risk and a:

$$r(n+1,\tau) = \mathbb{E}\left\{(a' X_T^{T-n} - x_{T+\tau})^2\right\} = \sigma_0 - 2a'\sigma_{(\tau)}^{(\tau+n)} + a'\Sigma_{(n+1)} a. \tag{7.6}$$

Let us consider a necessary minimization condition:

$$\nabla_a r(n+1,\tau) = 2\Sigma_{(n+1)} a - 2\sigma_{(\tau)}^{(\tau+n)} = \mathbf{0}_{n+1}.$$

Solving the above equation yields the point $a = a^*$ defined by (7.5). Since the matrix $\nabla_a^2 r(n+1,\tau) = 2\Sigma_{(n+1)}$ is positive-definite, the point (7.5) is the unique minimum of the risk function. Substituting (7.5) into (7.6) proves (7.4). □

Observe that if x_t is a Gaussian stationary random sequence, then by Theorem 4.1 we have that

$$r_0(n+1,\tau) = \mathbb{E}\left\{\mathbb{D}\{x_{T+\tau} | X_T^{T-n}\}\right\}$$

is the expectation of the conditional variance of the predicted random variable $x_{T+\tau}$ given a fixed history X_T^{T-n}.

Note that if $|\Sigma_{(n+1)}| = 0$, then the formulas (7.4), (7.5) cannot be used. Let us consider this singular case in more detail. Without loss of generality, assume

$$n = \min\left\{k : \left|\Sigma_{(k)}\right| \neq 0, \quad \left|\Sigma_{(k+1)}\right| = 0\right\}.$$

Since $\sigma_0 \neq 0$, we have $n \geq 1$. The properties of covariance matrices imply that if $\left|\Sigma_{(n+1)}\right| = 0$, and its submatrix $\Sigma_{(n)}$ is nonsingular, then there exists an n-vector $b = (b_i) \in \mathbb{R}^n$ such that for any time point $t \in \mathbb{Z}$ we have

$$x_t \overset{\text{a.s.}}{=} b' X_{t-1}^{t-n}, \tag{7.7}$$

i.e., any $n+1$ consecutive members of the random sequence are almost surely linearly dependent; due to stationarity, the coefficient vector b doesn't depend on t. By (7.7), we have a system of τ linear relations:

$$
\begin{aligned}
x_{T+\tau} &\overset{\text{a.s.}}{=} b' X_{T+\tau-1}^{T+\tau-n}, \\
x_{T+\tau-1} &\overset{\text{a.s.}}{=} b' X_{T+\tau-n-2}^{T+\tau-n-1}, \\
&\dots \\
x_{T+1} &\overset{\text{a.s.}}{=} b' X_{T}^{T-n+1},
\end{aligned} \tag{7.8}
$$

which allows us to obtain a linear expression for $x_{T+\tau}$ in $x_T, x_{T-1}, \dots, x_{T-n+1}$:

$$
x_{T+\tau} \overset{\text{a.s.}}{=} a'_* X_T^{T-n+1}. \tag{7.9}
$$

Here the coefficient vector $a_* = (a_{*i})$ is obtained from the vector b through step-by-step-ahead substitution of the linear relations in (7.8). For example:

- For $\tau = 1$, we have $a_* = b$;
- For $\tau = 2$,

$$
a_{*1} = b_1^2 + b_2, \quad a_{*2} = b_1 b_2 + b_3, \quad \dots, \quad a_{*,n-1} = b_1 b_{n-1} + b_n, \quad a_{*n} = b_1 b_n.
$$

From (7.9), (7.1), it follows that if the forecast coefficient vector is obtained from the relation

$$
a^* = \begin{pmatrix} a_* \\ \text{- - -} \\ 0 \end{pmatrix} \in \mathbb{R}^{n+1},
$$

then we have faultless forecasting, and consequently zero forecast risk:

$$
r_0(n+1, \tau) = 0. \tag{7.10}
$$

As noted above, the minimum risk $r_0(n+1, \tau)$ is a monotonous function:

$$
r_0(n+1, \tau) \le r_0(n, \tau), \quad n \ge 0.
$$

Let us analyze the size of the increments

$$
\Delta(n, \tau) = r_0(n, \tau) - r_0(n+1, \tau) \ge 0. \tag{7.11}
$$

Corollary 7.1. *If* $|\Sigma_{n+1}| \ne 0$, *then the risk increments* (7.11) *satisfy the formula*

$$
\Delta(n, \tau) = \frac{\left(\sigma_{\tau+n} - \left(\sigma_{(\tau)}^{(\tau+n-1)} \right)' \Sigma_{(n)}^{-1} \sigma_{(n)}^{(1)} \right)^2}{\sigma_0 - \left(\sigma_{(1)}^{(n)} \right)' \Sigma_{(n)}^{-1} \sigma_{(1)}^{(n)}} \ge 0. \tag{7.12}
$$

Proof. By (7.4), (7.11), we have

$$r_0(n+1,\tau) = \sigma_0 - Q_{n+1}, \qquad Q_{n+1} = \left(\sigma_{(\tau)}^{(\tau+n)}\right)' \Sigma_{(n+1)}^{-1} \sigma_{(\tau)}^{(\tau+n)},$$
$$r_0(n,\tau) = \sigma_0 - Q_n, \qquad Q_n = \left(\sigma_{(\tau)}^{(\tau+n-1)}\right)' \Sigma_{(n)}^{-1} \sigma_{(\tau)}^{(\tau+n-1)}, \tag{7.13}$$
$$\Delta(n,\tau) = Q_{n+1} - Q_n.$$

Let us rewrite the matrices included in the quadratic form Q_{n+1} in a block representation:

$$\Sigma_{(n+1)} = \left(\begin{array}{c|c} \Sigma_{(n)} & \sigma_{(n)}^{(1)} \\ \hline \left(\sigma_{(n)}^{(1)}\right)' & \sigma_0 \end{array}\right), \quad \sigma_{(\tau)}^{(\tau+n)} = \left(\begin{array}{c} \sigma_{(\tau)}^{(\tau+n-1)} \\ \hline \sigma_{\tau+n} \end{array}\right). \tag{7.14}$$

Taking into account (7.14) and the block matrix inversion formula [21], we obtain

$$\Sigma_{(n+1)}^{-1} = \left(\begin{array}{c|c} \left(\Sigma_{(n)} - \dfrac{1}{\sigma_0}\sigma_{(n)}^{(1)}\left(\sigma_{(n)}^{(1)}\right)'\right)^{-1} & -\dfrac{\Sigma_{(n)}^{-1}\sigma_{(n)}^{(1)}}{\sigma_0 - \left(\sigma_{(n)}^{(1)}\right)' \Sigma_{(n)}^{-1}\sigma_{(n)}^{(1)}} \\ \hline -\dfrac{\left(\Sigma_{(n)}^{-1}\sigma_{(n)}^{(1)}\right)'}{\sigma_0 - \left(\sigma_{(n)}^{(1)}\right)' \Sigma_{(n)}^{-1}\sigma_{(n)}^{(1)}} & \dfrac{1}{\sigma_0 - \left(\sigma_{(n)}^{(1)}\right)' \Sigma_{(n)}^{-1}\sigma_{(n)}^{(1)}} \end{array}\right),$$

where due to [22] we have

$$\left(\Sigma_{(n)} - \frac{1}{\sigma_0}\sigma_{(n)}^{(1)}\left(\sigma_{(n)}^{(1)}\right)'\right)^{-1} = \Sigma_{(n)}^{-1} + \frac{\Sigma_{(n)}^{-1}\sigma_{(n)}^{(1)}\left(\sigma_{(n)}^{(1)}\right)'\Sigma_{(n)}^{-1}}{\sigma_0 - \left(\sigma_{(1)}^{(n)}\right)' \Sigma_{(n)}^{-1}\sigma_{(1)}^{(n)}}.$$

Substituting this expression into (7.13), after equivalent transformations we obtain

$$Q_{n+1} = \left(\sigma_{(\tau)}^{(\tau+n-1)}\right)' \Sigma_{(n)}^{-1} \sigma_{(\tau)}^{(\tau+n-1)} + \frac{\left(\left(\sigma_{(\tau)}^{(\tau+n-1)}\right)' \Sigma_{(n)}^{-1}\sigma_{(n)}^{(1)}\right)^2}{\sigma_0 - \left(\sigma_{(1)}^{(n)}\right)' \Sigma_{(n)}^{-1}\sigma_{(1)}^{(n)}} -$$
$$- 2\frac{\left(\sigma_{(\tau)}^{(\tau+n-1)}\right)' \Sigma_{(n)}^{-1}\sigma_{(n)}^{(1)}\sigma_{\tau+n}}{\sigma_0 - \left(\sigma_{(1)}^{(n)}\right)' \Sigma_{(n)}^{-1}\sigma_{(1)}^{(n)}} + \frac{(\sigma_{\tau+n})^2}{\sigma_0 - \left(\sigma_{(1)}^{(n)}\right)' \Sigma_{(n)}^{-1}\sigma_{(1)}^{(n)}} =$$

$$= Q_n + \frac{\left(\sigma_{\tau+n} - \left(\sigma_{(\tau)}^{(\tau+n-1)}\right)' \Sigma_{(n)}^{-1} \sigma_{(n)}^{(1)}\right)^2}{\sigma_0 - \left(\sigma_{(1)}^{(n)}\right)' \Sigma_{(n)}^{-1} \sigma_{(1)}^{(n)}}.$$

Now (7.12) follows from (7.11), (7.13). □

Consider a special case of a stationary sequence—the autoregression time series of the first order AR(1) (as defined in Sect. 3.3):

$$x_t = \theta x_{t-1} + \xi_t, \quad t \in \mathbb{Z}, \tag{7.15}$$

where $\theta \in (-1, +1)$ is the autoregression coefficient, $\{\xi_t\}$ are jointly independent identically distributed random variables with zero expectations, $\mathbb{E}\{\xi_t\} = 0$, and variances $\mathbb{D}\{\xi_t\} = \sigma^2$.

Lemma 7.1. *An AR(1) time series defined by* (7.15) *has the following properties:*

$$\sigma_0 = \frac{\sigma^2}{1-\theta^2},$$

$$\sigma_{(\tau)}^{(\tau+n)} = \sigma_0 \begin{pmatrix} \theta^\tau \\ \theta^{\tau+1} \\ \vdots \\ \theta^{\tau+n} \end{pmatrix}; \text{ the matrix } \Sigma_{(n)} = \sigma_0 \begin{pmatrix} 1 & \theta & \theta^2 & \dots & \theta^{n-1} \\ \theta & 1 & \theta & \dots & \theta^{n-2} \\ \dots & \dots & \dots & \dots & \dots \\ \theta^{n-1} & \theta^{n-2} & \theta^{n-3} & \dots & 1 \end{pmatrix}$$

is a Toeplitz matrix; and the matrix

$$\Sigma_{(n)}^{-1} = \frac{1}{\sigma_0(1-\theta^2)} \begin{pmatrix} 1 & -\theta & 0 & \dots & 0 & 0 \\ -\theta & 1+\theta^2 & -\theta & \dots & 0 & 0 \\ \dots & \dots & \dots & \dots & \dots & \dots \\ 0 & 0 & 0 & \dots & 1+\theta^2 & -\theta \\ 0 & 0 & 0 & \dots & -\theta & 1 \end{pmatrix}$$

is a tridiagonal matrix.

Proof. The first three equalities have been proved in [2]. To prove the last equality, it is sufficient to verify the relation

$$\Sigma_{(n)} \Sigma_{(n)}^{-1} = \mathbf{I}_n.$$

□

Corollary 7.2. *For an AR(1) time series defined by* (7.15)*, we have*

$$\begin{aligned} a^* &= (\theta^\tau\, 0 \dots 0)', & \hat{x}^*_{T+\tau} &= \theta^\tau x_T, \\ r_0(n+1,\tau) &= \sigma^2 \frac{1-\theta^{2\tau}}{1-\theta^2}, & \Delta(n,\tau) &= 0. \end{aligned} \tag{7.16}$$

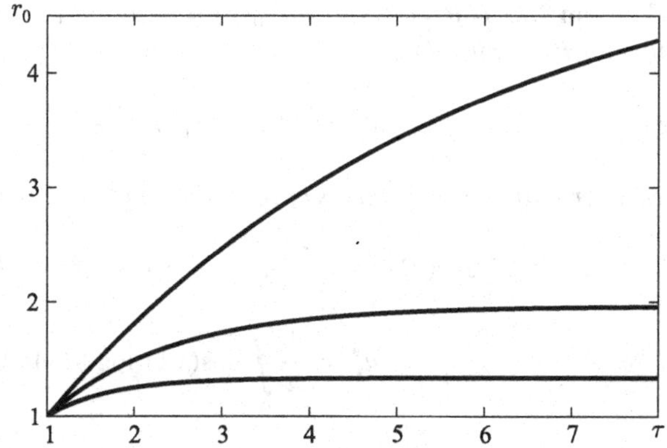

Fig. 7.1 Dependence of the minimum risk on the forecast depth τ, where $\theta = 0.9,\ 0.7,\ 0.5$ (from *top* to *bottom*)

Proof. Relations (7.16) follow from (7.4), (7.5), (7.12), and Lemma 7.1. □

In view of Lemma 7.1, the relations (7.16) show that despite the nonzero correlation of the predicted value $x_{T+\tau}$ and the history $\{x_T, x_{T-1}, \ldots, x_{T-n}\}$, the optimal forecast $\hat{x}^*_{T+\tau} = \theta^\tau x_T$ depends only on the one-step-back history (i.e., only on the observation x_T). Thus, the minimum forecast risk doesn't depend on n and is monotonous increasing as the forecast depth τ increases:

$$r_0(n+1,\tau) = \sigma^2 \frac{1-\theta^{2\tau}}{1-\theta^2}.$$

Figure 7.1 presents a plot of the minimum risk for $\theta \in \{0.5; 0.7; 0.9\}$, $\sigma^2 = 1$, and $\tau = 1, 2, \ldots, 8$.

In the most general case, Kolmogorov obtained an estimate for the limit value of the risk as the length of the observed history of the time series tends to infinity,

$$r_0(\tau) = \lim_{n\to+\infty} r_0(n+1,\tau),$$

by using the spectral density $S(\lambda)$ of the random sequence x_t. The spectral density and the covariance function σ_k are linked by a Fourier cosine transform:

$$\sigma_k = \frac{1}{\pi}\int_0^\pi \cos(k\lambda) S(\lambda) d\lambda, \quad k \in \mathbb{N}.$$

Let us introduce the notation

$$P = \frac{1}{\pi}\int_0^\pi \log S(\lambda) d\lambda.$$

Theorem 7.2. *If $P = -\infty$, then $r_0(\tau) = 0$ for all $\tau \geq 0$ (the singular case). If the integral P is finite, then*

$$r_0(\tau) = e^P(1 + r_1^2 + r_2^2 + \cdots + r_\tau^2),$$

where $\{r_k\}$ are the coefficients of the following power series expansion:

$$\exp(a_1^*\zeta + a_2^*\zeta^2 + \ldots) = 1 + r_1\zeta + r_2\zeta^2 + \ldots ,$$

$$a_k^* = \frac{1}{\pi}\int_0^\pi \cos(k\lambda)\, \log S(\lambda) d\lambda.$$

Theorem 7.2 was proved in [18]. Observe that if $S(\lambda) = 0$ on a set of positive Lebesgue measure, then $P = -\infty$ and we have the singular case.

7.2 Optimal Forecasting Under ARIMA Time Series Models

7.2.1 The General Method for Stationary Time Series

Let x_t, $t \in \mathbb{Z}$, be a stationary time series with a zero expectation, $\mathbb{E}\{x_t\} = 0$, the covariance function

$$\sigma_\tau \equiv \sigma_{-\tau} = \mathbb{E}\{x_t x_{t+\tau}\}, \quad \tau \in \mathbb{Z}, \quad \sigma_0 > 0,$$

and a spectral density $S(\lambda)$, $\lambda \in [-\pi, +\pi]$, which is defined by the Fourier cosine transform (3.8). The following theorem establishes the conditions that allow a representation of the stationary time series x_t in the form of a general linear process (3.18).

Theorem 7.3. *In the notation of Theorem 7.2, if*

$$P = \frac{1}{\pi}\int_0^\pi \log S(\lambda) d\lambda > -\infty,$$

then we have the following representation of x_t as a general linear process:

$$x_t = \sum_{j=0}^{+\infty} \gamma_j u_{t-j} \tag{7.17}$$

(in the mean square sense), where $\{u_t\}$ are uncorrelated random variables with zero expectations and finite variances $\mathbb{D}\{u_t\} = \sigma^2$, and $\{\gamma_j\}$ are uniquely determined coefficients such that

$$\gamma_0 = 1, \quad \sum_{j=0}^{+\infty} \gamma_j^2 < +\infty. \tag{7.18}$$

This theorem was proved in [2]. It is easy to see that (7.17) is a moving average model of infinite order, MA($+\infty$). In addition to the representation (7.17), a stationary time series can be written as an AR($+\infty$) autoregression model of infinite order. Let us introduce the following polynomials in a complex variable $z \in \mathbb{C}$:

$$\Gamma(z) = \sum_{j=0}^{+\infty} \gamma_j z^j, \quad B(z) = \frac{1}{\Gamma(z)} = \sum_{j=0}^{+\infty} \beta_j z^j. \tag{7.19}$$

Theorem 7.4. *Assume that the spectral density $S(\lambda)$ is bounded. Then the following $AR(+\infty)$-representation:*

$$u_t = \sum_{j=0}^{+\infty} \beta_j x_{t-j}, \quad t \in \mathbb{Z}; \quad \beta_0 = 1, \quad \sum_{j=0}^{+\infty} \beta_j^2 < +\infty; \tag{7.20}$$

holds if and only if

$$\Gamma(z) \neq 0, \quad |z| \leq 1,$$

and for $\varrho \to 1$ the integral

$$J = \int_0^{2\pi} \frac{1}{|\Gamma(\varrho e^{i\lambda})|^2} d\lambda$$

is bounded, i.e., if the function $1/\Gamma(z)$ belongs to the Hardy class H^2.

Theorem 7.4 was proved in [1].

Corollary 7.3. *The mean square optimal linear one-step-ahead ($\tau = 1$) forecast based on an infinite observed history $X_{-\infty}^T = (\ldots, x_{T-2}, x_{T-1}, x_T)'$ equals*

$$\hat{x}_{T+1} = -\sum_{j=1}^{+\infty} \beta_j x_{T+1-j} \tag{7.21}$$

and has the risk

$$r_0(+\infty, 1) = r_0(1) = \sigma^2. \tag{7.22}$$

Proof. From (7.20), for $t = T + 1$ we have

$$x_{T+1} = -\sum_{j=1}^{+\infty} \beta_j x_{T+1-j} + u_{T+1}. \tag{7.23}$$

By Theorem 4.1, the optimal forecasting statistic is defined as a conditional expectation, which by (7.23) can be written as

$$\hat{x}_{T+1} = \mathbb{E}\{x_{T+1} \mid x_T, x_{T-1}, \ldots\} = -\sum_{j=1}^{+\infty} \beta_j x_{T+1-j},$$

thus proving (7.21). The relation (7.22) follows from (7.21), (7.23), and the definition of the forecast risk. □

Corollary 7.4. *If a finite length-*$(n + 1)$ *history of a time series* $\{x_t\}$ *is observed,*

$$X_{T-n}^{T} = (x_{T-n}, \ldots, x_{T-1}, x_T)' \in \mathbb{R}^{n+1},$$

then the optimal linear forecast is the following truncation of (7.21)*:*

$$\tilde{x}_{T+1} = -\sum_{j=1}^{n+1} \beta_j x_{T+1-j}, \tag{7.24}$$

and the respective forecast risk equals

$$r_0(n + 1, 1) = \sigma^2 + \mathbb{E}\left\{\left(\sum_{j=n+2}^{+\infty} \beta_j x_{T+1-j}\right)^2\right\} \geq r_0(1). \tag{7.25}$$

Proof. From Theorem 4.1 and the total expectation formula, we have

$$\tilde{x}_{T+1} = \mathbb{E}\left\{x_{T+1} \mid X_{T-n}^{T}\right\} = \mathbb{E}\left\{\mathbb{E}\left\{x_{T+1} \mid X_{T-n}^{T}, X_{-\infty}^{T-n-1}\right\}\right\},$$

where the outer expectation is computed w.r.t. the complete history $X_{-\infty}^{T-n-1}$. Applying (7.23) and the fact that $\mathbb{E}\{x_t\} = 0$ yields (7.24). The relation (7.25) follows from (7.23), (7.24), and the definition of the forecast risk. □

Similarly to (7.21), (7.24), performing $\tau > 1$ repeated iterations of the relation (7.23) yields τ-step-ahead optimal forecasts for the general linear process. For example, taking $\tau = 2$ results in the two-step-ahead forecast

$$\hat{x}_{T+2} = \beta_1^2 x_T - \sum_{j=1}^{+\infty} (\beta_{j+1} - \beta_1 \beta_j) x_{T-j}.$$

Note that relations of the form (7.20) also hold for vector stationary time series (see Sect. 8.2).

7.2.2 Forecasting Under the AR(p) Model

From (3.15), the AR(p) model is defined by the following stochastic equation:

$$x_t - \alpha_1 x_{t-1} - \cdots - \alpha_p x_{t-p} = u_t, \quad 0 \leq \sum_{j=1}^{p} \alpha_j^2 < 1, \tag{7.26}$$

which is a special case of (7.20):

$$\beta_j = -\alpha_j, \; j = 1, 2, \ldots, p; \quad \beta_j = 0, \; j > p.$$

From Corollary 7.1, for $T > p$ the optimal forecast and its risk can be written, respectively, as

$$\hat{x}_{T+1} = \sum_{j=1}^{p} \alpha_j x_{T+1-j}, \quad r_0(T, 1) = \sigma^2. \tag{7.27}$$

Note that the optimal forecast (7.27) only depends on p previous observations X_{T+1-p}^T. It may seem that this forecast is rather inaccurate, since the forecast risk is equal to the variance of the innovation process σ^2. However, the value of the forecast risk must be considered in relation to the variance of the observed process:

$$R_0(T, 1) = \frac{r_0(T, 1)}{\mathbb{D}\{x_{T+1}\}}.$$

Take, for instance, $p = 1$, then $\mathbb{D}\{x_{T+1}\} = \sigma^2/(1 - \alpha_1^2)$ and $R_0(T + 1) = 1 - \alpha_1^2$. Thus, the relative risk of forecasting diminishes as the quantity $|\alpha_1|$ tends to 1, i.e., as the autoregression dependence becomes stronger (recall that, from the stationarity condition, we have $0 \leq |\alpha_1| \leq 1$).

A τ-step-ahead forecast is found by τ repeated applications of the recurrence relation (7.27):

$$\hat{x}_{T+\tau} = \sum_{j=1}^{p} (A^\tau)_{1j} \, x_{T+1-j}, \tag{7.28}$$

where the $(p \times p)$-matrix A is the companion matrix of the stochastic difference equation (7.26):

$$A = \begin{pmatrix} \alpha_1 & \alpha_2 & \alpha_3 & \dots & \alpha_{p-1} & \alpha_p \\ 1 & 0 & 0 & \dots & 0 & 0 \\ 0 & 1 & 0 & \dots & 0 & 0 \\ \vdots & \vdots & \vdots & \dots & \vdots & \vdots \\ 0 & 0 & 0 & \dots & 1 & 0 \end{pmatrix}.$$

As shown in [9], the forecast (7.28) can be computed by applying the recurrence relation for $j = 1, 2, \dots, \tau$:

$$\hat{x}_{T+j} = \alpha_1 \hat{x}_{T+j-1} + \alpha_2 \hat{x}_{T+j-2} + \dots + \alpha_p \hat{x}_{T+j-p}, \qquad (7.29)$$

where $\hat{x}_t = x_t$ for $t \leq T$.

7.2.3 Forecasting Under the MA(q) Model

From (3.16), the MA(q) model is defined by the following stochastic difference equation:

$$x_t = \gamma(B) u_t = u_t - \gamma_1 u_{t-1} - \dots - \gamma_q u_{t-q}, \quad t \in \mathbb{Z},$$

where the second expression is written in the operator form and B is the lag operator. Then the τ-step-ahead optimal forecast has the following form [9] for $\tau \leq q$:

$$\hat{x}_{T+\tau} = \frac{\gamma_\tau + \gamma_{\tau+1} B + \dots + \gamma_q B^{q-\tau}}{1 + \gamma_1 B + \dots + \gamma_q B^q} x_T, \qquad (7.30)$$

or, equivalently, $\hat{x}_{T+\tau} = \gamma_\tau \hat{\varepsilon}_T + \gamma_{\tau+1} \hat{\varepsilon}_{T-1} + \dots + \gamma_q \hat{\varepsilon}_{T+\tau-q}$, where $\{\hat{\varepsilon}_t\}$ is defined by the recurrence relation

$$\hat{\varepsilon}_t = x_t - \gamma_1 \hat{\varepsilon}_{t-1} - \gamma_2 \hat{\varepsilon}_{t-2} - \dots - \gamma_q \hat{\varepsilon}_{t-q}, \quad t = T - m + 1, \dots, T,$$

under the initial conditions

$$\hat{\varepsilon}_{T-m-q+1} = \hat{\varepsilon}_{T-m-q+2} = \dots = \hat{\varepsilon}_{T-m} = 0$$

for some sufficiently large m (optimality of the forecast is attained by passing to a limit as $m \to +\infty$).

Note that $\sigma_\tau = 0$ for $|\tau| > q$, i.e., the random variables $x_{T+\tau}$ and $\{x_T, x_{T-1}, \dots\}$ are uncorrelated. Thus, we have

$$\hat{x}_{T+\tau} \equiv 0, \quad \text{if } |\tau| > q. \tag{7.31}$$

By (7.31), in that case the forecast is equal to the expectation $\mathbb{E}\{x_{T+\tau}\} = 0$.

7.2.4 Forecasting Under the ARMA(p, q) Model

By (3.14), the ARMA(p, q) model is defined by the following stochastic difference equation:

$$\alpha(B)x_t = \gamma(B)u_t, \quad t \in \mathbb{Z},$$

$$\alpha(B) = 1 - \sum_{i=1}^{p} \alpha_i B^i, \qquad \gamma(B) = 1 - \sum_{j=1}^{q} \gamma_j B^j.$$

Let us represent this model in the AR$(+\infty)$ form:

$$u_t = \frac{\alpha(B)}{\gamma(B)} x_t, \quad t \in \mathbb{Z},$$

where the quotient $\alpha(B)/\gamma(B)$ is rewritten by expanding $1/\gamma(B)$ as an infinite power series and multiplying by the polynomial $\alpha(B)$. Now the general method of Sect. 7.2.1 can be applied to obtain the optimal forecast $\hat{x}_{T+\tau|T}$ for $x_{T+\tau}$ based on the observations X_1^T. This forecast can be computed recursively [9]:

$$\hat{x}_{T+s|T} = \begin{cases} \sum\limits_{i=1}^{p} \alpha_i \hat{x}_{T+s-i|T} - \sum\limits_{j=0}^{q} \gamma_j \hat{\varepsilon}_{T+s-j}, & s = 1, 2, \ldots, q; \\ \sum\limits_{i=1}^{p} \alpha_i \hat{x}_{T+s-i|T}, & s = q+1, q+2, \ldots, \tau, \end{cases} \tag{7.32}$$

where

$$\hat{x}_{t|T} = x_t \quad \text{for } t \le T, \quad \hat{\varepsilon}_t = x_t - \hat{x}_{t|t-1}. \tag{7.33}$$

7.2.5 Forecasting Under the ARIMA(p, d, q) Model

This model is defined by the stochastic difference equation (3.21). By this definition, a time series x_t fits the ARIMA(p, d, q) model if and only if the time series of order d differences

$$y_t = \Delta^d x_t, \quad t \in \mathbb{Z},$$

fits the ARMA(p, q) model. This leads to the following optimal forecasting algorithm ($T > d$):

1. Based on the observations $x_1, \ldots, x_T$, compute $T - d$ order d differences:

$$y_t = \Delta^d x_t, \quad t = d + 1, d + 2, \ldots, T;$$

2. Apply the results of Sect. 7.2.4 to construct optimal forecasts for the ARMA(p, q) model of the time series y_t:

$$\begin{aligned} \hat{y}_{T+\tau} &= f_\tau(y_{d+1}, \ldots, y_T), \\ \hat{y}_{T+\tau-1} &= f_{\tau-1}(y_{d+1}, \ldots, y_T), \\ &\cdots \\ \hat{y}_{T+d} &= f_{\tau-d}(y_{d+1}, \ldots, y_T); \end{aligned}$$

3. Based on the forecasts $\{\hat{y}_t\}$, compute the optimal forecast of $x_{T+\tau}$:

$$\hat{x}_{T+\tau} = g(\hat{y}_{T+\tau}, \hat{y}_{T+\tau-1}, \ldots, \hat{y}_{T+\tau-d}). \tag{7.34}$$

For example, in the ARIMA$(p, 1, q)$ model the optimal forecast is written as

$$\hat{x}_{T+\tau} = x_T + \sum_{i=1}^{\tau} \hat{y}_{T+i}.$$

7.3 Plug-In Forecasting Algorithms

7.3.1 Plug-In Forecasting Algorithms Based on Covariance Function Estimators

Section 7.1 presents an optimal forecasting algorithm (7.5) for stationary time series in a setting where the covariance function

$$\sigma_\tau = \mathbb{E}\left\{(x_t - \mathbb{E}\{x_t\})(x_{t+\tau} - \mathbb{E}\{x_{t+\tau}\})'\right\}, \quad \tau \in \mathbb{Z},$$

is known. In practice, this function usually remains unknown, and instead a consistent estimator $\hat{\sigma}_\tau$ constructed from the collected data is substituted into (7.5) in place of σ_τ. This yields a *plug-in forecasting algorithm based on an estimator of the covariance function* (we are considering the case $n = T - 1$):

$$\hat{x}_{T+\tau} = \hat{a}' X_T^1, \qquad \hat{a} = \hat{\Sigma}_{(n+1)}^{-1} \hat{\sigma}_{(\tau)}^{(\tau+n)},$$
$$\hat{\Sigma}_{n+1} = (\hat{\sigma}_{ij}) \in \mathbb{R}^{(n+1)\times(n+1)}, \qquad \hat{\sigma}_{ij} = \hat{\sigma}_{|i-j|}, \tag{7.35}$$
$$\hat{\sigma}_{(\tau)}^{(\tau+n)} = (\hat{\sigma}_\tau, \hat{\sigma}_{\tau+1}, \ldots, \hat{\sigma}_{\tau+n})'.$$

Estimation of σ_τ can be based on the following types of statistical data:

(a) The same time series $\{x_1, x_2, \ldots, x_T\}$ that is used in (7.35) to compute the forecast (unless stated otherwise, this case of "most meager" experimental data is assumed in this section);
(b) A different time series $\{\tilde{x}_1, \ldots, \tilde{x}_{T^0}\}$ of length T^0 independent of $\{x_1, \ldots, x_T\}$;
(c) Several time series independent of $\{x_1, \ldots, x_T\}$;
(d) Some combination of (a) and (b) or (c).

Let us present several statistical estimators $\hat{\sigma}_\tau$ which can be used in (7.35). The following two cases will be treated separately.

Case 1. $\mathbb{E}\{x_t\} = \mu$ is a priori known.
In this case, the sample covariance function

$$\hat{\sigma}_\tau = \frac{1}{T-|\tau|} \sum_{t=1}^{T-|\tau|} (x_t - \mu)(x_{t+|\tau|} - \mu), \quad \tau = 0, \pm 1, \ldots, \pm(T-1), \tag{7.36}$$

is a consistent unbiased estimator of σ_τ. If we also assume normality of the time series x_t, then the variance of this estimator equals [24]:

$$\mathbb{D}\{\hat{\sigma}_\tau\} = \frac{1}{T(1-|\tau|/T)^2} \sum_{m=-(T-|\tau|)+1}^{T-|\tau|-1} \left(1 - \frac{|m|+|\tau|}{T}\right) \left(\sigma_m^2 + \sigma_{m+\tau}\sigma_{m-\tau}\right), \tag{7.37}$$

and the following asymptotic relation holds as $T \to +\infty$:

$$\mathbb{D}\{\hat{\sigma}_\tau\} = \frac{1}{T-|\tau|} \sum_{m=-\infty}^{+\infty} \left(\sigma_m^2 + \sigma_{m+\tau}\sigma_{m-\tau}\right) + o\left(\frac{1}{T}\right).$$

Case 2. $\mathbb{E}\{x_t\} = \mu$ is a priori unknown.
In that case, the sample mean

$$\hat{\mu} = \overline{x} = \frac{1}{T} \sum_{t=1}^{T} x_t \tag{7.38}$$

is used as a consistent unbiased estimator of μ; the variance of the sample mean equals

$$\mathbb{D}\{\hat{\mu}\} = \frac{1}{T} \sum_{\tau=-(T-1)}^{T-1} (1 - |\tau|/T)\sigma_\tau;$$

and the following asymptotic equality holds as $T \to +\infty$:

$$\mathbb{D}\{\hat{\mu}\} = \frac{2\pi\sigma_0^2}{T} S(0) + o\left(\frac{1}{T}\right),$$

where $S(\lambda)$ is the spectral density.

A consistent asymptotically unbiased estimator of the covariance function for an unknown μ can be written as

$$\hat{\sigma}_\tau = \frac{1}{T - |\tau|} \sum_{t=1}^{T-|\tau|} (x_t - \overline{x})(x_{t+|\tau|} - \overline{x}), \quad \tau = 0, \pm 1, \ldots, \pm(T-1). \tag{7.39}$$

For $T \to \infty$, we have asymptotically

$$\mathbb{E}\{\hat{\sigma}_\tau\} = \sigma_\tau - \frac{2\pi\sigma_0^2}{T} S(0) + o\left(\frac{1}{T}\right).$$

Note that, under some fairly general conditions, the estimators (7.37)–(7.39) are asymptotically normally distributed as $T \to \infty$ [24].

7.3.2 *Plug-In Forecasting Algorithms Based on AR(p) Parameter Estimators*

From the results of Sect. 7.2.2, plug-in forecasting algorithms under the AR(p) model have the form defined by (7.28):

$$\hat{x}_{T+\tau} = \sum_{j=1}^{p} \left(\hat{A}^\tau\right)_{1j} x_{T+1-j},$$

where $\hat{A}$ is the following $(p \times p)$-matrix:

$$\hat{A} = \begin{pmatrix} \hat{\alpha}_1 & \hat{\alpha}_2 & \hat{\alpha}_3 & \ldots & \hat{\alpha}_{p-1} & \hat{\alpha}_p \\ 1 & 0 & 0 & \ldots & 0 & 0 \\ 0 & 1 & 0 & \ldots & 0 & 0 \\ \vdots & \vdots & \vdots & \vdots & \vdots & \vdots \\ 0 & 0 & 0 & 0 & 1 & 0 \end{pmatrix},$$

and $\hat{\alpha} = (\hat{\alpha}_i) \in \mathbb{R}^p$ is a column vector of consistent estimators of the unknown autoregression coefficients. The autoregression coefficient vector will be denoted as $\alpha = (\alpha_i)$.

Least squares estimators for the model parameters can be written as

$$\begin{aligned} \hat{\alpha} &= A^{-1} a_0, & \hat{\sigma}^2 &= \frac{1}{T} \sum_{t=1}^{T} (x_t - \hat{\alpha}' X_{t-1}^{t-p})^2, \\ A &= \frac{1}{T} \sum_{t=1}^{T} X_{t-1}^{t-p} (X_{t-1}^{t-p})', & a_0 &= \frac{1}{T} \sum_{t=1}^{T} x_t X_{t-1}^{t-p}. \end{aligned} \tag{7.40}$$

The above expressions assume knowledge of the history $X_0^{1-p} \in \mathbb{R}^p$; if this assumption isn't satisfied, then the lower bound of summation over t in the estimators (7.40) should be replaced by $t = p + 1$, and the multiplier $1/T$ should be replaced by $1/(T - p)$. Note that if the innovation process u_t is normal, then these least squares estimators are equal to the (conditional) likelihood estimators. This fact and the following theorem have been proved in [2].

Theorem 7.5. *Let x_t be a stationary AR(p) time series, and let $\{u_t\}$ be jointly independent. If, in addition, $\{u_t\}$ are either identically distributed or have bounded moments of order $2 + \varepsilon$ for some $\varepsilon > 0$,*

$$\mathbb{E}\{|u_t|^{2+\varepsilon}\} \leq m < +\infty,$$

then the least squares estimators (7.40) *are consistent for $T \to +\infty$:*

$$\hat{\alpha} \xrightarrow{\mathbf{P}} \alpha, \quad \hat{\sigma}^2 \xrightarrow{\mathbf{P}} \sigma^2.$$

If $\{u_t\}$ are also identically distributed, then $\hat{\alpha}$ is asymptotically normal as $T \to +\infty$:

$$\mathcal{L}\{\sqrt{T}(\hat{\alpha} - \alpha)\} \to N_p(\mathbf{0}_p, \sigma^2 F^{-1}),$$

$$F = \sum_{s=0}^{+\infty} A^s \Sigma A^s, \qquad \Sigma = \begin{pmatrix} \sigma^2 & 0 & \dots & 0 \\ 0 & 0 & \dots & 0 \\ \vdots & \vdots & \vdots & \vdots \\ 0 & 0 & \dots & 0 \end{pmatrix}.$$

Statistical estimators of the autoregression order p follow from either the Akaike information criterion (AIC) or the Bayesian information criterion (BIC) [24]:

$$\hat{p} = \arg\min \mathrm{AIC}(p), \quad \mathrm{AIC}(p) = T \log \hat{\sigma}^2(p) + 2(p+1);$$

$$\tilde{p} = \arg\min \mathrm{BIC}(p), \quad \mathrm{BIC}(p) = T \log \hat{\sigma}^2(p) - (T-p-1)\log\left(1 - \frac{p+1}{T}\right) + \\ + (p+1)\left(\log T + \log\left(\frac{1}{p+1}\left(\frac{\hat{\sigma}_x^2}{\hat{\sigma}^2} - 1\right)\right)\right), \tag{7.41}$$

where $\hat{\sigma}^2(p)$ is the remainder variance defined by (7.40) during the tuning stage of the AR(p) model, and

$$\hat{\sigma}_x^2 = \frac{1}{T}\sum_{t=1}^{T} x_t^2$$

is the sample variance (which, for an autoregression of order zero, can be considered as a special case of $\hat{\sigma}_0$).

Plug-in forecasting under vector autoregression models will be discussed in Chap. 8.

7.3.3 Plug-In Forecasting Algorithms Based on Parameter Estimation of MA(q) Models

As discussed in Sect. 7.2.3, plug-in forecasting algorithms under MA(q) models are defined by the relations (7.30), (7.31) after substituting consistent estimators $\hat{\gamma} = (\hat{\gamma}_1, \ldots, \hat{\gamma}_q)'$ in place of the unknown true values $\gamma = (\gamma_1, \ldots, \gamma_q)'$.

One of the following two methods is usually applied to construct the estimator $\hat{\gamma}$:

1. Numerical maximization of the likelihood function [2];
2. Durbin's approximation method [6].

Let us briefly describe the second method. As noted earlier, the MA(q) model allows a representation in the form AR($+\infty$):

$$u_t = \sum_{i=0}^{+\infty} \beta_i x_{t-i}. \tag{7.42}$$

Durbin [6] proposed to construct an estimator for γ based on a finite $(n+1)$th order approximation of the infinite sum (7.42):

$$u_t = \sum_{i=0}^{n} \beta_i x_{t-i}, \quad \beta_0 = 1, \tag{7.43}$$

where $n \in \mathbb{N}$ is a parameter defining the accuracy of the estimation algorithm. Now, similarly to Sect. 7.2.2, the sample covariance function $\hat{\sigma}_\tau$ will be used to construct an estimator for $\beta = (\beta_1, \dots, \beta_n)'$:

$$\hat{\beta} = -\hat{\Sigma}_{(n)}^{(-1)} \sigma_{(1)}^{(n)}, \quad \hat{\beta}_0 = 1. \tag{7.44}$$

This allows us to compute the estimator $\hat{\gamma}$ from the following relations:

$$\begin{aligned} &\hat{\gamma} = -C^{-1} g, \qquad C = (c_{ij}) \in \mathbb{R}^{q \times q}, \quad c_{ij} = \sum_{u=0}^{n-|i-j|} \hat{\beta}_u \hat{\beta}_{u+|i-j|}, \\ &g = (g_k) \in \mathbb{R}^q, \quad g_k = \sum_{u=0}^{k} \hat{\beta}_u \hat{\beta}_{u+k}, \end{aligned} \tag{7.45}$$

where $i, j = 1, \dots, q, k = 1, \dots, q$.

The order q can be estimated by AIC or BIC information criteria defined by (7.41), where p is replaced by q, and the remainder variance can be found from (7.43):

$$\hat{\sigma}^2 = \frac{1}{T-n} \sum_{t=n+1}^{T} \left(\sum_{i=0}^{n} \hat{\beta}_i x_{t-i} \right)^2. \tag{7.46}$$

It is recommended to compute the estimators (7.44)–(7.46) for several increasing values of n until the approximation (7.43) is verified to be sufficiently accurate.

7.3.4 Plug-In Forecasting Algorithms Based on ARMA(p, q) Parameter Estimators

In Sect. 7.2.4, it was shown that a plug-in forecasting algorithm under the ARMA(p, q) model is defined by the relations (7.32), (7.33), where the vectors of unknown true coefficients $\alpha = (\alpha_1, \dots, \alpha_p)' \in \mathbb{R}^p$, $\gamma = (\gamma_1, \dots, \gamma_q)' \in \mathbb{R}^q$ are replaced by the vectors of their consistent estimators $\hat{\alpha} = (\hat{\alpha}_i) \in \mathbb{R}^p$, $\hat{\gamma} = (\hat{\gamma}_j) \in \mathbb{R}^q$.

Let us present a method to construct the estimators $\hat{\alpha}$, $\hat{\gamma}$ based on maximization of the conditional likelihood function $L(\alpha, \gamma)$ under a normality assumption on the innovation process ε_t [4, 9]:

$$L(\alpha, \gamma) = -\frac{T-p}{2} \log(2\pi) - \frac{T-p}{2} \log(\sigma^2) - \frac{1}{2\sigma^2} Q(\alpha, \gamma),$$

$$Q(\alpha, \gamma) = \sum_{t=p+1}^{T} u_t^2,$$

where $u_p = u_{p-1} = \cdots = u_{p-q+1} = 0$, and for $t \geq p+1$ the values $\{u_t\}$ are computed recursively from the stochastic difference equation defining the ARMA model:

$$u_t = x_t - \alpha_1 x_{t-1} - \cdots - \alpha_p x_{t-p} + \gamma_1 u_{t-1} + \cdots + \gamma_q u_{t-q}.$$

Maximization of the likelihood function in α, γ is equivalent to minimization of Q:

$$Q(\hat{\alpha}, \hat{\gamma}) = \min_{\alpha,\gamma} Q(\alpha, \gamma);$$

this procedure is performed numerically [9]. Subsequent maximization of the likelihood function in σ^2 leads to the estimator

$$\hat{\sigma}^2_{(p,q)} = \frac{1}{T-p} Q(\hat{\alpha}, \hat{\gamma}).$$

This statistic can also be used in information criteria (7.41), where p is replaced by $p+q$.

Note that there exists a modification of Durbin's procedure [2] (see Sect. 7.3.3) to estimate the parameters of the ARMA(p, q) model.

7.3.5 *Plug-In Forecasting Algorithms Based on ARIMA(p, d, q) Parameter Estimators*

From the results of Sect. 7.2.5, a plug-in forecasting algorithm for a non-stationary ARIMA(p, d, q) model is defined by (7.34), where the vectors of unknown true coefficients $\alpha = (\alpha_i) \in \mathbb{R}^p$, $\gamma = (\gamma_j) \in \mathbb{R}^q$ are replaced by their consistent estimators $\hat{\alpha} = (\hat{\alpha}_i) \in \mathbb{R}^p$, $\hat{\gamma} = (\hat{\gamma}_j) \in \mathbb{R}^q$. Assume that the integration order $d \in \mathbb{N}$ is known, then the estimators $\hat{\alpha}$, $\hat{\gamma}$ can be constructed by using the plug-in approach, as summarized below:

1. From the observations $x_1, \ldots, x_T$, compute $T - d$ differences of order d:

$$y_t = \Delta^d x_t, \quad t = d+1, d+2, \ldots, T;$$

2. Applying the results of Sect. 7.3.4, identify an ARMA(p, q) model describing the time series $\{y_t\}$; this step yields parameter estimators $\hat{\alpha}$, $\hat{\gamma}$, $\hat{\sigma}^2$;
3. Substitute the estimators $\hat{\alpha}$, $\hat{\gamma}$, $\hat{\sigma}^2$ in place of the unknown parameters in (7.34).

The integration parameter d can be estimated by one of the methods from [9].

7.4 Robustness Under Parametric Model Specification Errors

7.4.1 The General Case

Consider a problem of forecasting a future time series element $x_{T+\tau} \in \mathbb{R}^1$ based on observing the history $X_1^T = (x_1, x_2, \ldots, x_T)' \in \mathbb{R}^T$. The mean square optimal forecast, by Theorem 4.1, can be represented in the following general form:

$$\hat{x}^*_{T+\tau} = f_0(X_1^T; \theta^0) = \mathbb{E}_{\theta^0}\left\{x_{T+\tau} \mid X_1^T\right\}, \tag{7.47}$$

and its risk equals

$$\begin{aligned} r_0(\tau) &= \mathbb{E}_{\theta^0}\left\{(\hat{x}^*_{T+\tau} - x_{T+\tau})^2\right\} = \\ &= \mathbb{E}_{\theta^0}\left\{\mathbb{D}_{\theta^0}\left\{x_{T+\tau} \mid X_1^T\right\}\right\}, \end{aligned} \tag{7.48}$$

where $\theta^0 \in \mathbb{R}^m$ is a parameter vector of the undistorted time series model.

A parametric time series model specification error occurs due to the uncertainty of the true value of θ^0, which necessitates an application of the plug-in approach:

$$\hat{x}_{T+\tau} = f_0(X_1^T; \theta), \tag{7.49}$$

where

$$\theta = \theta^0 + \Delta\theta \in \mathbb{R}^m, \tag{7.50}$$

and $\Delta\theta \in \mathbb{R}^m$ is the parameter vector defining the specification error. In establishing the asymptotic properties of forecasting under misspecification errors, we are going to distinguish between the following two cases:

Case 1. *Absolute specification error* defined as

$$|\Delta\theta| = \sqrt{(\Delta\theta)'(\Delta\theta)} \le \varepsilon, \tag{7.51}$$

where $\varepsilon \ge 0$ is a given upper bound of the distortion level.

Case 2. *Relative specification error:*

$$\frac{|\Delta\theta|}{|\theta^0|} \le \varepsilon. \tag{7.52}$$

Theorem 7.6. *If the plug-in forecasting algorithm* (7.49) *is used under the specification error* (7.50)*, the function $f_0(\cdot)$ is twice differentiable, the expectation*

$$\mathbb{E}_{\theta^0}\left\{\nabla_\theta f_0\left(X_1^T; \theta^0\right)\left(\nabla_{\theta^0} f_0\left(X_1^T; \theta^0\right)\right)'\right\} \tag{7.53}$$

is finite, and for all $\bar{\theta}$ in an ε-neighborhood U_{θ^0} of the point θ^0 defined by (7.51) *or* (7.52), *the expectations*

$$\mathbb{E}_{\theta^0}\left\{\frac{\partial^2 f_0(X_1^T;\bar{\theta})}{\partial\bar{\theta}_i\partial\bar{\theta}_j}\frac{\partial^2 f_0(X_1^T;\bar{\theta})}{\partial\bar{\theta}_k\partial\bar{\theta}_l}\right\}, \quad i,j,k,l \in \{1,\dots,m\}, \tag{7.54}$$

are bounded, then for $\varepsilon \to 0$ the forecast risk satisfies the asymptotic expansion

$$\begin{aligned} r_\varepsilon(\tau) &= r_0(\tau) + \lambda'\Delta\theta + O(\varepsilon^2), \\ \lambda &= 2\mathbb{E}_{\theta^0}\left\{\left(f_0(X_1^T;\theta^0) - x_{T+\tau}\right)\nabla_{\theta^0} f_0(X_1^T;\theta^0)\right\} \in \mathbb{R}^m. \end{aligned} \tag{7.55}$$

Proof. Let us write a first order Taylor's expansion of the function (7.49) w.r.t. θ in an ε-neighborhood of θ^0, expressing the remainder term in the Lagrange form:

$$f_0(X_1^T;\theta) = f_0(X_1^T;\theta^0) + \left(\nabla_{\theta^0} f_0(X_1^T;\theta^0)\right)'\Delta\theta + (\Delta\theta)'\,\nabla^2_{\bar{\theta}} f_0(X_1^T;\bar{\theta})\Delta\theta,$$

where $\bar{\theta} \in U_{\theta^0}$. This expansion, together with (7.48) and the boundedness of the expectations (7.53), (7.54), proves (7.55). □

7.4.2 Stationary Time Series Forecasting Under Misspecification of Covariance Functions

Let us consider a special case where the time series x_t is stationary, has a zero expectation and a covariance function $\sigma^0_\tau = \mathbb{E}\{x_t x_{t+\tau}\}$, $\tau \in \mathbb{Z}$. Then the optimal linear forecast is a special case of (7.47) defined by Theorem 7.1 for $n = T - 1$:

$$\hat{x}^*_{T+\tau} = \left(a^*\right)' X_T^1, \quad a^* = \left(\Sigma^0_{(T)}\right)^{-1}\sigma^{0(\tau+T-1)}_{(\tau)}, \tag{7.56}$$

$$\Sigma^0_T = \begin{pmatrix} \sigma^0_0 & \sigma^0_1 & \cdots & \sigma^0_{T-1} \\ \sigma^0_1 & \sigma^0_0 & \cdots & \sigma^0_{T-2} \\ \vdots & \vdots & \cdots & \vdots \\ \sigma^0_{T-1} & \sigma^0_{T-2} & \cdots & \sigma^0_0 \end{pmatrix}, \quad \sigma^{0(\tau+T-1)}_{(\tau)} = \begin{pmatrix} \sigma^0_\tau \\ \sigma^0_{\tau+1} \\ \vdots \\ \sigma^0_{\tau+T-1} \end{pmatrix}.$$

Assume that the m-vector ($m = \tau + T$)

$$\theta^0 = (\sigma^0_0, \sigma^0_1, \dots, \sigma^0_{\tau+T-1})'$$

of covariance function values [this θ^0 influences the vector a^* of the optimal forecast coefficients (7.56)] contains specification errors:

$$\theta = (\sigma_0, \sigma_1, \dots, \sigma_{\tau+T-1})', \quad \sigma_k = \sigma^0_k + \Delta\sigma_k, \quad k \in \{0, 1, \dots, \tau + T - 1\}. \tag{7.57}$$

Once again, we define the ε-neighborhood of the allowed values of θ by (7.51) or (7.52). The plug-in forecasting algorithm is based on estimation of the parameter vector θ and has the following form:

$$\hat{x}_{T+\tau} = a' X_T^1, \quad a = \Sigma_{(T)}^{-1} \sigma_{(\tau)}^{(\tau+T-1)}, \tag{7.58}$$

where the matrices

$$\Sigma_{(T)}, \quad \sigma_{(\tau)}^{(\tau+T-1)}$$

are constructed similarly to (7.56) with σ_k^0 replaced by σ_k.

Let us introduce the following notation:

$$\Delta\Sigma_{(T)} = \begin{pmatrix} \Delta\sigma_0 & \Delta\sigma_1 & \cdots & \Delta\sigma_{T-1} \\ \Delta\sigma_1 & \Delta\sigma_0 & \cdots & \Delta\sigma_{T-2} \\ \vdots & \vdots & \cdots & \vdots \\ \Delta\sigma_{T-1} & \Delta\sigma_{T-2} & \cdots & \Delta\sigma_0 \end{pmatrix}, \quad \Delta\sigma_{(\tau)}^{(\tau+T-1)} = \begin{pmatrix} \Delta\sigma_{(\tau)} \\ \Delta\sigma_{(\tau+1)} \\ \vdots \\ \Delta\sigma_{(\tau+T-1)} \end{pmatrix}. \tag{7.59}$$

Theorem 7.7. *The forecast risk of the plug-in forecasting algorithm* (7.58) *under the specification errors in the covariances* (7.57), (7.51), (7.52) *satisfies the following asymptotic expansion as* $\varepsilon \to 0$*:*

$$\begin{aligned} r_\varepsilon(\tau) =& r_0(\tau) + \left(\Delta\Sigma_{(T)}\left(\Sigma_{(T)}^0\right)^{-1}\sigma_{(\tau)}^{0(\tau+T-1)} - \Delta\sigma_{(\tau)}^{(\tau+T-1)}\right)'\left(\Sigma_{(T)}^0\right)^{-1} \times \\ &\times \left(\Delta\Sigma_{(T)}\left(\Sigma_{(T)}^0\right)^{-1}\sigma_{(\tau)}^{0(\tau+T-1)} - \Delta\sigma_{(\tau)}^{(\tau+T-1)}\right) + O(\varepsilon^3). \end{aligned} \tag{7.60}$$

Proof. Applying (7.57), (7.59), and simplifying yields

$$\Delta a = a - a^* = \left(\Sigma_{(T)}^0\right)^{-1}\left(-\Delta\Sigma_{(T)}\left(\Sigma_{(T)}^0\right)^{-1}\sigma_{(\tau)}^{0(\tau+T-1)} + \Delta\sigma_{(\tau)}^{(\tau+T-1)}\right) + O(\varepsilon^2)\mathbf{1}_T. \tag{7.61}$$

From (7.56), (7.58) we can obtain

$$\begin{aligned} r_\varepsilon(\tau) &= \mathbb{E}\left\{\left((a^* + \Delta a)' X_T^1 - x_{T+\tau}\right)^2\right\} = r_0 + (\Delta a)'\Sigma_T^0 \Delta a + \\ &+ 2\left((a^*)'\mathbb{E}\{X_T^1 (X_T^1)'\} - \mathbb{E}\{x_{T+\tau}(X_T^1)'\}\right)\Delta a = r_0(\tau) + (\Delta a)'\Sigma_T^0 \Delta a. \end{aligned}$$

Substituting (7.61) into this expression yields (7.60). □

It is easy to see that the asymptotic expansion (7.60) is more accurate than the general result of Sect. 7.4.1: the risk increment due to misspecification is of the second order w.r.t. ε. Also note that this risk increment grows as the matrix $\Sigma_{(T)}$ becomes ill-conditioned—in particular, as the dependence between the elements of the time series increases.

7.4.3 Forecasting of AR(p) Time Series Under Misspecification of Autoregression Coefficients

Let us consider another special case of the setting introduced in Sect. 7.4.1, where the predicted time series $\{x_t\}$ follows a stationary AR(p) model (see Sect. 7.2.2 for the definition):

$$x_t = \theta_1^0 x_{t-1} + \cdots + \theta_p^0 x_{t-p} + u_t, \quad t \in \mathbb{Z}. \tag{7.62}$$

In that case, the optimal forecasting statistic is defined by (7.28). Due to prior uncertainty, the autoregression coefficients $\{\theta_i^0\}$ are subject to specification errors $\{\Delta\theta_i\}$, and thus the forecast is based not on the true model parameters, but on their distorted values:

$$\theta_i = \theta_i^0 + \Delta\theta_i, \quad i = 1, 2, \ldots, p. \tag{7.63}$$

The plug-in forecasting algorithm is based on estimation of the vector θ and can be written as follows:

$$\hat{x}_{T+\tau} = \sum_{i=1}^{p} (A^\tau)_{1i} x_{T+1-i}, \tag{7.64}$$

$$A = \begin{pmatrix} \theta_1 & \theta_2 & \theta_3 & \cdots & \theta_{p-1} & \theta_p \\ 1 & 0 & 0 & \cdots & 0 & 0 \\ 0 & 1 & 0 & \cdots & 0 & 0 \\ \vdots & \vdots & \vdots & \vdots & \vdots & \vdots \\ 0 & 0 & 0 & \cdots & 1 & 0 \end{pmatrix}.$$

To obtain a formula for the risk of the forecasting statistic (7.64), we are going to need the following auxiliary result.

Lemma 7.2. *Under the AR(p) model defined by* (7.62)*, the following representation holds:*

$$X_{t+\tau}^{t+\tau-p+1} = A_0^\tau X_t^{t-p+1} + \sum_{k=0}^{\tau-1} A_0^k U_{t+\tau-k}, \quad \tau \in \mathbb{N}, \ t \in \mathbb{Z}, \tag{7.65}$$

where $U_t = (u_t \vdots \mathbf{0}_{p-1})' \in \mathbb{R}^p$, *the column vector* X_m^n *is defined by* (7.56), *and* A_0 *is a companion* $(p \times p)$*-matrix derived from* (7.64) *by taking* $\theta = \theta^0$.

Proof. In the above notation, due to (7.62) we have

$$X_{t+\tau}^{t+\tau-p+1} = A_0 X_{t+\tau-1}^{t+\tau-p} + U_{t+\tau}.$$

Applying this recursion $\tau - 1$ times yields (7.65). □

Let $(A_0^k)_{1\cdot}$ denote the 1st row of a $(p \times p)$-matrix A_0^k.

Corollary 7.5. *The following representation holds:*

$$x_{T+\tau} = (A_0^\tau)_{1\cdot} X_T^{T-p+1} + \sum_{k=0}^{\tau-1} (A_0^k)_{11} u_{T+\tau-k}. \tag{7.66}$$

Proof. To prove the corollary, it is sufficient to assume $t = T$ in (7.65), remembering that the 2nd, 3rd, ..., pth components of the vector U_t are zeros. □

Theorem 7.8. *Under the specification errors* (7.63), (7.51), (7.52) *in the* $AR(p)$ *autoregression coefficients, the* τ*-step-ahead forecast risk equals*

$$r_\varepsilon(\tau) = r_0(\tau) + (A^\tau - A_0^\tau)_{1\cdot} \Sigma_{(p)} \left((A^\tau - A_0^\tau)_{1\cdot}\right)', \tag{7.67}$$

where

$$r_0(\tau) = \sigma^2 \left(1 + \sum_{k=1}^{\tau-1} \left((A_0^k)_{11}\right)^2\right) \tag{7.68}$$

is the τ*-step-ahead autoregression forecast risk under prior knowledge of autoregression coefficients (i.e., the minimum possible risk).*

Proof. By (7.64), (7.66) we have:

$$\hat{x}_{T+\tau} - x_{T+\tau} = (A^\tau)_{1\cdot} X_T^{T-p+1} - \left((A_0^\tau)_{1\cdot} X_T^{T-p+1} + \sum_{k=0}^{\tau-1} (A_0^k)_{11} u_{T+\tau-k}\right) =$$

$$= (A^\tau - A_0^\tau)_{1\cdot} X_T^{T-p+1} - \sum_{k=0}^{\tau-1} (A_0^k)_{11} u_{T+\tau-k}.$$

Let us substitute this expression for the random forecast error in the risk functional:

$$r_\varepsilon(\tau) = \mathbb{E}\left\{\left((A^\tau - A_0^\tau)_{1\cdot} X_T^{T-p+1} - \sum_{k=0}^{\tau-1} (A_0^k)_{11} u_{T+\tau-k}\right)^2\right\} =$$

$$= (A^\tau - A_0^\tau)_{1\cdot} \mathbb{E}\left\{X_T^{T-p+1} (X_T^{T-p+1})'\right\} (A^\tau - A_0^\tau)'_{1\cdot} -$$

$$- 2\sum_{k=0}^{\tau-1}(A_0^k)_{11}(A^\tau - A_0^\tau)_{1\cdot}\mathbb{E}\{X_T^{T-p+1}u_{T+\tau-k}\}+$$

$$+\sum_{k,l=0}^{\tau-1}(A_0^k)_{11}(A_0^l)_{11}\mathbb{E}\{u_{T+\tau-k}u_{T+\tau-l}\}.$$

From the stationarity of the time series, we have $\mathbb{E}\{X_T^{T-p+1}(X_T^{T-p+1})'\} = \Sigma_{(p)}$. Then the model properties (7.62) imply $\mathbb{E}\{u_t u_{t'}\} = \sigma^2\delta_{tt'}$, $\mathbb{E}\{x_t u_{t''}\} = 0$ for $t < t''$. Substituting this result into the previous expression leads to (7.67), (7.68). □

The second nonnegative summand in the formula (7.67) is the risk increment due to the specification error $\Delta\theta = \theta - \theta^0$ in the autoregression coefficients.

Corollary 7.6. *Under a specification error $\Delta\theta$ in the AR(p) model, the τ-step-ahead forecast risk satisfies the following asymptotic expansion:*

$$r_\varepsilon(\tau) = r_0(\tau) + (\Delta\theta)' B_\tau \Sigma_{(p)} B_\tau' \Delta\theta + O(|\Delta\theta|^3), \tag{7.69}$$

where B_τ is a $(p \times p)$-matrix defined as

$$B_\tau = \sum_{k=1}^{\tau-1}(A_0^k)_{11}(A_0^{\tau-k})'.$$

Proof. Let us denote $\Delta A = A - A_0$ and write a matrix expansion

$$(A_0 + \Delta A)^\tau = A_0^\tau + \sum_{k=1}^{\tau-1} A_0^k \Delta A A_0^{\tau-k} + O(||\Delta A||^2),$$

which yields the following asymptotic expansion of the first row of A^τ:

$$(A^\tau)_{1\cdot} = (A_0^\tau)_{1\cdot} + \Delta\theta' B_\tau + O(|\Delta\theta|^2).$$

Here we have used the fact that every row of the matrix ΔA, except for the first row, contains only zeros. Substituting $(A^\tau - A_0^\tau)_{1\cdot}$ into (7.67) leads to (7.69). □

Corollary 7.7. *Under the AR(p) specification error* (7.51)*, the following asymptotic expansions hold for the guaranteed risk and the risk instability coefficient:*

$$r_+(\tau) = r_0(\tau) + \varepsilon^2\lambda_{\max}(B_\tau \Sigma_{(p)} B_\tau') + O(\varepsilon^3),$$

$$\kappa(\tau) = \frac{r_+(\tau) - r_0(\tau)}{r_0(\tau)} = \varepsilon^2\frac{\lambda_{\max}(B_\tau \Sigma_{(p)} B_\tau')}{r_0(\tau)} + O(\varepsilon^3),$$

where $\lambda_{\max}(C)$ is the maximum eigenvalue of a matrix C.

Proof. By (7.69) and well-known extremal properties of matrix eigenvalues [22], we have

$$r_+(\tau) = \max_{|\Delta\theta|\leq\varepsilon} r_\varepsilon(\tau) = r_0(\tau) + \max_{(\Delta\theta)'\Delta\theta\leq\varepsilon^2} \left((\Delta\theta)' B_\tau \Sigma_{(p)} B_\tau'\right) + O(\varepsilon^3) =$$

$$= r_0(\tau) + \varepsilon^2 \lambda_{\max}(B_\tau \Sigma_{(p)} B_\tau) + O(\varepsilon^3). \qquad \square$$

Corollary 7.8. *Under the specification error* (7.52), *we have*

$$r_+(\tau) = r_0(\tau) + \varepsilon^2 |\theta^0|^2 \lambda_{\max}(B_\tau \Sigma_{(p)} B_\tau') + O(\varepsilon^3),$$

$$\kappa(\tau) = \varepsilon^2 |\theta^0|^2 \lambda_{\max}(B_\tau \Sigma_{(p)} B_\tau') / r_0(\tau) + O(\varepsilon^3).$$

Note that for $\tau = 1$, i.e., in the case of one-step-ahead forecasting, the remainder terms in the expansion (7.69) and the expansions proved by Corollaries 7.7, 7.8 become zeros.

Also note that the covariance matrix $\Sigma_{(p)}$ in Theorem 7.7 and its corollaries can be computed by applying one of the following three procedures:

1. Solving a Yule–Walker system of equations [2] in covariances $\{\sigma_0, \sigma_1, \ldots, \sigma_{p-1}\}$ which make up $\Sigma_{(p)}$;
2. Evaluating the sum of a converging matrix series

$$\Sigma_{(p)} = \sum_{k=0}^{\infty} A_0^k \Sigma (A_0^k)' = \sigma^2 \sum_{k=0}^{\infty} (A_0^k)_{\cdot 1}((A_0^k)_{\cdot 1})',$$

   where the $(p \times p)$-matrix Σ is made up of zero elements, except for $\sigma_{11} = \sigma^2$;
3. Using the following iterative procedure [2]:

$$\Sigma_{(p)}^{(i)} = \Sigma + A_0 \Sigma_{(p)}^{(i-1)} A_0', \quad i = 1, 2, \ldots; \quad \Sigma_{(p)}^{(0)} ::= \Sigma.$$

For instance, taking $\tau = 1$, $p = 1$ leads to an explicit form of (7.69):

$$r_\varepsilon(1) = \sigma^2 + \sigma^2 \frac{(\Delta\theta)^2}{1 - (\theta^0)^2}, \quad \kappa(1) = \frac{(\Delta\theta)^2}{1 - (\theta^0)^2}.$$

The condition $\kappa(1) \leq \delta$ leads to the following expression for the δ-admissible level of the specification error:

$$|\Delta\theta| \leq \varepsilon_+(\delta) = \sqrt{(1 - (\theta^0)^2)\delta}\,,$$

where $\delta > 0$; thus, the effect of specification errors becomes the largest as $|\theta^0|$ approaches 1.

7.5 Robustness Under Functional Innovation Process Distortions in the Mean Value

Let us discuss a common situation where the autoregression model (7.62) is affected by functional distortions in the mean value of the innovation process:

$$x_t = (\theta^0)' X_{t-1}^{t-p} + \tilde{u}_t, \quad \tilde{u}_t = u_t + \mu_t, \quad t \in \mathbb{N}, \tag{7.70}$$

where $\theta^0 = (\theta_i^0) \in \mathbb{R}^p$ is a column vector of p hypothetical autoregression coefficients; $X_{t-1}^{t-p} = (x_{t-1}, x_{t-2}, \ldots, x_{t-p})' \in \mathbb{R}^p$; $\{u_t\}$ are i.i.d. random variables with zero expectations $\mathbb{E}\{u_t\} = 0$ and finite variances $\mathbb{D}\{u_t\} = \sigma^2$; an unknown deterministic function $\mu_t \in \mathbb{R}^1$ defines the distortion; $\tilde{u}_t$ is the distorted innovation process. The hypothetical model (7.62) is obtained from (7.70) by taking $\mu_t \equiv 0$. Since the time series (7.70) is, in general, non-stationary, the initial condition will be chosen as $X_0^{-p+1} = \mathbf{0}_p$.

Let us study the effect of distortion $\{\mu_t\}$ combined with specification errors $\Delta\theta = \theta - \theta^0$ in the autoregression coefficients on the risk of the traditional forecasting statistic (7.64).

We are going to use the following notation: let $\mathbf{1}_{p(1)}$ be a unit column p-vector with the first element equal to 1 and the remaining elements equal to zero, and let

$$a(k,\tau) = \left((A^\tau - A_0^\tau)A_0^k\right)_{11} \in \mathbb{R}^1, \tag{7.71}$$

where A_0, A are companion $(p \times p)$-matrices, defined in Sect. 7.4.

Theorem 7.9. *Under an autoregression model* (7.70) *of order p in the presence of functional distortion and specification error* $\Delta\theta = \theta - \theta^0$, *the forecasting statistic* (7.64) *has the risk*

$$r_\varepsilon(\tau) = r_0(\tau) + \sigma^2 \sum_{k=0}^{T-1} a^2(k,\tau) + \left(\sum_{k=0}^{T-1} a(k,\tau)\mu_{T-k} - \sum_{k=0}^{\tau-1} (A_0^k)_{11}\mu_{T+\tau-k}\right)^2. \tag{7.72}$$

Proof. Similarly to the proof of Lemma 7.2, in the above notation (7.70) yields that

$$x_{T+\tau} = (A_0^\tau)_{1\cdot} X_T^{T-p+1} + \sum_{k=0}^{\tau-1} (A_0^k)_{11} u_{T+\tau-k} + \sum_{k=0}^{\tau-1} (A_0^k)_{11} \mu_{T+\tau-k},$$

$$\hat{x}_{T+\tau} = (A^\tau)_{1\cdot} X_T^{T-p+1}.$$

Now, applying the same reasoning as in the proof of Theorem 7.7, we can obtain

$$r_\varepsilon(\tau) = (A^\tau - A_0^\tau)_{1\cdot}\, \mathbb{E}\left\{X_T^{T-p+1}\left(X_T^{T-p+1}\right)'\right\}(A^\tau - A_0^\tau)'_{1\cdot} + \\ + \left(\sum_{k=0}^{\tau-1}(A_0^k)_{11}\mu_{T+\tau-k}\right)^2 + \sigma^2\sum_{k=0}^{\tau-1}\left((A_0^k)_{11}\right)^2 - \\ - 2(A^\tau - A_0^\tau)_{1\cdot}\, \mathbb{E}\{X_T^{T-p+1}\}\sum_{k=0}^{\tau-1}(A_0^k)_{11}\mu_{T+\tau-k}. \tag{7.73}$$

Similarly to Lemma 7.2, we have the following linear representation based on the initial state $X_0^{-p+1} = \mathbf{0}_p$:

$$X_T^{T-p+1} = A_0^T X_0^{-p+1} + \sum_{k=0}^{T-1} A_0^k U_{T-k} + \sum_{k=0}^{T-1} A_0^k \mu_{T-k}\cdot \mathbf{1}_{p(1)} = \\ = \sum_{k=0}^{T-1} u_{T-k}A_0^k\mathbf{1}_{p(1)} + \sum_{k=0}^{T-1}\mu_{T-k}A_0^k\mathbf{1}_{p(1)}.$$

Taking into account the identity $A_0^k\mathbf{1}_{p(1)} \equiv (A_0^k)_{\cdot 1}$, we obtain

$$X_T^{T-p+1} = \sum_{k=0}^{T-1} u_{T-k}(A_0^k)_{\cdot 1} + \sum_{k=0}^{T-1}\mu_{T-k}(A_0^k)_{\cdot 1}.$$

By the properties of $\{u_t\}$, this yields the following equalities:

$$\mathbb{E}\left\{X_T^{T-p+1}\right\} = \sum_{k=0}^{T-1}\mu_{T-k}(A_0^k)_{\cdot 1},$$

$$\mathbb{E}\left\{X_T^{T-p+1}(X_T^{T-p+1})'\right\} = \sigma^2\sum_{k=0}^{T-1}(A_0^k)_{\cdot 1}(A_0^k)'_{\cdot 1} + \left(\sum_{k=0}^{T-1}\mu_{T-k}(A_0^k)_{\cdot 1}\right)(\cdot)'.$$

Substituting them into (7.73), applying (7.71), (7.68), and performing equivalent transformations yields

$$r_\varepsilon(\tau) = r_0(\tau) + \sigma^2\sum_{k=0}^{T-1}a^2(k,\tau) + \left(\sum_{k=0}^{T-1}a(k,\tau)\mu_{T-k}\right)^2 + \\ + \left(\sum_{k=0}^{\tau-1}(A_0^k)_{11}\mu_{T+\tau-k}\right)^2 - 2\left(\sum_{k=0}^{T-1}a(k,\tau)\mu_{T-k}\right)\left(\sum_{k=0}^{\tau-1}(A_0^k)_{11}\mu_{T+\tau-k}\right),$$

proving (7.72). □

The right-hand side of (7.72) is composed of three summands. The first is the forecast risk for the undistorted hypothetical model, the second is the risk increment due to the specification error, the third is the risk increment due to both the distortion in the mean value of the innovation process and the specification errors. Note that the second summand is different from the risk increment obtained in Theorem 7.7 since the process in (7.70) is non-stationary with the initial state $X_0^{-p+1} = \mathbf{0}_p$.

Now let us evaluate the guaranteed forecast risk.

Theorem 7.10. *Under the conditions of Theorem 7.9, assuming that the functional distortion in the mean value of the innovation process in* (7.70) *is bounded in the* l_2*-norm,*

$$T^{-1}\sum_{t=1}^{T}\mu_t^2 \le \varepsilon_{(1)}^2, \quad \tau^{-1}\sum_{t=T+1}^{T+\tau}\mu_t^2 \le \varepsilon_{(2)}^2, \tag{7.74}$$

where $\varepsilon_{(1)}, \varepsilon_{(2)} \ge 0$ *are the respective distortion levels for the base (observation) interval and the forecast interval, the guaranteed forecast risk is equal to*

$$r_+(\tau) = r_0(\tau) + \sigma^2\sum_{k=0}^{T-1}a^2(k,\tau) + \left(\varepsilon_{(1)}\sqrt{T\sum_{k=0}^{T-1}a^2(k,\tau)} + \varepsilon_{(2)}\sqrt{\tau\sum_{k=0}^{\tau-1}\left((A_0^k)_{11}\right)^2}\right)^2. \tag{7.75}$$

Proof. By (7.72), finding the guaranteed risk under the conditions of the theorem can be reduced to solving the following optimization problem:

$$\left(\sum_{k=0}^{T-1}a(k,\tau)\mu_{T-k} - \sum_{k=0}^{\tau-1}(A_0^k)_{11}\mu_{T+\tau-k}\right)^2 \to \max_{\{\mu_t\}} \tag{7.76}$$

under the conditions (7.74). This is a quadratic programming problem with a quadratic separable objective function and quadratic restrictions. Separability of the objective function, together with the special form of the restrictions (7.74), allows us to reduce (7.76) to a pair of independent optimization problems of the following form:

$$f(z) = b'z \to \max_z, \quad z'z \le \varepsilon^2,$$

where

$$b = (b_i) \in R^N, \quad z = (z_i) \in \mathbb{R}^N, \quad \varepsilon > 0.$$

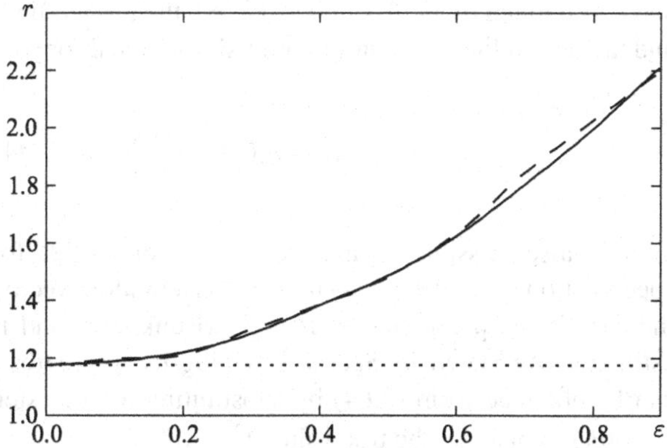

Fig. 7.2 Forecast risk for different distortion levels ε

Since $b'z = |b||z|\cos(\alpha)$, where α is the angle between vectors b and z, the maximum is equal to

$$\max f(z) = |b|\sqrt{d}|b| = \varepsilon|b|^2$$

and is attained at $z^* = \varepsilon|b|$. Using this result to solve (7.76) yields (7.75). □

Corollary 7.9. *Under specification error* $\Delta\theta$ *and functional distortion in the mean value* $\{\mu_t\}$ *satisfying* (7.74)*, the risk instability coefficient of the* τ*-step-ahead autoregression forecast* (7.64) *based on* T *observations equals*

$$\kappa(\tau) = \frac{r_+(\tau) - r_0(\tau)}{r_0(\tau)} = \frac{\sigma^2}{r_0(\tau)} \sum_{k=0}^{T-1} a^2(k,\tau) +$$

$$+ \frac{1}{r_0(\tau)} \left(\varepsilon_{(1)} \sqrt{T^{-1} \sum_{k=0}^{T-1} a^2(k,\tau)} + \varepsilon_{(2)} \sqrt{\tau^{-1} \sum_{k=0}^{\tau-1} \left((A_0^k)_{11}\right)^2} \right)^2 \geq 0.$$

Results of Monte-Carlo simulations performed by the author [15, 16] were in line with the obtained theoretical results.

Figure 7.2 presents results of simulations based on the model (7.70) with

$$p = 2, \quad \theta^0 = (0.3, 0.4)', \quad \theta = (0.4, 0.5)', \quad \sigma^2 = 1,$$

$$\mu_t = \varepsilon \sin(t), \quad \varepsilon_{(1)} = \varepsilon_{(2)} = \varepsilon, \quad T = 40, \quad \tau = 2.$$

For each simulated distortion level, 10^4 Monte-Carlo simulation rounds were performed. The solid line shows the theoretical dependence between the risk $r_\varepsilon(\tau)$

and ε as defined by (7.72), (7.71), (7.68), the dashed line is the experimental curve, and the dotted line is the minimum risk (obtained for $\mu_t \equiv 0$):

$$r_{\min} = r_0(\tau) + \sigma^2 \sum_{k=o}^{T-1} a^2(k,\tau).$$

It is also possible to use the expression (7.72) for the forecast risk if the specification error $\Delta\theta = \theta - \theta^0 \in \mathbb{R}^p$ is a random vector. For instance, this may be the case if the parameter θ^0 is a priori unknown and is replaced by a statistical estimator. As stated in Sect. 7.2, a plug-in autoregression forecasting algorithm can be obtained from (7.64) by substituting a (consistent) least squares estimator $\hat{\theta} = (\hat{\theta}_i)$ in place of the true value $\theta^0 = (\theta_i^0)$:

$$\hat{x}_{T+\tau} = (\hat{A}^\tau)_{1\cdot} X_T^{T-p+1}, \tag{7.77}$$

$$\hat{A} = \left(\begin{array}{c:c} \multicolumn{2}{c}{\hat{\theta}'} \\ \hdashline I_{p-1} & \mathbf{0}_{p-1} \end{array}\right), \quad \hat{\theta} = \left(\sum_{t=p+1}^{T} X_{t-1}^{t-p} (X_{t-1}^{t-p})'\right)^{-1} \sum_{t=p+1}^{T} x_t X_{t-1}. \tag{7.78}$$

The nonlinear dependence of the forecast $\hat{x}_{T+\tau}$ on the observed time series $X = (x_1, \ldots, x_T)' \in \mathbb{R}^T$ and the dependence between $\hat{\theta}$ and X_T^{T-p+1}, which in turn leads to a dependence between the factors in (7.77), make it extremely difficult to evaluate the robustness of the forecasting statistic (7.77) explicitly. Thus, we are going to restrict ourselves to a simplified case where computation of the least squares estimate (7.78) is based on an auxiliary time series

$$\tilde{X} = (\tilde{x}_1, \ldots, \tilde{x}_{\tilde{T}}) \in \mathbb{R}^{\tilde{T}}$$

of length $\tilde{T}$, which is independent from the time series X and is unaffected by distortion [this is the case (b) of Sect. 7.3.1]:

$$\hat{\theta} = \left(\sum_{t=p+1}^{\tilde{T}} \tilde{X}_{t-1}^{t-p} (\tilde{X}_{t-1}^{t-p})'\right)^{-1} \sum_{t=p+1}^{\tilde{T}} \tilde{x}_t \tilde{X}_{t-1}^{t-p}. \tag{7.79}$$

This situation arises if the forecasting process is divided into two stages. In the first stage (the training stage), a training sample $\tilde{X}$ is used to construct a parameter estimator $\tilde{\theta}$ defined in (7.79). In the second stage (forecasting stage), a forecast $\hat{x}_{T+\tau}$ defined by (7.77) is computed based on the observations X. Note that the results of a single training stage can be used to construct multiple forecasts.

Theorem 7.11. *Under the* $AR(p)$ *model* (7.70), *assuming functional distortion in the mean value* $\{\mu_t\}$ *satisfying the conditions* (7.74) *for* $\varepsilon_{(1)} = \varepsilon_+$, $\varepsilon_{(2)} = 0$ *(i.e.,*

$\mu_t = 0$ *for* $t > T$), *consider a one-step-ahead* ($\tau = 1$) *plug-in forecasting statistic* (7.77), (7.79), *where* $\tilde{X}$ *is an undistorted time series independent of* X. *Defining the mean square error matrix of the estimator* $\hat{\theta}$ *as*

$$\mathbb{E}\{(\hat{\theta} - \theta^0)(\hat{\theta} - \theta^0)'\} = (\tilde{T})^{-1} F,$$

the guaranteed forecast risk is equal to

$$\tilde{r}_+(1) = \sigma^2 + \frac{\sigma^2}{\tilde{T}} \sum_{t=1}^{T} (A_0^{t-1})'_{\cdot 1} F (A_0^{t-1})_{\cdot 1} + \varepsilon_+^2 \frac{T}{\tilde{T}} \lambda_{\max}(G_T), \tag{7.80}$$

where $\lambda_{\max}(G_T)$ *is the maximum eigenvalue of the* $(T \times T)$*-matrix* $G_T = (g_{ij})$*:*

$$g_{ij} = (A_0^{i-1})'_{\cdot 1} F (A_0^{j-1})_{\cdot 1}, \quad i, j = 1, \ldots, T. \tag{7.81}$$

Proof. Let us apply (7.72), (7.68), (7.71) for $\varepsilon_{(1)} = \varepsilon_+$, $\varepsilon_{(2)} = 0$, $\tau = 1$:

$$a(k,1) = \left((A - A_0) A_0^k\right)_{11} = (\hat{\theta} - \theta^0)' (A_0)^k_{\cdot 1};$$

$$r_\varepsilon(1) = \sigma^2 + \sigma^2 \sum_{k=0}^{T-1} \left((\hat{\theta} - \theta^0)'(A_0^k)_{\cdot 1}\right)^2 + \left(\sum_{k=0}^{T-1} (\hat{\theta} - \theta^0)'(A_0^k)_{\cdot 1} \mu_{T-k}\right)^2.$$

The mean value of this risk w.r.t. the distribution of $\hat{\theta}$, under the conditions of the theorem and in the notation (7.81), can be written as

$$\mathbb{E}\{r_\varepsilon(1)\} = \sigma^2 + \frac{\sigma^2}{\tilde{T}} \sum_{k=0}^{T-1} (A_0^k)'_{\cdot 1} F (A_0^k)_{\cdot 1} + \frac{1}{\tilde{T}} \sum_{i,j=0}^{T-1} \mu_{T-i} \mu_{T-j} g_{ij}.$$

Maximizing this quadratic function in $\{\mu_1, \ldots, \mu_T\}$ under the quadratic condition

$$T^{-1} \sum_{t=1}^{T} \mu_t^2 \leq \varepsilon_+^2$$

by using Lagrange multipliers leads to (7.80). □

Corollary 7.10. *Under the conditions of Theorem 7.10, the instability coefficient of the forecast risk is equal to*

$$\kappa(1) = \frac{1}{\tilde{T}} \sum_{t=1}^{T} (A_0^{t-1})'_{\cdot 1} F (A_0^{t-1})_{\cdot 1} + \varepsilon_+^2 \frac{T}{\tilde{T}} \frac{\lambda_{\max}(G_T)}{\sigma^2}.$$

7.6 Robustness of Autoregression Forecasting Under Heteroscedasticity of the Innovation Process

7.6.1 The Mathematical Model

Let us consider the following rather common setting: an AR(p) autoregression model is complicated by heteroscedasticity (inhomogeneity w.r.t. the variances) of the innovation process. In other words, let us assume functional distortion of the innovation process in the variance [cf. (7.70)]:

$$x_t = (\theta^0)' X_{t-1}^{t-p} + \tilde{u}_t, \quad \tilde{u}_t = \mu_t u_t, \quad t \in \mathbb{Z}, \tag{7.82}$$

where $\theta^0 = (\theta_i^0) \in \mathbb{R}^p$ is a column vector of p hypothetical autoregression coefficient values;

$$X_{t-1}^{t-p} = (x_{t-1}, \dots, x_{t-p})' \in \mathbb{R}^p;$$

random variables $\{u_t\}$ are i.i.d. and follow the $N(0, \sigma^2)$ normal distribution law;

$$0 \le \mu_t \le \varepsilon, \quad t \in \mathbb{Z}, \tag{7.83}$$

is an unknown deterministic sequence defining the distortion; $\tilde{u}_t$ is the distorted innovation process. The expression (7.82) is reduced to the hypothetical model for $\mu_t \equiv 1$. Note that, in general, heteroscedasticity implies non-stationarity of the time series (7.82):

$$\mathbb{D}\{\tilde{u}_t\} = \sigma^2 \mu_t^2, \quad t \in \mathbb{Z}.$$

We are going to assume that the hypothetical AR(p) model is stable, i.e., every root of the equation

$$x^p - \theta_1 x^{p-1} - \dots - \theta_p = 0$$

lies within the unit circle.

The model (7.82) can be represented as a first order p-variate vector autoregression VAR(1) (see Sect. 3.5.2):

$$X_t^{t-p+1} = A_0 X_{t-1}^{t-p} + \mu_t U_t, \quad t \in \mathbb{Z}, \tag{7.84}$$

where

$$X_{t-1}^{t-p} = \begin{pmatrix} x_{t-1} \\ \vdots \\ x_{t-p} \end{pmatrix} \in \mathbb{R}^p, \quad U_t = \begin{pmatrix} u_t \\ 0 \\ \vdots \\ 0 \end{pmatrix} \in \mathbb{R}^p, \quad A_0 = \left(\begin{array}{c:c} \multicolumn{2}{c}{\theta^0} \\ \hdashline I_{p-1} & \mathbf{0}_{p-1} \end{array} \right) \in \mathbb{R}^{p \times p}.$$

In this notation, the maximum absolute eigenvalue of the matrix A_0 satisfies the stability condition above:

$$\lambda_{\max}(A_0) < 1. \tag{7.85}$$

Let us study the influence of distortion $\{\mu_t\}$ on the forecast risk of the traditional forecasting statistic (7.64), also accounting for the specification error $\Delta\theta = \theta - \theta^0$ in the definition of autoregression coefficients.

7.6.2 Presence of a Specification Error

As in Sect. 7.5, we are going to consider the situation where the autoregression coefficient vector θ^0 is known up to a certain deterministic error $\Delta\theta = \theta - \theta^0$, and the plug-in forecasting statistic for $x_{T+\tau}$ based on observations $\{x_1, \ldots, x_T\}$ is defined from the vector $\theta = \theta^0 + \Delta\theta$:

$$\hat{x}_{T+\tau} = (\hat{X}_{T+\tau}^{T+\tau-p+1})_1, \quad \hat{X}_{T+\tau}^{T+\tau-p+1} = A^\tau X_T^{T-p+1}, \quad A = \left(\begin{array}{c|c} \multicolumn{2}{c}{\theta'} \\ \hline I_{p-1} & \mathbf{0}_{p-1} \end{array}\right). \tag{7.86}$$

As before, forecasting performance of (7.86) will be characterized by the mean square forecast risk in a matrix form

$$R = (r_{ij}) = R(\theta^0, \theta, T, \tau) = \mathbb{E}\left\{\left(\hat{X}_{T+\tau}^{T+\tau-p+1} - X_{T+\tau}^{T+\tau-p+1}\right)(\cdot)'\right\} \tag{7.87}$$

and the scalar risk

$$r = r(\theta^0, \theta, T, \tau) = r_{11} = \mathbb{E}\left\{(\hat{x}_{T+\tau} - x_{T+\tau})^2\right\} \geq 0. \tag{7.88}$$

Let us use the following notation: $\mathbf{1}_{p(1)} = (1, 0 \ldots 0)' \in \mathbb{R}^p$ is a unit basis vector, $(A_0^k)_{\cdot 1} \in \mathbb{R}^p$ is the first column of the matrix A_0^k, and

$$S = \sigma^2 \sum_{t=0}^{+\infty} (A_0^t)_{\cdot 1} (A_0^t)'_{\cdot 1} \mu_{T-t}^2 \in \mathbb{R}^{p \times p}, \tag{7.89}$$

$$S_{T\tau} = \sigma^2 \sum_{t=0}^{\tau-1} (A_0^t)_{\cdot 1} (A_0^t)'_{\cdot 1} \mu_{T+\tau-t}^2 \in \mathbb{R}^{p \times p}. \tag{7.90}$$

Theorem 7.12. *In the AR(p) model with heteroscedasticity defined by* (7.82), (7.83), (7.85) *under specification error* $\Delta\theta$, *the matrix forecast risk* (7.87) *of the forecasting statistic* (7.86) *is equal to*

$$R(\theta^0, \theta, T, \tau) = S_{T\tau} + (A^\tau - A_0^\tau)S(A^\tau - A_0^\tau)' \tag{7.91}$$

for $T \geq p$.

Proof. Using the methods of [2], the following equalities can be obtained from (7.82), (7.86):

$$X_{T+\tau}^{T+\tau-p+1} = \sum_{k=0}^{+\infty} A_0^k \mu_{T+\tau-k} U_{T+\tau-k}, \quad \hat{X}_{T+\tau}^{T+\tau-p+1} = A^\tau \sum_{k=0}^{+\infty} A_0^k \mu_{T-k} U_{T-k},$$

$$\hat{X}_{T+\tau}^{T+\tau-p+1} - X_{T+\tau}^{T+\tau-p+1} = \sum_{k=0}^{\tau-1} A_0^k \mu_{T+\tau-k} U_{T+\tau-k} + \sum_{l=0}^{+\infty} (A^\tau - A_0^\tau) A_0^l \mu_{T-l} U_{T-l}. \tag{7.92}$$

Substituting (7.92) in the definition of the mean square risk (7.88) and accounting for the independence of $\{u_t\}$, we obtain, after equivalent matrix transformations, the following expression for the risk:

$$R(\theta^0, \theta, T, \tau) = \mathbb{E}\left\{\left(\sum_{k=0}^{\tau-1} A_0^k \mu_{T+\tau-k} U_{T+\tau-k}\right)(\cdot)'\right\} +$$

$$+ \mathbb{E}\left\{\left(\sum_{l=0}^{+\infty} (A^\tau - A_0^\tau) A_0^l \mu_{T-l} U_{T-l}\right)(\cdot)'\right\} =$$

$$= \sigma^2 \sum_{k=0}^{\tau-1} \mu_{T+\tau-k}^2 (A_0^k \mathbf{1}_{p(1)})(\cdot)' + \sigma^2 \sum_{l=0}^{+\infty} \mu_{T-l}^2 \left((A^\tau - A_0^\tau) A_0^l \mathbf{1}_{p(1)}\right)(\cdot)'.$$

In the notation (7.89), the previous expression is equivalent to (7.91). □

Corollary 7.11. *Under the conditions of Theorem 7.12, the scalar risk of forecasting the future value* $x_{T+\tau}$ *equals*

$$r(\theta^0, \theta, T, \tau) = \sigma^2 \sum_{t=0}^{\tau-1} ((A_0^t)_{11})^2 \mu_{T+\tau-t}^2 + \sigma^2 \sum_{t=0}^{+\infty} \left(\left((A^\tau - A_0^\tau) A_0^t\right)_{11}\right)^2 \mu_{T-t}^2. \tag{7.93}$$

Proof. It suffices to apply (7.88)–(7.91) and certain well-known matrix properties. □

In the expression for the risk (7.93), the first summand is due to heteroscedasticity in the forecasting interval, and the second summand is due to both heteroscedasticity and the specification error for time points $s \leq T$.

Corollary 7.12. *Under the conditions of Theorem 7.12, the minimum value of the scalar risk over $\theta \in \mathbb{R}^p$ is equal to*

$$r_{\min}(\theta^0, T, \tau) = r(\theta^0, \theta^0, T, \tau) = \sigma^2 \sum_{t=0}^{\tau-1} ((A_0^t)_{11})^2 \mu^2_{T+\tau-t} \cdot$$

Now let us investigate the behavior of the risk under heteroscedasticity in a setting of asymptotically small specification errors $\Delta\theta = \theta - \theta^0$ [14]:

$$|\Delta\theta| = \sqrt{(\theta - \theta^0)'(\theta - \theta^0)} \leq \gamma, \quad \gamma \to 0. \tag{7.94}$$

Theorem 7.13. *Under the conditions of Theorem 7.12 and in the asymptotics* (7.94)*, the following expansion holds for the scalar risk of forecasting* (7.88)*:*

$$r(\theta^0, \theta, T, \tau) = r_{\min}(\theta^0, T, \tau) + (\Delta\theta)'\beta(T, \tau)\Delta\theta + O(\gamma^3), \tag{7.95}$$

where β is a quadratic form defined by the following symmetric positive-definite matrix:

$$\beta(T, \tau) = (\beta_{jk}(T, \tau)) = \left(\sum_{i=0}^{\tau-1} (A_0^i)_{11} A_0^{\tau-i-1}\right) S \left(\sum_{i=0}^{\tau-1} (A_0^i)_{11} A_0^{\tau-i-1}\right)' .$$

Proof. As in the proof of Corollary 7.6, substitute a matrix expansion of $(A_0 + \Delta A)^\tau$ into (7.93). □

Corollary 7.13. *Under the conditions of Theorem 7.13, the guaranteed forecast risk can be written as*

$$r_+(\theta^0, T, \tau, \gamma) = r_{\min}(\theta^0, T, \tau) + \gamma^2 \lambda_{\max}(\beta(T, \tau)) + O(\gamma^3), \tag{7.96}$$

where $\lambda_{\max}(C)$ is the maximum eigenvalue of a matrix C.

Proof. Applying some well-known extremal properties of eigenvalues [22] allows us to construct an argument similar to the proof of Corollary 7.7. □

From the main term of the expansion (7.96) and Definition 4.7, we can obtain the δ-admissible ($\delta > 0$) distortion level for the vector of autoregression coefficients:

$$\gamma_+(\delta) = \sqrt{\delta \frac{r_{\min}(\theta^0, T, \tau)}{\lambda_{\max}(\beta(T, \tau))}}.$$

Theorems 7.12, 7.13, together with their corollaries, allow us to study another commonly encountered specification error—incorrectly defined autoregression order. Assume that the observed time series x_t is described by an AR(p^0) autoregression model of true order $p^0 \in \mathbb{N}$, but the forecast (7.86) is based on an AR(p) model of order $p \in \mathbb{N}$ different from p^0. To be more specific, let us assume $p < p^0$ (underestimated autoregression order; the case of overestimation, where $p > p^0$, can be studied similarly). Let us split the matrices into blocks (the numbers outside the brackets denote block dimensions):

$$\theta^0 = \begin{matrix} p \\ p^0 - p \end{matrix} \begin{pmatrix} \theta^0_{(1)} \\ \hdashline \theta^0_{(2)} \end{pmatrix}, \quad \theta = \begin{matrix} p \\ p^0 - p \end{matrix} \begin{pmatrix} \theta_{(1)} \\ \hdashline \theta_{(p^0-p)} \end{pmatrix}, \quad \Delta\theta = \begin{matrix} p \\ p^0 - p \end{matrix} \begin{pmatrix} \theta_{(1)} - \theta^0_{(1)} \\ \hdashline -\theta^0_{(2)} \end{pmatrix},$$

$$\beta(T, \tau) = \begin{matrix} \\ p \\ p^0 - p \end{matrix} \begin{matrix} \quad p \qquad p^0 - p \\ \begin{pmatrix} \beta_{(11)} & \beta_{(12)} \\ \beta_{(21)} & \beta_{(22)} \end{pmatrix} \end{matrix}.$$

Corollary 7.14. *Under the conditions of Theorem 7.13, assuming underestimated regression order, the scalar forecast risk satisfies the asymptotic expansion*

$$r(\theta^0, \theta, T, \tau) = r_{\min}(\theta^0, T, \tau) + (\theta^0_{(2)})' \beta_{(22)}(T, \tau)\theta^0_{(2)} + O(\gamma^3); \tag{7.97}$$

and the specification error is δ-admissible ($\delta > 0$) if

$$(\theta^0_{(2)})' \beta_{(22)}(T, \tau)\theta^0_{(2)} \leq \delta\, r_{\min}(\theta^0, T, \tau). \tag{7.98}$$

Proof. Substituting block representations of the matrices $\Delta\theta$ and $\beta(T, \tau)$ into (7.95) leads to (7.97). The relation (7.98) follows directly from (7.97). □

7.6.3 Least Squares Estimation of θ^0

Now consider a situation where the parameter vector θ^0 of an AR(p) model with heteroscedasticity (7.82) is a priori unknown, and an estimator is constructed for θ^0 based on T observations $x_1, \ldots, x_T$. Traditionally, a least squares estimator is used to estimate θ^0 [2]. The estimator is constructed under the following assumptions:

$$T > p, \qquad \left| \sum_{t=p}^{T-1} X_t^{t-p+1} \left(X_t^{t-p+1} \right)' \right| \neq 0,$$

and it has the form

$$\hat{\theta} = \begin{pmatrix} \hat{\theta}_{(1)} \\ \vdots \\ \hat{\theta}_p \end{pmatrix} = \left(\frac{1}{T-p} \sum_{t-p}^{T-1} X_t^{t-p+1} (X_t^{t-p+1})' \right)^{-1} \frac{1}{T-p} \sum_{t=p}^{T-1} x_{t+1} X_t^{t-p+1} \in \mathbb{R}^p. \tag{7.99}$$

Let us start by studying the consistency and asymptotic normality of the least squares estimator (7.99) under heteroscedasticity. We are going to need the following auxiliary result [13, 15].

Lemma 7.3. *Under the $AR(p)$ model with heteroscedasticity* (7.82)*, if the conditions* (7.83), (7.85)*, and an additional asymptotic condition on μ_t,*

$$\exists c \in (0, +\infty) \ : \ \frac{1}{T} \sum_{t=1}^{T} \mu_t^2 \underset{T \to +\infty}{\longrightarrow} c, \tag{7.100}$$

are satisfied, then the following mean square convergences hold as $T \to +\infty$:

$$\frac{1}{T-p} \sum_{t=p}^{T-1} \mu_{t+1} u_{t+1} X_t^{t-p+1} \xrightarrow{\text{m.s.}} \mathbf{0}_p,$$

$$\frac{1}{T-p} \sum_{t=p}^{T-1} X_t^{t-p+1} (X_t^{t-p+1})' \xrightarrow{\text{m.s.}} cF, \qquad F = \sigma^2 \sum_{i=0}^{+\infty} (A_0^i)_{\cdot 1} (A_0^i)'_{\cdot 1} \ .$$

Theorem 7.14. *Under the conditions of Lemma 7.3, the least squares estimator of the autoregression coefficients* (7.99) *is consistent in probability:*

$$\hat{\theta} \underset{T \to +\infty}{\xrightarrow{\mathbf{P}}} \theta^0.$$

Proof. From (7.82), (7.99) we have the following representation for the difference between the estimator and the vector of true autoregression coefficients:

$$\hat{\theta} - \theta^0 = \left(\frac{1}{T-p} \sum_{t=p}^{T-1} X_t^{t-p+1} (X_t^{t-p+1})' \right)^{-1} \frac{1}{T-p} \sum_{T-p}^{T-1} \mu_{t+1} u_{t+1} X_t^{t-p+1}. \tag{7.101}$$

Then, by Lemma 7.3 and properties of convergent random sequences [25], we have $\hat{\theta} - \theta^0 \xrightarrow{\mathbf{P}} \mathbf{0}_p$, proving the theorem. □

Theorem 7.15. *Under the conditions of Lemma 7.3, assume that, in addition to* (7.100), *we have*

$$\forall \tau \in \mathbb{N} \quad \exists c_\tau \in (0, +\infty) \ : \quad \frac{1}{T}\sum_{t=1}^{T} \mu_t^2 \mu_{t+\tau}^2 \underset{T\to+\infty}{\longrightarrow} c_\tau.$$

Then the least squares estimator for the autoregression coefficients (7.99) *is asymptotically normal:*

$$\sqrt{T}\left(\hat{\theta} - \theta^0\right) \underset{T\to+\infty}{\xrightarrow{\mathbf{D}}} \eta, \qquad \mathcal{L}\{\eta\} = \mathcal{N}_p\left(\mathbf{0}_p, \frac{\sigma^2}{c^2} F^{-1} F_c F^{-1}\right),$$

where

$$F_c = \sum_{i=0}^{+\infty} c_{i+1} (A_0^i)_{\cdot 1} (A_0^i)'_{\cdot 1} \in \mathbb{R}^{p\times p}.$$

Proof. From Lemma 7.3 and Theorem 7.14, a proof can be constructed similarly to the case where the model is free from distortion [2]. □

Note that in the absence of heteroscedasticity we have $\mu_t \equiv 1$, $c_\tau \equiv c = 1$, $F_c = F$, and Theorems 7.14, 7.15 are reduced to known results [2].

Let us study the risk of a plug-in forecasting algorithm using the least squares estimator (7.99) for the autoregression coefficients:

$$\hat{x}_{T+\tau} = \left(\hat{A}^\tau X_T^{T-p+1}\right)_1, \quad \hat{A} = \left(\begin{array}{c:c} \multicolumn{2}{c}{\hat{\theta}'} \\ \hdashline I_{p-1} & \mathbf{0}_{p-1} \end{array}\right). \tag{7.102}$$

Theorem 7.16. *Under the conditions of Lemma 7.3, the forecast risk of the plug-in algorithm* (7.102) *satisfies the following asymptotic expansion as* $T \to +\infty$:

$$r = \sigma^2 \sum_{t=0}^{\tau-1} \mu_{T+\tau-t}^2 \left((A^t)_{11}\right)^2 + O\left(T^{-1}\right). \tag{7.103}$$

Proof. The expansion (7.103) follows from the representation (7.101) for the deviation $\hat{\theta} - \theta^0$ of the least squares estimator from the true value, the expression for the random forecast error

$$\hat{x}_{T+\tau} - x_{T+\tau} = \left(\hat{X}_{T+\tau}^{T+\tau-p+1} - X_{T+\tau}^{T+\tau-p+1}\right)_1 =$$
$$= \left((\hat{A}^\tau - A_0^\tau)X_T^{T-p+1} + \mu_{T+1}A_0^{\tau-1}U_{T+1} + \cdots + \mu_{T+\tau}U_{T+\tau}\right)_1,$$

and Lemma 7.3. □

Observe that the leading term of the asymptotic expansion established by Theorem 7.16 is the minimum risk value under heteroscedasticity for an a priori known θ^0 (cf. Corollary 7.12):

$$r = r_{\min}(\theta^0, T, \tau) + O\left(\frac{1}{T}\right).$$

This shows that for $T \to +\infty$ we have

$$r - r_{\min}(\theta^0, T, \tau) = O\left(\frac{1}{T}\right) \to 0.$$

Thus, under the above conditions, we can say that the plug-in forecasting algorithm (7.102), (7.99) is robust under heteroscedasticity.

7.7 Robustness of Autoregression Time Series Forecasting Under IO-Outliers

As mentioned in Sect. 4.3, outliers are a very common type of distortions. Autoregression time series are affected by two types of outliers: *IO (innovation outliers)* and *AO (additive outliers)* [20,23].

Let us start by considering IO distortion. In that case, the observed time series is subject to stochastic distortion of the innovation process:

$$x_t = (\theta^0)' X_{t-1}^{t-p} + \xi_t u_t, \quad t \in \mathbb{Z}, \tag{7.104}$$

where $\theta^0 = (\theta_i^0) \in \mathbb{R}^p$ is a column vector of p hypothetical autoregression coefficients; $\{u_t\}$ are jointly independent identically distributed $\mathcal{N}(0, \sigma^2)$ normal random variables; $\{\xi_t\}$ are jointly independent identically distributed discrete random variables independent of $\{u_t\}$ which assume one of the following two values:

$$\xi_t \in \{1, \sqrt{K}\}, \quad \mathbb{P}\{\xi_t = \sqrt{K}\} = \varepsilon, \quad \mathbb{P}\{\xi_t = 1\} = 1 - \varepsilon, \tag{7.105}$$

where $K \in (1, K_+]$ is the outlier magnitude, $\varepsilon \in [0, \varepsilon_+]$ is the probability that an outlier is observed.

It follows from (7.104), (7.105) that if at a time point $t = t'$ we have $\xi_{t'} = 1$, then no outlier has been observed: the innovation process equals $\tilde{u}_{t'} = u_{t'}$ and has the variance $\mathbb{D}\{\tilde{u}_{t'}\} = \sigma^2$; if at a certain other time point $t = t''$ we have $\xi_{t''} = \sqrt{K}$, then we have observed an outlier $\tilde{u}_{t''} = \sqrt{K} u_{t''}$ with the variance which is K times higher than hypothetical value: $\mathbb{D}\{\tilde{u}_{t''}\} = K\sigma^2 > \sigma^2$. In applications, it is usually said that we are dealing with outliers in the variance, and typically we have $K \gg 1$, $\varepsilon_+ \leq 0.1$.

Comparing (7.104), (7.105) with (7.82), (7.83), note that the difference between these models lies in the boundedness of the multiplier $\mu_t = \xi_t$ in (7.82). Also note that modifying the model (7.70) similarly to (7.104) leads to a model of outliers in the mean value.

Influence of IO distortion on forecast risk can be characterized by applying the results of Sect. 7.6. First, let us consider the case where the forecasting statistic defined by (7.86) is based on the autoregression coefficient vector $\theta = (\theta_i) \in \mathbb{R}^p$ which is distorted by a deterministic error: $\Delta\theta = \theta - \theta^0 \in \mathbb{R}^p$.

Theorem 7.17. *In the $AR(p)$ model under IO distortion* (7.104), (7.105), *assuming that the stability condition* (7.85) *is satisfied, and a specification error $\Delta\theta$ is present, the matrix risk* (7.87) *of the forecasting statistic* (7.86) *equals*

$$R(\theta^0, \theta, T, \tau) = \sigma^2(1 + \varepsilon(K-1)) \left(\sum_{t=0}^{\tau-1} (A_0^t)_{\cdot 1} (A_0^t)'_{\cdot 1} + \right.$$
$$\left. + (A^\tau - A_0^\tau) \sum_{t=0}^{+\infty} (A_0^t)_{\cdot 1} (A_0^t)'_{\cdot 1} (A^\tau - A_0^\tau)' \right) \tag{7.106}$$

if the observation length satisfies the condition $T \geq p$.

Proof. By (7.87) and the total expectation formula, we have

$$R(\theta^0, \theta, T, \tau) = \mathbb{E}\left\{\mathbb{E}\left\{\left(\hat{X}_{T+\tau}^{T+\tau-p+1} - X_{T+\tau}^{T+\tau-p+1}\right)(\cdot)' \mid \{\xi_{T+\tau}, \xi_{T+\tau-1}, \dots\}\right\}\right\}.$$

Here the conditional expectation for fixed

$$\{\xi_{T+\tau}, \xi_{T+\tau-1}, \dots\}$$

is found by applying Theorem 7.12, i.e., replacing $\{\mu_{T+\tau}, \mu_{T+\tau-1}, \dots\}$ in the expressions (7.89), (7.91) by, respectively, $\{\xi_{T+\tau}, \xi_{T+\tau-1}, \dots\}$. From (7.91), computing the unconditional expectation yields

$$R(\theta^0, \theta, T, \tau) = \mathbb{E}\{S_{T\tau}\} + (A^\tau - A_0^\tau)\mathbb{E}\{S\}(A^\tau - A_0^\tau)'. \tag{7.107}$$

By (7.105), we have

$$\mathbb{E}\{\xi_t^2\} = 1 + \varepsilon(K-1) \geq 1, \quad t \in \mathbb{Z}.$$

Substituting the above into (7.89), (7.107) yields (7.106). □

Corollary 7.15. *Under the conditions of Theorem 7.17, the scalar risk of forecasting the future value $x_{T+\tau}$ equals*

$$r(\theta^0,\theta,T,\tau)=\sigma^2(1+\varepsilon(K-1))\left(\sum_{t=0}^{\tau-1}\left((A_0^t)_{11}\right)^2+\sum_{t=0}^{+\infty}\left(((A^\tau-A_0^\tau)A_0^t)_{11}\right)^2\right).$$

Proof. It is sufficient to apply the definition of the scalar risk (7.88), the expression for the matrix risk (7.106), and certain well-known matrix properties. □

As earlier, let us introduce the following notation for the risk under the hypothetical model without outliers ($\varepsilon=0$) or specification errors ($\Delta\theta=\mathbf{0}_p$):

$$r_0(T,\tau)=\sigma^2\sum_{t=0}^{\tau-1}\left((A_0^t)_{11}\right)^2>0. \tag{7.108}$$

Corollary 7.16. *Under the conditions of Theorem 7.17, the scalar risk of forecasting $x_{T+\tau}$ can be represented as follows:*

$$\begin{aligned}r(\theta^0,\theta,T,\tau)&=r_0(T,\tau)+\varepsilon(K-1)r_0(T,\tau)+\\&+\sigma^2\sum_{t=0}^{+\infty}\left(\left(\left(A^\tau-A_0^\tau\right)A_0^t\right)_{11}\right)^2+\sigma^2\varepsilon(K-1)\sum_{t=0}^{+\infty}\left(\left(\left(A^\tau-A_0^\tau\right)A_0^t\right)_{11}\right)^2.\end{aligned} \tag{7.109}$$

Proof. Applying Corollary 7.15 and the expression (7.108) proves this statement. □

The representation (7.109) of the forecast risk contains four summands. The first is the hypothetical risk (in the absence of distortion), the second is due to outliers (it is proportional to the outlier probability and the variance increment $K-1$), the third is due to specification error $\Delta\theta$, and the fourth results from joint influence of the outliers and the specification error.

Let us study the case of small specification errors (7.94).

Consider the following positive-definite $(p\times p)$-matrix:

$$B(T,\tau)=\left(\sum_{i=0}^{\tau-1}(A_0^i)_{11}A_0^{\tau-i-1}\right)\sum_{t=0}^{+\infty}(A_0^t)_{\cdot 1}(A_0^t)'_{\cdot 1}\left(\sum_{i=0}^{\tau-1}(A_0^i)_{11}A_0^{\tau-i-1}\right)'. \tag{7.110}$$

Theorem 7.18. *Under the conditions] of Theorem 7.17, in the asymptotics* (7.94), *the scalar forecast risk for $x_{T+\tau}$ satisfies the following expansion:*

$$r(\theta^0, \theta, T, \tau) = r_0(T, \tau) + \varepsilon(K-1)r_0(T, \tau) + \sigma^2(\Delta\theta)' B(T, \tau)\Delta\theta + \\ + \sigma^2\varepsilon(K-1)\Delta\theta' B(T, \tau)\Delta\theta + O(\gamma^3). \quad (7.111)$$

Proof. It suffices to use a matrix expansion for $A^\tau = (A_0 + \Delta A)^\tau$ in (7.109) and to apply the notation (7.110). □

Corollary 7.17. *Under the conditions of Theorem 7.18, the guaranteed forecast risk satisfies the following asymptotic expansion:*

$$r_+(\theta^0, T, \tau, \gamma, \varepsilon) = (1 + \varepsilon(K-1))\big(r_0(T, \tau) + \gamma^2 \lambda_{\max}(B(T, \tau))\big) + O(\gamma^3), \quad (7.112)$$

and the risk instability coefficient can be written as

$$\kappa(\theta^0, T, \tau, \gamma, \varepsilon) = \varepsilon(K-1) + (1 + \varepsilon(K-1))\gamma^2 \frac{\lambda_{\max}(B(T, \tau))}{r_0(T, \tau)} + O(\gamma^3), \quad (7.113)$$

where $\lambda_{\max}(C)$ is the maximum eigenvalue of a matrix C.

Proof. As in the proof of Corollary 7.7, the relation (7.112) is obtained from maximizing the expression (7.111) in $\Delta\theta$ by applying extremal properties of eigenvalues. Now (7.113) follows from the definition of the risk instability coefficient,

$$\kappa(\theta^0, T, \tau, \gamma, \varepsilon) = \frac{r_+(\theta^0, T, \tau, \gamma, \varepsilon) - r_0(T, \tau)}{r_0(T, \tau)},$$

and the relation (7.112). □

Setting some fixed (critical) level $\delta > 0$ for the risk instability coefficient, discarding the last term of (7.113), and solving the inequality $\kappa \leq \delta$ in γ leads to an expression for the δ-admissible level of the specification error $|\Delta\theta|$ in the autoregression coefficient vector θ:

$$|\Delta\theta| \leq \gamma_+(\delta) = \sqrt{\frac{(\delta - \varepsilon(K-1))_+}{1 + \varepsilon(K-1)} \frac{r_0(T, \tau)}{\lambda_{\max}(B(T, \tau))}}, \quad (7.114)$$

where $(x)_+ = \{x, \ x > 0; \ 0, \ x \leq 0\}$ is the so-called ramp function of x.

It should be noted that dependence of the risk (7.109), the guaranteed risk (7.112), the risk instability coefficient (7.113), and the δ-admissible specification error level (7.114) on the outlier probability ε and the variance increment $K-1$ can be expressed in terms of a single auxiliary variable

$$\nu = \varepsilon(K-1) \geq 0. \quad (7.115)$$

This means, for example, that a 10 % proportion of outliers for $K = 4$ results in the same risk increment as a 1 % outlier proportion with $K = 31$. Solving the inequality $\kappa \leq \delta$ in the new variable (7.115) for a fixed specification error level γ yields the δ-admissible value of ν:

$$\nu \leq \nu_+(\delta) = \left(\delta - \gamma^2 \frac{\lambda_{\max}(B(T,\tau))}{r_0(T,\tau)}\right)_+ \left(1 + \gamma^2 \frac{\lambda_{\max}(B(T,\tau))}{r_0(T,\tau)}\right)^{-1}.$$

Let us proceed by considering a setting where the true parameters θ^0 of an AR(p) model under IO distortion (7.104), (7.105) are a priori unknown. The parameter vector θ^0 is estimated by using the least squares method (7.99) for the T recorded observations $x_1, \ldots, x_T$; the forecasting statistic is defined by (7.102).

By (7.104), (7.105), as well as independence and identical distribution of $\{\xi_t\}$, the strong law of large numbers is satisfied for $\{\xi_t^2\}$ [19]:

$$\frac{1}{T}\sum_{t=1}^{T} \xi_t^2 \xrightarrow[T\to+\infty]{\text{a.s.}} c = 1 + \varepsilon(K-1) \geq 1.$$

Thus, the condition (7.100) holds almost surely. Therefore, we can apply the result of Lemma 7.3:

$$\frac{1}{T-p}\sum_{t=p}^{T-1} \xi_{t+1} u_{t+1} X_t^{t-p+1} \xrightarrow{\text{m.s.}} \mathbf{0}_p,$$

$$\frac{1}{T-p}\sum_{t=p}^{T-1} X_t^{t-p+1} (X_t^{t-p+1})' \xrightarrow{\text{m.s.}} cF, \quad F = \sigma^2 \sum_{i=0}^{+\infty} (A_0^i)_{\cdot 1} (A_0^i)'_{\cdot 1}\,.$$

This, in turn, proves the consistency of the least squares estimator $\hat{\theta}$ under IO distortion defined by (7.105):

$$\hat{\theta} \xrightarrow[T\to+\infty]{\mathbf{P}} \theta^0.$$

Using this fact, an asymptotic expansion of the risk for $\varepsilon \to 0$, $T \to +\infty$ can be constructed similarly to Theorem 7.16:

$$r = (1 + \varepsilon(K-1)) r_0(T,\tau) + O(T^{-1}) + o(\varepsilon). \tag{7.116}$$

The expansion (7.116) leads to asymptotic expansions for the guaranteed risk:

$$r_+(T,\tau) = (1 + \varepsilon_+(K-1)) r_0(T,\tau) + O(T^{-1}) + o(\varepsilon),$$

and for the risk instability coefficient:

$$\kappa(T,\tau) = \varepsilon_+(K-1) + O(T^{-1}) + o(\varepsilon).$$

The last expansion yields an approximation for the δ-admissible proportion (or probability) of outliers:

$$\varepsilon_+(\delta) = \delta/(K-1).$$

From (7.116), we can also see that under IO distortion the forecasting statistic (7.102), (7.101) is consistent for $T \to +\infty$: $r_+(T,\tau) - r_0(T,\tau) \to 0$.

7.8 Robustness of Autoregression Time Series Forecasting Under AO Outliers

The $AR(p)$ model of a time series $x_t$under additive outliers (AO distortion) is defined by the following three stochastic equations [10, 11, 20]:

$$\begin{aligned} x_t &= y_t + h_t, \\ h_t &= \xi_t v_t, \\ y_t &= (\theta^0)' Y_{t-1}^{t-p} + u_t, \quad t \in \mathbb{Z}, \end{aligned} \tag{7.117}$$

where $\theta^0 = (\theta_i^0) \in \mathbb{R}^p$ is a column vector of p hypothetical autoregression coefficients describing the time series y_t; random variables $\{u_t\}$ are jointly independent and have identical $\mathcal{N}(0,\sigma^2)$ normal distributions; $\{\xi_t\}$ are independent Bernoulli random variables,

$$\xi_t \in \{0,1\}, \quad \mathbb{P}\{\xi_t = 1\} = 1 - \mathbb{P}\{\xi_t = 0\} = \varepsilon;$$

$\{v_t\}$ are jointly independent $\mathcal{N}(0, K\sigma^2)$ normal random variables independent of $\{\xi_t\}$, $\{u_t\}$. The value $\varepsilon \in [0, \varepsilon_+]$ is the outlier probability, and $K \in (1, K_+]$ is the outlier magnitude.

The third equation of (7.117) defines the undistorted hypothetical model of the unobservable autoregression time series y_t, and the observed time series x_t is defined by the first equation of (7.117). The time series x_t is an additive mixture of the unobservable elements y_t and the distortion process

$$h_t = \xi_t v_t, \quad t \in \mathbb{Z}. \tag{7.118}$$

From (7.117) and the definition of ξ_t, each observation is free of distortion with probability $\mathbb{P}\{h_t = 0\} = \mathbb{P}\{\xi_t = 0\} = 1 - \varepsilon$. If $\xi_t = 1$, an additive outlier is observed:

$$x_t = y_t + v_t.$$

Note that these outliers are also outliers in the variance (see Sect. 7.7) since their variance is increased compared to undistorted observations:

$$\mathcal{L}\{x_t \mid \xi_t = 1\} = \mathcal{N}\big(0, \mathbb{D}\{y_t\} + K\sigma^2\big).$$

Consider the problem of forecasting the future value $y_{T+\tau}$ based on an observed time series $x_1, \ldots, x_T$.

Let us introduce the notation

$$H_T^{T-p+1} = \begin{pmatrix} h_T \\ h_{T-1} \\ \vdots \\ h_{T-p+1} \end{pmatrix}, \quad Y_{T+\tau}^{T+\tau-p+1} = \begin{pmatrix} y_{T+\tau} \\ y_{T+\tau-1} \\ \vdots \\ y_{T+\tau-p+1} \end{pmatrix} \in \mathbb{R}^p.$$

From (7.117), (7.118), and the results of Sect. 7.2, we are going to construct an optimal forecast for $Y_{T+\tau}^{T+\tau-p+1}$, written as the following conditional expectation:

$$\hat{Y}_{T+\tau}^{*T+\tau-p+1} = \mathbb{E}\left\{Y_{T+\tau}^{T+\tau-p+1} \mid X_T^1\right\}. \tag{7.119}$$

We are going to require the following auxiliary result on properties of p-variate normal densities $n_p(x \mid \mu, \Sigma)$.

Lemma 7.4. *For any $p \in \mathbb{N}$, $(p \times p)$-matrices Σ_1, Σ_2, and points $x_{(1)}, x_{(2)} \in \mathbb{R}^p$, the following identity is satisfied:*

$$\begin{aligned} n_p(x_{(1)} \mid \mathbf{0}_p, \Sigma_1) n_p(x_{(2)} \mid x_{(1)}, \Sigma_2) &\equiv n_p(x_{(2)} \mid \mathbf{0}_p, \Sigma_1 + \Sigma_2) \times \\ &\times n_p\left(x_{(1)} \mid \Sigma_1(\Sigma_1+\Sigma_2)^{-1}x_{(2)},\ \Sigma_1 - \Sigma_1(\Sigma_1+\Sigma_2)^{-1}\Sigma_1\right). \end{aligned} \tag{7.120}$$

Proof. Consider a composite normal $(2p)$-vector $X = (X'_{(1)} \vdots X'_{(2)})' \in \mathbb{R}^{2p}$, where $\mathcal{L}\{X_{(1)}\} = \mathcal{N}_p(\mathbf{0}_p, \Sigma_1)$, and the conditional distribution of $X_{(2)}$ for a fixed $X_{(1)} = x_{(1)}$ is normal, $\mathcal{L}\{X_{(2)} \mid X_{(1)} = x_{(1)}\} = \mathcal{N}_p(x_{(1)}, \Sigma_2)$. Then the left-hand side of (7.120) is the joint probability density of X:

$$p_{X_{(1)},X_{(2)}}(x_{(1)}, x_{(2)}) = p_{X_{(1)}}(x_{(1)}) p_{X_{(2)}|X_{(1)}}(x_{(2)} \mid x_{(1)}). \tag{7.121}$$

One can verify directly that the joint $(2p)$-variate distribution of X has the following form:

$$\mathcal{L}\left\{\begin{pmatrix} X_{(1)} \\ --- \\ X_{(2)} \end{pmatrix}\right\} = \mathcal{N}_{2p}\left(\begin{pmatrix} \mathbf{0}_p \\ -- \\ \mathbf{0}_p \end{pmatrix}, \left(\begin{array}{c|c} \Sigma_1 & \Sigma_1 \\ \hline \Sigma_1 & \Sigma_1 + \Sigma_2 \end{array}\right)\right). \tag{7.122}$$

Similarly to (7.121), multiplication theorem for probability density functions yields

$$p_{X_{(1)},X_{(2)}}(x_{(1)},x_{(2)}) = p_{X_{(2)}}(x_{(2)})\,p_{X_{(1)}|X_{(2)}}(x_{(1)} \mid x_{(2)}). \tag{7.123}$$

Applying the properties of marginal and conditional normal probability densities to (7.122) leads to:

$$p_{X_{(2)}}(x_{(2)}) = n_p(x_{(2)} \mid \mathbf{0}_p, \Sigma_1 + \Sigma_2),$$
$$p_{X_{(1)}|X_{(2)}}(x_{(1)} \mid x_{(2)}) = n_p\left(x_{(1)} \mid \Sigma_1(\Sigma_1+\Sigma_2)^{-1}x_{(2)},\ \Sigma_1 - \Sigma_1(\Sigma_1+\Sigma_2)^{-1}\Sigma_1\right).$$

Substituting the above expressions into (7.120) yields (7.124). □

Let us introduce the following notation: $V = \{0,1\}$; the set V_p is composed of binary p-vectors $J = (j_1,\ldots,j_p)'$, $j_k \in V$, $k = 1,\ldots,p$; $D_p(J)$ denotes a diagonal $(p \times p)$-matrix with diagonal entries $j_1,\ldots,j_p$; the norm $|J| = \sum_{k=1}^{p} j_k$ is the Hamming weight of a binary vector J; the mean square matrix $\mathbb{E}\{Y_T^{T-p+1}(Y_T^{T-p+1})'\}$ is denoted as Σ_p.

Finally, denote

$$B_\varepsilon = B_\varepsilon(X_T^{T-p+1}) = \sum_{J\in V_p}\left(\varepsilon^{|J|}(1-\varepsilon)^{p-|J|}n_p\left(X_T^{T-p+1} \mid \mathbf{0}_p, \Sigma_p + K\sigma^2 D_p(J)\right)\times\right.$$
$$\left.\times K\sigma^2 D_p(J)(\Sigma_p + K\sigma^2 D_p(J))^{-1}\right)\times$$
$$\times\left(\sum_{J\in V_p}\varepsilon^{|J|}(1-\varepsilon)^{p-|J|}n_p\left(X_T^{T-p+1} \mid \mathbf{0}_p, \Sigma_p + K\sigma^2 D_p(J)\right)\right)^{-1}. \tag{7.124}$$

Theorem 7.19. *In the AR*(p) *model with AO distortion* (7.117), *the mean square optimal forecast of the p-vector* $Y_{T+\tau}^{T+\tau-p+1}$ *for* $\tau \geq p$ *based on the observations* $x_1,\ldots,x_T$ *is nonlinear:*

$$\hat{Y}_{T+\tau}^{*T+\tau-p+1} = A^\tau\left(\mathbf{I}_p - B_\varepsilon\left(X_T^{T-p+1}\right)\right)X_T^{T-p+1}, \tag{7.125}$$

where A_0 *is the companion matrix* (7.84).

Proof. From (7.117), (7.118), in the above notation, by Lemma 7.2, we have

$$\begin{aligned} Y_{T+\tau}^{T+\tau-p+1} &= A_0^\tau Y_T^{T-p+1} + \sum_{k=0}^{\tau-1} A_0^k U_{T+\tau-k}, \\ Y_T^{T-p+1} &= X_T^{T-p+1} - H_T^{T-p+1}. \end{aligned} \tag{7.126}$$

Taking into account the independence of $\{u_{T+1}, u_{T+\tau}\}$ from X_T^1, let us use (7.119), (7.126) to compute the conditional expectation:

$$\begin{aligned}\hat{Y}_{T+\tau}^{*T+\tau-p+1} &= A_0^\tau \mathbb{E}\left\{X_T^{T-p+1} - H_T^{T-p+1} \mid X_T^{T-p+1}, X_{T-p}^1\right\} = \\ &= A_0^\tau \left(X_T^{T-p+1} - \mathbb{E}\left\{H_T^{T-p+1} \mid X_T^{T-p+1}\right\}\right).\end{aligned} \tag{7.127}$$

Here we also use the fact that $\{h_T, h_{T-1}, \dots, h_{T-p+1}\}$ and $\{x_{T-p}, x_{T-p-1}, \dots, x_1\}$ are independent.

Let us compute the auxiliary expectation in (7.127):

$$\begin{aligned}\overline{H}_T^{T-p+1} &= \mathbb{E}\left\{H_T^{T-p+1} \mid X_T^{T-p+1}\right\} = \\ &= \int_{\mathbb{R}^p} z p_{H_T^{T-p+1} \mid X_T^{T-p+1}}\left(z \mid X_T^{T-p+1}\right) dz,\end{aligned} \tag{7.128}$$

where the conditional probability density equals

$$p_{H_T^{T-p+1} \mid X_T^{T-p+1}}\left(z \mid X_T^{T-p+1}\right) = \frac{p_{H_T^{T-p+1}, X_T^{T-p+1}}\left(z, X_T^{T-p+1}\right)}{p_{X_T^{T-p+1}}\left(X_T^{T-p+1}\right)}. \tag{7.129}$$

By (7.126), we have

$$X_T^{T-p+1} = Y_T^{T-p+1} + D_p\left(\Xi_T^{T-p+1}\right) V_T^{T-p+1}, \tag{7.130}$$

and applying the total expectation formula yields

$$p_{X_T^{T-p+1}}\left(X_T^{T-p+1}\right) = \sum_{J \in V_p} \mathbb{P}\left\{\Xi_T^{T-p+1} = J\right\} p_{Y_T^{T-p+1} + D_p(J) V_T^{T-p+1}}\left(X_T^{T-p+1}\right).$$

Since $\mathcal{L}\{Y_T^{T-p+1}\} = \mathcal{N}_p(\mathbf{0}_p, \Sigma_p)$, $\mathcal{L}\{V_T^{T-p+1}\} = \mathcal{N}_p(\mathbf{0}_p, K\sigma^2 \mathbf{I}_p)$, and the random variables Y_T^{T-p+1}, V_T^{T-p+1} are independent, from the properties of linear transformations of normal random vectors [3] we have

$$p_{Y_T^{T-p+1} + D_p(J) V_T^{T-p+1}}\left(X_T^{T-p+1}\right) = n_p\left(X_T^{T-p+1} \mid \mathbf{0}_p, \Sigma_p + K\sigma^2 D_p(J)\right).$$

This allows us to write

$$p_{X_T^{T-p+1}}\left(X_T^{T-p+1}\right) = \sum_{J \in V_p} \varepsilon^{|J|}(1-\varepsilon)^{p-|J|} n_p\left(X_T^{T-p+1} \mid \mathbf{0}_p, \Sigma_p + K\sigma^2 D_p(J)\right). \tag{7.131}$$

From (7.130) and the second equation of (7.126), by a similar argument, we have

$$p_{H_T^{T-p+1}, X_T^{T-p+1}}\left(z, X_T^{T-p+1}\right) = p_{H_T^{T-p+1}}(z) p_{Y_T^{T-p+1}}\left(X_T^{T-p+1} - z\right) =$$
$$= \sum_{J \in V_p} \varepsilon^{|J|}(1-\varepsilon)^{p-|J|} n_p\left(z \mid \mathbf{0}_p, K\sigma^2 D_p(J)\right) n_p\left(X_T^{T-p+1} \mid z, \Sigma_p\right).$$

Let us rewrite this relation by using Lemma 7.4 in the notation

$$\Sigma_1 ::= K\sigma^2 D_p(J), \quad \Sigma_2 ::= \Sigma_p, \quad x_{(1)} ::= z, \quad x_{(2)} ::= X_T^{T-p+1},$$

and substitute the result, together with (7.131), into (7.129). This yields

$$p_{H_T^{T-p+1}|X_T^{T-p+1}}\left(z \mid X_T^{T-p+1}\right) =$$
$$= \left(\sum_{J \in_p} \varepsilon^{|J|}(1-\varepsilon)^{p-|J|} n_p\left(X_T^{T-p+1} \mid \mathbf{0}_p, \Sigma_p + K\sigma^2 D_p(J)\right)\right)^{-1} \times$$
$$\times \sum_{J \in V_p} \varepsilon^{|J|}(1-\varepsilon)^{p-|J|} n_p\left(X_T^{T-p+1} \mid \mathbf{0}_p, \Sigma_p + K\sigma^2 D_p(J)\right) \times$$
$$\times n_p\Big(z \mid K\sigma^2 D_p(J)\left(\Sigma_p + K\sigma^2 D_p(J)\right)^{-1} X_T^{T-p+1},$$
$$K\sigma^2 D_p(J)\left(\mathbf{I}_p - K\sigma^2\left(\Sigma_p + K\sigma^2 D_p(J)\right)^{-1} D_p(J)\right)\Big).$$

By well-known properties of multivariate normal probability densities [3], substituting this expression into (7.128) proves (7.124), (7.125). □

Note that the expression for the conditional probability density H_T^{T-p+1} given a fixed p-step-back history X_T^{T-p+1} that was used to prove Theorem 7.19 can be rewritten as a mixture of 2^p normal p-variate densities:

$$p_{H_T^{T-p+1}|X_T^{T-p+1}}\left(z \mid X_T^{T-p+1}\right) = \sum_{J \in V_p} q_J n_p(z \mid \mu_J, \Xi_J), \quad z \in \mathbb{R}^p, \tag{7.132}$$

$$\mu_J = K\sigma^2 D_p(J)\left(\Sigma_p + K\sigma^2 D_p(J)\right)^{-1} X_T^{T-p+1},$$

$$\Xi_J = K\sigma^2 D_p(J)\left(\mathbf{I}_p - K\sigma^2\left(\Sigma_p + K\sigma^2 D_p(J)\right)^{-1} D_p(J)\right),$$

$$q_J = \frac{\varepsilon^{|J|}(1-\varepsilon)^{p-|J|} n_p\left(X_T^{T-p+1} \mid \mathbf{0}_p, \Sigma_p + K\sigma^2 D_p(J)\right)}{\sum_{J \in V_p} \varepsilon^{|J|}(1-\varepsilon)^{p-|J|} n_p\left(X_T^{T-p+1} \mid \mathbf{0}_p, \Sigma_p + K\sigma^2 D_p(J)\right)} \in [0, 1],$$

where $\sum_{J \in V_p} q_J \equiv 1$.

Corollary 7.18. *The minimum conditional mean square matrix risk for the optimal forecasting statistic* (7.125) *given fixed history* X_T^1 *is equal to*

$$R_\varepsilon^\star(\theta^0, T, \tau) = \sigma^2 \sum_{k=0}^{\tau-1} (A_0^k)_{\cdot 1}((A_0^k)_{\cdot 1})' + \\ + A_0^\tau \left(\sum_{J \in V_p} q_J(\mu_J \mu_J' + \Xi_J) - \left(\sum_{J \in V_p} q_J \mu_J \right) \left(\sum_{J \in V_p} q_J \mu_J \right)' \right) (A_0^\tau)'. \quad (7.133)$$

Proof. Using (7.125), (7.126), let us obtain an expression for the random forecast error vector:

$$\hat{Y}_{T+\tau}^{\star T+\tau-p+1} - Y_{T+\tau}^{T+\tau-p+1} = A_0^\tau H_T^{T-p+1} - A_0^\tau B_\varepsilon X_T^{T-p+1} - \sum_{k=0}^{\tau-1} A_0^k U_{T+\tau-k}.$$

Applying model assumptions (7.117) together with the representation (7.132) and taking into account the fixed history X_T^1 leads to an expression for the conditional matrix risk:

$$R_\varepsilon^\star(\theta^0, T, \tau) = \mathbb{E}\left\{ \left(\hat{Y}_{T+\tau}^{\star T+\tau-p+1} - Y_{T+\tau}^{T+\tau-p+1} \right) (\cdot)' \mid X_T^{T-p+1}, X_{T-p}^1 \right\} = \\ = A_0^\tau B_\varepsilon X_T^{T-p+1} (X_T^{T-p+1})' B_\varepsilon' (A_0^\tau)' + \\ + A_0^\tau \mathbb{E}\left\{ H_T^{T-p+1} (H_T^{T-p+1})' \mid X_T^{T-p+1} \right\} + \\ + \sigma^2 \sum_{k=0}^{\tau-1} (A_0^k)_{\cdot 1}((A_0^k)_{\cdot 1})' - 2 A_0^\tau B_\varepsilon X_T^{T-p+1} (B_\varepsilon X_T^{T-p+1})' (A_0^\tau)'. \quad (7.134)$$

From (7.132), we have

$$\mathbb{E}\left\{ H_T^{T-p+1} (H_T^{T-p+1})' \mid X_T^{T-p+1} \right\} = \sum_{J \in V_p} q_J (\mu_J \mu_J' + \Xi_J),$$

$$\mathbb{E}\left\{ H_T^{T-p+1} \mid X_T^{T-p+1} \right\} = \sum_{J \in V_p} q_J \mu_J = B_\varepsilon X_T^{T-p+1}.$$

Substituting the above expressions into (7.134) and performing equivalent transformations proves (7.133). □

Corollary 7.19. *An optimal one-step-ahead,* $\tau = 1$, *forecasting statistic for a first order autoregression model* ($p = 1$) *can be written as*

$$\hat{y}_{T+1} = \theta^0 x_T \left(1{-}\varepsilon \left(1 + \frac{1}{K^{-1}\left(1-(\theta^0)^2\right)}\right)^{-1} \times \right.$$

$$\left. \times \left(\varepsilon + (1-\varepsilon)\frac{n_1\left(x_T \mid 0, \sigma^2/\left(1-(\theta^0)^2\right)\right)}{n_1\left(x_T \mid 0, \sigma^2\left(K + (1-(\theta^0)^2)\right)\right)}\right)^{-1}\right).$$

Proof. It suffices to assume $p = 1$, $\tau = 1$ in (7.124), (7.125) and to apply the equality

$$\Sigma_p = 1/\left(1-(\theta^0)^2\right). \qquad \square$$

For an $AR(p)$ hypothetical model without distortion ($H_1^T \equiv 0$), the autoregression statistic

$$\tilde{Y}_{T+\tau}^{T+\tau-p+1} = A_0^\tau X_T^{T-p+1} \tag{7.135}$$

is optimal, as established in Sect. 7.2. Let us compare the properties of forecasting statistics (7.125) and (7.135).

Corollary 7.20. *As the outlier probability tends to zero ($\varepsilon \to 0$), the following asymptotic expansion holds for the optimal forecasting statistic:*

$$\hat{Y}_{T+\tau}^{*T+\tau-p+1} = \tilde{Y}_{T+\tau}^{T+\tau-p+1} + O(\varepsilon)\mathbf{1}_p. \tag{7.136}$$

Proof. Since for $J = \mathbf{0}_p$ the matrix $D_p(\mathbf{0}_p)$ is a zero $(p \times p)$-matrix, the numerator in (7.124) is of the order $O(\varepsilon)$, and the denominator is of the order $O(1)$. Applying this fact to (7.125) and using the notation (7.135) proves (7.136). $\square$

Note that if $K \to 0$, which implies

$$\mathbb{E}\{H_T^{T-p+1}\} \to \mathbf{0}_p,$$

then we obtain a result similar to (7.136):

$$\hat{Y}_{T+\tau}^{*T+\tau-p+1} - \tilde{Y}_{T+\tau}^{T+\tau-p+1} \to \mathbf{0}_p.$$

In applications, the values ε, K and the distribution $\{v_t\}$ are usually unknown, and therefore it is impossible to construct the optimal forecasting statistic (7.125). Instead, the autoregression forecasting statistic (7.135) is used. Let us evaluate the robustness of this statistic under AO distortion.

Theorem 7.20. *In the $AR(p)$ model under AO distortion* (7.117)*, let us construct an autoregression forecasting statistic* (7.135) *to predict $Y_{T+\tau}^{T+\tau-p+1}$ for $\tau \geq p$. Then the matrix risk of forecasting is equal to*

$$R_\varepsilon(\theta^0, T, \tau) = \sigma^2 \sum_{k=0}^{\tau-1} (A_0^k)_{\cdot 1}((A_0^k)_{\cdot 1})' + \varepsilon K \sigma^2 A_0^\tau (A_0^\tau)'. \tag{7.137}$$

Proof. From (7.125), (7.135) we can obtain an expression for the random forecast error vector:

$$\tilde{Y}_{T+\tau}^{T+\tau-p+1} - Y_{T+\tau}^{T+\tau-p+1} = A_0^\tau H_T^{T-p+1} - \sum_{k=0}^{\tau-1} A_0^k U_{T+\tau-k}.$$

Substituting this expression in the formula for the matrix risk,

$$R_\varepsilon(\theta^0, T, \tau) = \mathbb{E}\left\{\left(\tilde{Y}_{T+\tau}^{T+\tau-p+1} - Y_{T+\tau}^{T+\tau-p+1}\right)(\cdot)'\right\},$$

and calculating the mean value under the model assumptions (7.117) yields the formula (7.137). □

Corollary 7.21. *Under the conditions of Theorem 7.20, the risk of forecasting the future value $y_{T+\tau}$ equals*

$$r_\varepsilon(\theta^0, T, \tau) = \sigma^2 \sum_{k=0}^{\tau-1} \left((A_0^k)_{11}\right)^2 + \varepsilon K \sigma^2 \sum_{j=1}^{p} \left((A_0^\tau)_{1j}\right)^2. \tag{7.138}$$

Proof. The relation (7.138) follows from (7.137) and the definition of the scalar risk:

$$r_\varepsilon(\theta^0, T, \tau) = \mathbb{E}\{(\tilde{y}_{T+\tau} - y_{T+\tau})^2\} = \left(R_\varepsilon(\theta^0, T, \tau)\right)_{11}.$$ □

From (7.138) and (7.108), we can see that the forecast risk can be represented as follows:

$$r_\varepsilon(\theta^0, T, \tau) = r_0(T, \tau) + \varepsilon K \sigma^2 \sum_{j=1}^{p} \left((A_0^\tau)_{1j}\right)^2,$$

where $r_0(T, \tau)$ is the risk under the undistorted hypothetical model ($\varepsilon = 0$).

Similarly to Theorem 7.20 and its corollaries, we can evaluate the forecast risk under a specification error $\Delta\theta = \theta - \theta^0$. Then the statistic (7.135) is replaced by the following forecasting statistic:

$$\tilde{Y}_{T+\tau}^{T+\tau-p+1} = A^\tau X_T^{T-p+1}, \tag{7.139}$$

where the matrix A is defined by (7.86).

As earlier, the covariance matrix of order p is defined as

$$\Sigma_p = \mathbb{E}\left\{Y_T^{T-p+1}(Y_T^{T-p+1})'\right\}.$$

Theorem 7.21. *In the $AR(p)$ model under AO distortion and specification error $\Delta\theta = \theta - \theta^0$, $\gamma = |\Delta\theta| \to 0$, let us construct an autoregression forecasting statistic (7.139) for predicting $Y_{T+\tau}^{T+\tau-p+1}$, $\tau \geq p$. Then the matrix risk of forecasting satisfies the following asymptotic expansion:*

$$R_\varepsilon(\Delta\theta, \theta^0, T, \tau) = \sigma^2 \sum_{k=0}^{\tau-1} (A_0^k)_{\cdot 1}((A_0^k)_{\cdot 1})' + \varepsilon K \sigma^2 \Bigg((A_0^\tau)(A_0^\tau)' +$$

$$+ \sum_{k=0}^{\tau-1} \left(A_0^\tau (A_0^{\tau-k-1})' \Delta\theta ((A_0^k)_{\cdot 1})' + (A_0^k)_{\cdot 1} (\Delta\theta)' A_0^{\tau-k-1} (A_0^\tau)' \right) \Bigg) + O(\gamma^2) \mathbf{1}_{p\times p}. \tag{7.140}$$

Proof. As in the proof of Theorem 7.20, from (7.126) and (7.139) we have

$$\tilde{Y}_{T+\tau}^{T+\tau-p+1} - Y_{T+\tau}^{T+\tau-p+1} = A^\tau H_{T+}^{T-p+1} - \sum_{k=0}^{\tau-1} A_0^k U_{T+\tau-k} + (A^\tau - A_0^\tau) Y_T^{T-p+1}.$$

Substituting this result into the expression for the matrix risk and taking an average under the model assumptions (7.117) yields

$$R_\varepsilon(\Delta\theta, \theta^0, T, \tau) = \sigma^2 \sum_{k=0}^{\tau-1} (A_0^k)_{\cdot 1}((A_0^k)_{\cdot 1})' + (A^\tau - A_0^\tau) \Sigma_p (A^\tau - A_0^\tau)' + \varepsilon K \sigma^2 A^\tau (A^\tau)'. \tag{7.141}$$

By (7.86), we have

$$A = A_0 + \Delta A, \quad \Delta A = \begin{pmatrix} (\Delta\theta)' \\ \hdashline \mathbf{0}_{(p-1)\times p} \end{pmatrix} \in \mathbb{R}^{p\times p}.$$

As in the proof of Corollary 7.6, we obtain

$$(A_0 + \Delta A)^\tau = A_0^\tau + \sum_{k=0}^{\tau-1} A_0^k \Delta A A_0^{\tau-k-1} + O(||\Delta A||^2),$$

and, from well-known matrix properties, we can write

$$A_0^k \Delta A = \left((A_0^k)_{\cdot 1} \;\vdots\; (A_0^k)_{22} \right) \begin{pmatrix} (\Delta\theta)' \\ \hdashline \mathbf{0}_{(p-1)\times p} \end{pmatrix} = (A_0^k)_{\cdot 1} (\Delta\theta)'.$$

Substituting these relations into (7.141) and performing equivalent matrix transformations proves (7.140). □

Corollary 7.22. *Under the conditions of Theorem 7.20, for $\gamma = |\Delta\theta| \to 0$ the risk of forecasting the future value $y_{T+\tau}$ equals*

$$r_\varepsilon(\Delta\theta, \theta^0, T, \tau) = r_\varepsilon(\theta^0, T, \tau) + 2\varepsilon K\sigma^2 \sum_{k=0}^{\tau-1} (A_0^k)_{11} \sum_{j=1}^{p} \left(A_0^\tau (A_0^{\tau-k-1})'\right)_{1j} \Delta\theta_j + O(\gamma^2),$$

where the risk $r_\varepsilon(\theta^0, T, \tau)$ is defined by (7.138).

Proof. The proof is similar to the proof of Corollary 7.21. □

The Corollaries 7.21, 7.22 allow us to calculate the guaranteed risk r_+ and the δ-admissible level of outlier probability $\varepsilon_+(\delta)$ assuming that the specification error is bounded from above, $|\Delta\theta| \leq \gamma$.

To conclude the section, let us mention that an asymptotic expansion for the risk of a plug-in forecasting statistic based on the least squares estimator $\hat{\theta}$ has been obtained in [11]; this result will not be reproduced here due to space considerations.

7.9 Robustness of Autoregression Forecasting Under Bilinear Distortion

7.9.1 Introduction

Although most processes in the real world are nonlinear, they are often modeled by simpler and better-studied linear stochastic equations. When such linearization is applied, it is quite important to quantify the effect of nonlinearity on the mathematical results obtained under a linear model. This section introduces a nonlinear modification of the linear time series autoregression model—the bilinear (BL) model [7,8]. Bilinear models have many potential applications, however, most of the available results are related to macroeconomic and financial time series. For example, Buyers and Peel [5] have established bilinearity of currency exchange rates, and Terdik [26] used the bilinear model to study the dynamics of the S&P 500 stock index.

Let us give a formal definition of the BL model, establish some of its properties, and analyze the robustness of autoregression forecasting under bilinear distortion BL$(p, 0, 1, 1)$ by evaluating the respective mean square forecast risk.

7.9.2 The Bilinear Model and Its Stationarity Conditions

Let us consider a time series $x_t^0 \in \mathbb{R}$, $t \in \mathbb{Z}$, defined over a probability space $(\Omega, \mathcal{F}, \mathbb{P})$ by an AR(p) linear autoregression model [2] of order p, $p \in \mathbb{N}$:

$$x_t^0 = \sum_{j=1}^{p} \alpha_j x_{t-j}^0 + u_t, \quad t \in \mathbb{Z}, \tag{7.142}$$

where $\{\alpha_j\}_{j=1}^p$ are autoregression coefficients, $\{u_t\}_{t\in\mathbb{Z}}$ are i.i.d. normal random observation errors with zero expectations and finite variances $\mathbb{D}\{u_t\} = \sigma^2$. Under this model, the time series is strictly stationary [2, 7] if we have

$$\varrho(A) < 1, \tag{7.143}$$

where $\varrho(A)$ is the spectral radius of the matrix $A \in \mathbb{R}^{p\times p}$ defined from $\{\alpha_j\}_{j=1}^p$ as follows:

$$A = \begin{pmatrix} -\alpha_1 & -\alpha_2 & \dots & -\alpha_{p-1} & -\alpha_p \\ 1 & 0 & \dots & 0 & 0 \\ 0 & 1 & \dots & 0 & 0 \\ \dots & \dots & \dots & \dots & \dots \\ 0 & 0 & \dots & 1 & 0 \end{pmatrix}.$$

It is said [7, 8] that a time series $x_t \in \mathbb{R}$, $t \in \mathbb{Z}$, defined over $(\Omega, \mathfrak{F}, \mathbb{P})$ satisfies a bilinear model $BL(p, 0, 1, 1)$ if the following bilinear equation holds:

$$x_t = \sum_{j=1}^{p} \alpha_j x_{t-j} + \beta x_{t-1} u_{t-1} + u_t, \quad t \in \mathbb{Z}, \tag{7.144}$$

where β is the bilinearity coefficient. If $\beta = 0$, then the model (7.144) is equivalent to (7.142): $x_t \equiv x_t^0$. If we have

$$\rho\left(A \otimes A + \sigma^2 B \otimes B\right) < 1, \tag{7.145}$$

then the time series (7.144) is strictly stationary [26]; here $\otimes$ is the Kronecker matrix product, and the elements of the matrix $B \in \mathbb{R}^{p\times p}$ are zeros except for $B(1,1) = \beta$.

7.9.3 First and Second Order Moments in Stationary Bilinear Time Series Models

It is known [2] that the expectation of a process defined by a stationary autoregression model (7.142) equals zero: $\mathbb{E}\{x_t^0\} = 0$, and that the expectation of a stationary bilinear process (7.144) can be written as

$$\mathbb{E}\{x_t\} = \beta\sigma^2 \left(1 - \sum_{j=1}^{p} \alpha_j\right)^{-1}.$$

Assuming stationarity, let us introduce the following notation for the respective second order moment functions of the random processes (7.142) and (7.144):

$$c_0(s) ::= \mathbb{E}\{x_t^0 x_{t-s}^0\}, \quad c(s) ::= \mathbb{E}\{x_t x_{t-s}\}, \quad s \in \mathbb{Z}.$$

Moments of stationary AR(p) processes (7.142) satisfy the Yule–Walker system of equations [2]:

$$c_0(0) = \sum_{j=1}^{p} \alpha_j c_0(j) + \sigma^2,$$

$$c_0(s) = \sum_{j=1}^{p} \alpha_j c_0(j-s), \quad s = 1, \ldots, p\ ,$$

or, in a matrix form,

$$W c_0 = \sigma^2 i_1,$$

where the matrix W is $W = W_1 + \mathbf{I}_{p+1} + W_2 \in \mathbb{R}^{(p+1)\times(p+1)}$; the moment vector c_0 is defined as $c_0 = (c_0(0), \ldots, c_0(p))' \in \mathbb{R}^{p+1}$; the matrix $\mathbf{I}_{p+1} \in \mathbb{R}^{(p+1)\times(p+1)}$ is the identity matrix; $i_1 = (1, 0, \ldots, 0) \in \mathbb{R}^{p+1}$ represents the first column of the matrix $\mathbf{I}_{p+1}$; and

$$W_1 = \begin{pmatrix} 0 & -\alpha_1 & -\alpha_2 & \ldots & -\alpha_p \\ 0 & -\alpha_2 & -\alpha_3 & \ldots & 0 \\ \ldots & \ldots & \ldots & \ldots & \ldots \\ 0 & -\alpha_p & 0 & \ldots & 0 \\ 0 & 0 & 0 & \ldots & 0 \end{pmatrix} \in \mathbb{R}^{(p+1)\times(p+1)},$$

$$W_2 = \begin{pmatrix} 0 & 0 & \ldots & 0 & 0 \\ -\alpha_1 & 0 & \ldots & 0 & 0 \\ \ldots & \ldots & \ldots & \ldots & \ldots \\ -\alpha_{p-1} & -\alpha_{p-2} & \ldots & 0 & 0 \\ -\alpha_p & -\alpha_{p-1} & \ldots & -\alpha_1 & 0 \end{pmatrix} \in \mathbb{R}^{(p+1)\times(p+1)}.$$

If the matrix W is nonsingular, then we have $c_0 = W^{-1} i_1$.

Theorem 7.22. *Under the bilinear model* (7.144), (7.145), *the moments of the time series satisfy the following relations:*

$$(1-\beta^2\sigma^2)c(0) = \sum_{l=1}^{p} \alpha_j c(j) + \sigma^2 + \beta^2\sigma^4 + \beta^2\sigma^4(1+\alpha_1)\left(1-\sum_{j=1}^{p}\alpha_j\right)^{-1},$$

$$c(1) = \sum_{j=1}^{p} \alpha_j c(j-1) + 2\beta^2\sigma^4\left(1-\sum_{j=1}^{p}\alpha_j\right)^{-1},$$

$$c(s) = \sum_{j=1}^{p} \alpha_j c(j-s) + \beta^2\sigma^4\left(1-\sum_{j=1}^{p}\alpha_j\right)^{-1}, \quad s = 2,\ldots,p.$$

Proof. Let us multiply every term in the stochastic difference equation (7.144) by x_{t-s} and compute the expectations

$$\mathbb{E}\{x_t x_{t-s}\} = \sum_{j=1}^{p} \alpha_j \mathbb{E}\{x_{t-j}x_{t-s}\} + \beta\mathbb{E}\{x_{t-1}x_{t-s}u_{t-1}\} + \mathbb{E}\{x_{t-s}u_t\}, \quad s = 0,\ldots,p. \tag{7.146}$$

For $s = 0$, the above expression becomes

$$\mathbb{E}\{x_t^2\} = \sum_{j=1}^{p} \alpha_j \mathbb{E}\{x_{t-j}x_t\} + \beta\mathbb{E}\{x_{t-1}x_t u_{t-1}\} + \mathbb{E}\{x_t u_t\}.$$

From (7.144), it follows that $\mathbb{E}\{x_t u_t\} = \sigma^2$. To compute $\mathbb{E}\{x_{t-1}x_t u_{t-1}\}$, let us multiply each term of (7.144) by $x_{t-1}u_{t-1}$:

$$\mathbb{E}\{x_t x_{t-1} u_{t-1}\} = \alpha_1 x_t \mathbb{E}\{x_{t-1}^2 u_{t-1}\} + \\ + \sum_{j=2}^{p} \alpha_j \mathbb{E}\{x_{t-j}x_{t-1}u_{t-1}\} + \beta\mathbb{E}\{x_{t-1}^2 u_{t-1}^2\} + \mathbb{E}\{u_t x_{t-1} u_{t-1}\}.$$

The independence between the elements of the time series and the future observation errors, as well as the uncorrelatedness of the errors, implies that $\mathbb{E}\{x_{t-1}u_{t-1}u_t\} = 0$. Now it is easy to show that

$$\mathbb{E}\{x_{t-k}x_t u_t\} = \beta\sigma^4\left(1-\sum_{j=1}^{p}\alpha_j\right)^{-1}, \quad k = 1,\ldots,p,$$

$$\mathbb{E}\{x_t^2 u_t\} = 2\beta\sigma^4\left(1-\sum_{j=1}^{p}\alpha_j\right)^{-1}.$$

Let us obtain an expression for $\mathbb{E}\{x_{t-1}^2 u_{t-1}^2\}$ by squaring the relation (7.144) and calculating the expectation

$$\mathbb{E}\{x_t^2\} = \mathbb{E}\left\{\left(\sum_{j=1}^{p} \alpha_j x_{t-j} + \beta x_{t-1} u_{t-1}\right)^2\right\} + \mathbb{E}\{u_t^2\}.$$

Similarly, let us compute $\mathbb{E}\{x_t^2 u_t^2\}$ by using the normality of $\{u_t\}_{t\in\mathbb{Z}}$:

$$\mathbb{E}\{x_t^2 u_t^2\} = \mathbb{E}\left\{\left(\sum_{j=1}^{p} \alpha_j x_{t-j} + \beta x_{t-1} u_{t-1}\right)^2\right\} \mathbb{E}\{u_t^2\} + \mathbb{E}\{u_t^4\} = \sigma^2 \mathbb{E}\{x_t^2\} + 2\sigma^4$$

since we have $\mathbb{E}\{u_t^2\} = \sigma^2$, $\mathbb{E}\{u_t^3\} = 0$, $\mathbb{E}\{u_t^4\} = 3\sigma^4$. Then we can write

$$\mathbb{E}\{x_t x_{t-1} u_{t-1}\} = \beta\sigma^4 (1 + \alpha_1)\left(1 - \sum_{j=1}^{p} \alpha_j\right)^{-1} + \beta\sigma^4 + \beta\sigma^2 \mathbb{E}\{x_t^2\}.$$

Substituting the above expression into (7.146) for $s = 0$ and applying the notation proves the first equality stated in the theorem, and the remaining relations follow from (7.146) and the above expressions for $\mathbb{E}\{x_t^2 u_t\}$ and $\mathbb{E}\{x_t x_{t-k} u_t\}$ with $k \geq 1$.

□

Let us rewrite the relations of Theorem 7.22 in a matrix form and obtain their asymptotic expansions as $\beta \to 0$. We are going to use the following notation:

$$\mathbf{1}_{p+1} = (1, 1, \ldots, 1)' \in \mathbb{R}^{p+1} \quad \text{and} \quad \mathbf{1}_{(p+1)\times(p+1)} \in \mathbb{R}^{(p+1)\times(p+1)}$$

denote, respectively, a $(p+1)$-vector and a $(p+1) \times (p+1)$-matrix of ones, and f is a column vector defined as

$$f = \left(1 + (1+\alpha_1)\left(1 - \sum_{j=1}^{p} \alpha_j\right)^{-1}, \; 2\left(1 - \sum_{j=1}^{p} \alpha_j\right)^{-1}, \right.$$
$$\left. \left(1 - \sum_{j=1}^{p} \alpha_j\right)^{-1}, \ldots, \left(1 - \sum_{j=1}^{p} \alpha_j\right)^{-1}\right)' \in \mathbb{R}^{p+1}.$$

Corollary 7.23. *Under the conditions of Theorem 7.22, the following matrix relation holds:*

$$(W - \beta^2\sigma^2 i_1 i_1')\, c = \sigma^2 i_1 + \beta^2\sigma^4 f.$$

If $|W| \neq 0$, $\beta^2\sigma^2 W^{-1}(1,1) \neq 1$, *then we also have*

$$c = W^{-1}\big(\mathbf{I}_{p+1} - \beta^2\sigma^2 i_1 i_1' W^{-1}\big)^{-1}(\sigma^2 i_1 + \beta^2\sigma^4 f). \qquad (7.147)$$

Theorem 7.23. *Under the stationary bilinear model* (7.144), (7.145), *assuming that* $|W| \neq 0$, *the second order moment vector satisfies the following asymptotic expansion as* $\beta \to 0$*:*

$$c = \sigma^2 W^{-1} i_1 + \beta^2\sigma^4 (W^{-1} f + W^{-1}(1,1) W^{-1} i_1) + o(\beta^2)\mathbf{1}_{p+1}.$$

Proof. By the conditions of the theorem, we have

$$\big(\mathbf{I}_{p+1} - \beta^2\sigma^2 i_1 i_1' W^{-1}\big)^{-1} =$$
$$= \mathbf{I}_{p+1} + \beta^2\sigma^2 i_1 i_1' W^{-1} + \big(\beta^2\sigma^2 i_1 i_1' W^{-1}\big)^2 \big(\mathbf{I}_{p+1} - \beta^2\sigma^2 i_1 i_1' W^{-1}\big)^{-1}.$$

Let us fix $\{\alpha_j\}_{j=1}^{p}$ (consequently, the matrix W) and the variance σ^2. There exists a sufficiently small β_0 such that $\beta^2\sigma^2 W^{-1}(1,1) \neq 1$ for all $\beta < \beta_0$; then we have the following expansion:

$$\big(\mathbf{I}_{p+1} - \beta^2\sigma^2 i_1 i_1' W^{-1}\big)^{-1} = \mathbf{I}_{p+1} + \beta^2\sigma^2 i_1 i_1' W^{-1} + o(\beta^2)\,\mathbf{1}_{(p+1)\times(p+1)}.$$

Substituting this expression into (7.147) proves the theorem. □

Corollary 7.24. *Under the conditions of Theorem 7.23, the following asymptotic expansion holds for* $c(0) = \mathbb{E}\{x_t^2\}$ *as* $\beta \to 0$*:*

$$c(0) = \sigma^2 W^{-1}(1,1) + \beta^2\sigma^4\big(i_1' W^{-1} f + (W^{-1}(1,1))^2\big) + o(\beta^2). \qquad (7.148)$$

7.9.4 Robustness of Autoregression Forecasting Under Bilinear Distortion

Under a bilinear model, let a p-segment $x_{T+1}, \ldots, x_{T+p}$ of future time series elements be forecast based on an observation $X = \{x_1, \ldots, x_T\}$ of length T, $T > p$. We are going to construct an autoregression forecast [2, 12] which is mean square optimal under an AR(p) linear model with a priori known parameters and depends only on the p previous elements of the time series: $\{x_{T-p+1}, \ldots, x_T\}$. Let us investigate the influence of the bilinear term in (7.144) on forecasting performance.

Lemma 7.5. *Stochastic difference equations* (7.142), (7.144) *can be rewritten in the following equivalent form:*

$$S_1 X_{T+1}^{0T+p} = S_2 X_{T-p+1}^{0T} + U_{T+1}^{T+p}, \tag{7.149}$$

$$S_1 X_{T+1}^{T+p} = S_2 X_{T-p+1}^{T} + \beta \langle XU \rangle_T^{T+p-1} + U_{T+1}^{T+p}, \tag{7.150}$$

where $S_1, S_2 \in \mathbb{R}^{p \times p}$,

$$S_1 = \begin{pmatrix} 1 & -\alpha_1 & \dots & -\alpha_{p-1} \\ 0 & 1 & \dots & -\alpha_{p-2} \\ \dots & \dots & \dots & \dots \\ 0 & 0 & \dots & 1 \end{pmatrix}, \quad S_2 = \begin{pmatrix} \alpha_p & 0 & \dots & 0 \\ \alpha_{p-1} & \alpha_p & \dots & 0 \\ \dots & \dots & \dots & \dots \\ \alpha_1 & \alpha_2 & \dots & \alpha_p \end{pmatrix},$$

$$X_n^{0m} = (x_m^0, \dots, x_n^0)' \in \mathbb{R}^{m-n+1}, \qquad X_n^m = (x_m, \dots, x_n)' \in \mathbb{R}^{m-n+1},$$

$$U_n^m = (u_m, \dots, u_n)' \in \mathbb{R}^{m-n+1}, \quad \langle XU \rangle_n^m = (x_m u_m, \dots, x_n u_n)' \in \mathbb{R}^{m-n+1}.$$

Proof. The relations (7.149), (7.150) follow from (7.142), (7.144) and the notation. □

Thus, from the vector of known values of the time series $X_1^T \in \mathbb{R}^p$ we need to construct a vector of estimators for the future values $\hat{X}_{T+1}^{T+p} = (\hat{x}_{T+\tau}) \in \mathbb{R}^p$, where each $\hat{x}_{T+\tau}$ is a τ-step-ahead forecast, $\tau = 1, \dots, p$. For a quantitative characterization of forecasting performance, we are going to calculate the mean square matrix risk, which is defined as

$$R = (R(i,j))_{i,j=1}^p = \mathbb{E}\left\{ \left(X_{T+1}^{T+p} - \hat{X}_{T+1}^{T+p}\right)\left(X_{T+1}^{T+p} - \hat{X}_{T+1}^{T+p}\right)' \right\},$$

and a collection of scalar "local" risks

$$r(\tau) = R(p-\tau+1, p-\tau+1) = \mathbb{E}\left\{ (x_{T+\tau} - \hat{x}_{T+\tau})^2 \right\}, \quad \tau = 1, \dots, p.$$

Under the autoregression model (7.142), (7.149), the autoregression forecast $\hat{X}_{T+1}^{0T+p}$ is derived from the following relations [2, 12]:

$$S_1 \hat{X}_{T+1}^{0T+p} = S_2 X_{T-p+1}^{0T}, \quad \hat{X}_{T+1}^{0T+p} = S_1^{-1} S_2 X_{T-p+1}^{0T}. \tag{7.151}$$

Theorem 7.24. *In the stationary* $AR(p)$ *model* (7.142), (7.143), *the autoregression forecast* (7.151) *is unbiased:*

$$\mathbb{E}\{\hat{X}_{T+1}^{0T+p} - X_{T+1}^{0T+p}\} = \mathbf{0}_p \in \mathbb{R}^p,$$

its matrix risk equals $R_0 = \sigma^2(S_1'S_1)^{-1}$ *and is minimal, i.e., for an arbitrary forecasting statistic and its matrix risk* R_0^**, the matrix difference* $(R_0^* - R_0) \in \mathbb{R}^{p\times p}$ *is positive-semidefinite.*

Proof. By (7.149), (7.151), we have

$$S_1\left(X_{T+1}^{0T+p} - \hat{X}_{T+1}^{0T+p}\right) = U_{T+1}^{T+p}, \tag{7.152}$$

and consequently

$$S_1\mathbb{E}\{X_{T+1}^{0T+p} - \hat{X}_{T+1}^{0T+p}\} = \mathbf{0}_p.$$

Nonsingularity of S_1 proves the unbiasedness of the forecast. From (7.152) and the definition of the forecast risk, we have $S_1R_0S_1^{-1} = \sigma^2\mathbf{I}_p$. Since $|S_1| \neq 0$, we can write $R_0 = \sigma^2(S_1S_1')^{-1}$.

Let us prove optimality of the forecast (7.151). Consider an arbitrary forecasting statistic $\hat{X}_{T+1}^{0*T+p} \in \mathbb{R}^p$ and compute its risk:

$$\begin{aligned}R_0^* &= \mathbb{E}\left\{(\hat{X}_{T+1}^{0*T+p} - \hat{X}_{T+1}^{0T+p})(\hat{X}_{T+1}^{0*T+p} - \hat{X}_{T+1}^{0T+p})'\right\} + \\ &+ \mathbb{E}\left\{(\hat{X}_{T+1}^{0*T+p} - \hat{X}_{T+1}^{0T+p})(\hat{X}_{T+1}^{0T+p} - X_{T+1}^{0T+p})'\right\} + \\ &+ \mathbb{E}\left\{(\hat{X}_{T+1}^{0T+p} - X_{T+1}^{0T+p})(\hat{X}_{T+1}^{0*T+p} - \hat{X}_{T+1}^{0T+p})'\right\} + R_0.\end{aligned}$$

By (7.152), we have

$$X_{T+1}^{0*T+p} - \hat{X}_{T+1}^{0T+p} = S_1^{-1}U_{T+1}^{T+p},$$

and the second and third summands are equal to zero since a forecasting statistic doesn't depend on future observation errors $\{u_{T+1}, \ldots, u_{T+p}\}$. This proves that the matrix $R_0^* - R_0$ is positive-semidefinite. □

Corollary 7.25. *Under the conditions of Theorem 7.24, the local forecast risks* (7.151) *can be written as*

$$r(\tau) = \sigma^2 \sum_{j=1}^{\tau} \left(S_1^{-1}(1, j)\right)^2, \quad \tau = 1, \ldots, p.$$

Let us assume that the bilinearity coefficient β in the stationary model (7.144), (7.145) is sufficiently small to continue using an autoregression forecast statistic similar to (7.151):

$$S_1\hat{X}_{T+1}^{T+p} = S_2X_{T-p+1}^{T}, \quad \hat{X}_{T+1}^{T+p} = S_1^{-1}S_2X_{T-p+1}^{T}. \tag{7.153}$$

Lemma 7.6. *In the stationary bilinear model* (7.144), (7.145), *the forecast* (7.153) *is biased:*

$$\mathbb{E}\{X_{T+1}^{T+p} - \hat{X}_{T+1}^{T+p}\} = \beta\sigma^2\mathbf{1}_p \neq \mathbf{0}_p \in \mathbb{R}^p.$$

Proof. From (7.150), (7.153), we have

$$X_{T+1}^{T+p} - \hat{X}_{T+1}^{T+p} = \beta S_1^{-1}\langle XU\rangle_T^{T+p-1} + S_1^{-1}U_{T+1}^{T+p}.$$

This leads to the equality

$$\mathbb{E}\left\{X_{T+1}^{T+p} - \hat{X}_{T+1}^{T+p}\right\} = \beta\sigma^2\mathbf{1}_p.$$

Since $|S_1| \neq 0$, we have

$$\mathbb{E}\left\{X_{T+1}^{T+p} - \hat{X}_{T+1}^{T+p}\right\} \neq \mathbf{0}_p,$$

proving the bias. □

Theorem 7.25. *In the stationary bilinear model* (7.144), (7.145), *the mean square matrix risk of the autoregression forecast* (7.153) *equals*

$$R = \sigma^2(S_1'S_1)^{-1} + \beta^2\sigma^2c(0)(S_1)^{-1}(S_1')^{-1} + \beta^2\sigma^4(S_1)^{-1}(\mathbf{1}_{p\times p} + \mathbf{I}_p)(S_1')^{-1}+$$
$$+ 2\beta^2\sigma^4\left(1 - \sum_{j=}^{p}\alpha_j\right)^{-1}(S_1)^{-1}(\mathbf{I}_p^{-1,0,1} - \mathbf{I}_p)(S_1')^{-1}, \quad (7.154)$$

where $\mathbf{I}_p^{-1,0,1}$ *is a tridiagonal* $(p \times p)$*-matrix with the elements lying on the three diagonals equal to 1.*

Proof. From (7.150) and (7.153), we have

$$S_1\left(X_{T+1}^{T+p} - \hat{X}_{T+1}^{T+p}\right) = \beta\langle XU\rangle_T^{T+p-1} + U_{T+1}^{T+p},$$

and

$$S_1\mathbb{E}\left\{\left(\hat{X}_{T+1}^{T+p} - X_{T+1}^{T+p}\right)\left(X_{T+1}^{T+p} - \hat{X}_{T+1}^{T+p}\right)'\right\}S_1' = S_1RS_1' =$$
$$= \beta^2\mathbb{E}\left\{\langle XU\rangle_T^{T+p-1}(\langle XU\rangle_T^{T+p-1})'\right\} + \beta\mathbb{E}\left\{\langle XU\rangle_T^{T+p-1}(U_{T+1}^{T+p})'\right\} +$$
$$+ \beta\mathbb{E}\left\{U_{T+1}^{T+p}(\langle XU\rangle_T^{T+p-1})'\right\} + \mathbb{E}\left\{U_{T+1}^{T+p}(U_{T+1}^{T+p})'\right\}.$$

Calculating the matrices in the right-hand side of this relation elementwise, using the intermediate results from the proof of Theorem 7.22, and applying the nonsingularity condition $|S_1| \neq 0$ proves (7.154). □

Corollary 7.26. *Under the conditions of Theorem 7.25, the local risks equal*

$$r(\tau) = \sigma^2 \sum_{j=1}^{\tau} \left(S_1^{-1}(1,j)\right)^2 + \beta^2\sigma^2 c(0) \sum_{j=1}^{\tau} \left(S_1^{-1}(1,j)\right)^2 + \\ + \beta^2\sigma^4 \left(\sum_{j=1}^{\tau} \left(S_1^{-1}(1,j)\right)^2 + \left(\sum_{j=1}^{\tau} S_1^{-1}(1,j) \right)^2 + \right. \\ \left. +4\left(1 - \sum_{j=1}^{p} \alpha_j\right)^{-1} \sum_{j=1}^{\tau-1} S_1^{-1}(1,j) S_1^{-1}(1,j+1) \right), \quad \tau = 1, \ldots, p. \tag{7.155}$$

Corollary 7.27. *If $|W| \neq 0$, then asymptotically as $\beta \to 0$ we have:*

$$R = \sigma^2 (S_1' S_1)^{-1} + \beta^2 \sigma^4 \left(W^{-1}(1,1)(S_1)^{-1}(S_1')^{-1} + (S_1')^{-1}(\mathbf{1}_{p\times p} + \mathbf{I}_p)(S_1')^{-1} + \right. \\ \left. +2\left(1 - \sum_{j=1}^{p} \alpha_j\right)^{-1} (S_1)^{-1}(\mathbf{I}_p^{-1,0,1} - \mathbf{I}_p)(S_1')^{-1} \right) + o(\beta^2)\mathbf{1}_{p\times p}, \tag{7.156}$$

$$r(\tau) = \sigma^2 \sum_{j=1}^{\tau} (S_1^{-1}(1,j))^2 + \beta^2\sigma^4 \left((W^{-1}(1,1) + 1) \sum_{j=1}^{\tau} (S_1^{-1}(1,j))^2 + \right. \\ \left. + \left(\sum_{j=1}^{\tau} S_1^{-1}(1,j) \right)^2 + 4\left(1 - \sum_{j=1}^{p} \alpha_j\right)^{-1} \sum_{j=1}^{\tau-1} S_1^{-1}(1,j) S_1^{-1}(1,j+1) \right) + \\ + o(\beta^2), \quad \tau = 1, \ldots, p. \tag{7.157}$$

Proof. Substitute (7.150) into (7.156) and (7.157). □

7.9.5 Robustness Analysis of Autoregression Forecasting

Assume that the bilinear distortion level β lies in a small interval $[-\beta_+, \beta_+]$, $\beta_+ \geq 0$. Robustness of the autoregression forecasting statistic (7.153) will be evaluated by using the robustness indicators defined in Sect. 4.4:

- Risk instability coefficient $\kappa(\tau)$:

$$\kappa(\tau) = \frac{r_+(\tau) - r_0(\tau)}{r_0(\tau)},$$

where $r_+(\tau) = \sup\{r(\tau) : \beta \in [-\beta_+, \beta_+]\}$ is the guaranteed forecast risk, and

$$r_0(\tau) = \sigma^2 \sum_{j=1}^{\tau} \left((S_1)^{-1}(1, j)\right)^2$$

is the minimum risk value (obtained under the undistorted model, $\beta = 0$);
- δ-admissible distortion level $\beta^+(\delta, \tau)$:

$$\beta^+(\delta, \tau) = \sup\{\beta_+ : \kappa(\tau) \le \delta\}.$$

Smaller values of $r_+(\tau)$, $\kappa(\tau)$ and larger values of $\beta^+(\delta, \tau)$ correspond to higher robustness under bilinear distortion.

Theorem 7.26. *In a stationary bilinear model* (7.144), (7.145), *if* $|W| \neq 0$, $\beta_+ \to 0$, *then*

$$\kappa(\tau) = \beta_+^2 \sigma^2 \left(\sum_{j=1}^{\tau} \left(S_1^{-1}(1, j)\right)^2\right)^{-1} \left(\left(W^{-1}(1,1) + 1\right) \sum_{j=1}^{\tau} \left(S_1^{-1}(1, j)\right)^2 + \right.$$
$$\left. + \left(\sum_{j=1}^{\tau} S_1^{-1}(1, j)\right)^2 + 4\left(1 - \sum_{j=1}^{p} \alpha_j\right)^{-1} \sum_{j=1}^{\tau-1} S_1^{-1}(1, j) S_1^{-1}(1, j+1)\right) + o(\beta_+^2).$$

Proof. Substitute (7.157) into the expression for $\kappa(\tau)$. □

Theorem 7.26 yields an approximation for $\beta^+(\delta, \tau)$:

$$\beta^+(\delta, \tau) \approx \delta^{1/2} (\sigma^2)^{-1/2} \left(\sum_{j=1}^{\tau} (S_1^{-1}(1, j))^2\right)^{1/2} \times$$
$$\times \left((W^{-1}(1,1) + 1) \sum_{j=1}^{\tau} (S_1^{-1}(1, j))^2 + \left(\sum_{j=1}^{\tau} S_1^{-1}(1, j)\right)^2 + \right.$$
$$\left. +4\left(1 - \sum_{j=1}^{p} \alpha_j\right)^{-1} \sum_{j=1}^{\tau-1} S_1^{-1}(1, j) S_1^{-1}(1, j+1)\right)^{-1/2}. \tag{7.158}$$

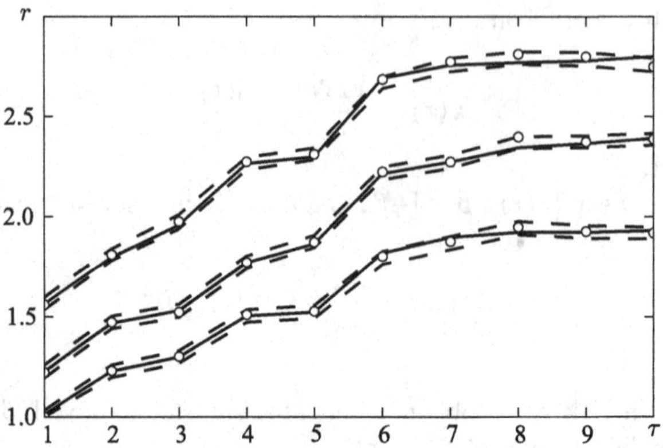

Fig. 7.3 The exact risk, its point and 95 % confidence interval estimates for $\beta = 0; 0.2; 0.3$

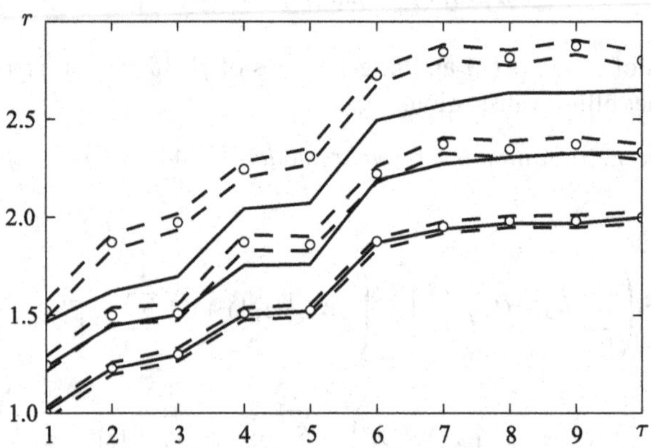

Fig. 7.4 The asymptotic risk, its point and 95 % confidence interval estimates for $\beta = 0; 0.2; 0.3$

7.9.6 Numerical Results

Let us consider a bilinear model BL(10,0,1,1) of order $p = 10$:

$$x_t = -0.5x_{t-1} - 0.5x_{t-2} + 0.1x_{t-3} + 0.2x_{t-4} - 0.2x_{t-5} - 0.1x_{t-6} - 0.2x_{t-7} +$$
$$+ 0.2x_{t-8} + 0.1x_{t-9} - 0.1x_{t-10} + \beta x_{t-1}u_{t-1} + u_t, \qquad \mathcal{L}\{u\}_t = \mathcal{N}(0, 1).$$

A total of $N = 1{,}000$ Monte-Carlo simulation rounds were performed, and point estimates, as well as $(1 - \varepsilon)$-confidence intervals, were computed for the forecast risk (7.155). In Figs. 7.3 and 7.4, these estimates of the forecast risk are plotted against forecast depth τ; circles indicate point estimates and dashed lines are the 95 % confidence limits; solid lines are the exact (Fig. 7.3) and asymptotic (Fig. 7.4) values of risk defined by the formulas (7.155), (7.157). Figures 7.5 and 7.6 give a

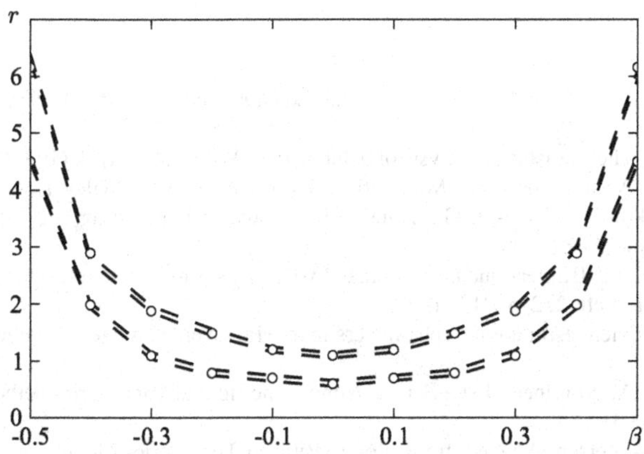

Fig. 7.5 The exact risk, its point and 95 % confidence interval estimates for $\tau = 1; 5$

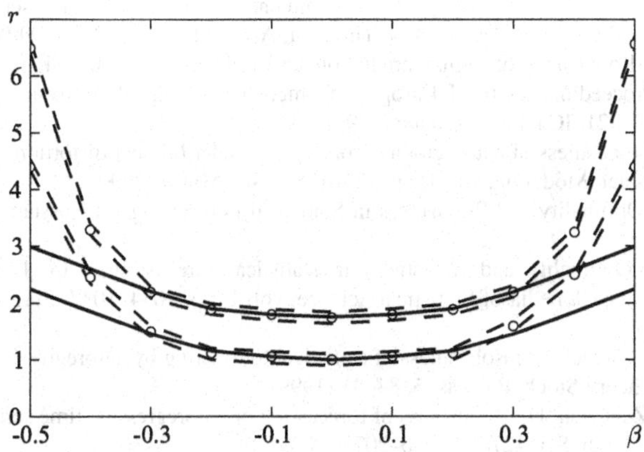

Fig. 7.6 The asymptotic risk, its point and 95 % confidence interval estimates for $\tau = 1; 5$

Table 7.1 Critical levels of bilinear distortion

δ \ τ	1	2	5
0.1	0.141	0.171	0.163
0.5	0.316	0.383	0.364
1.0	0.447	0.542	0.515

similar presentation of the dependence between the forecast risk and the bilinearity coefficient β.

Figures 7.5 and 7.6 show that as the bilinearity level β increases, the risk of the autoregression forecasting statistic, starting with some level β_*, is rapidly increasing; in this case the autoregression forecast becomes unusable.

Table 7.1 presents δ-admissible distortion levels $\beta^+(\delta, \tau)$ computed from (7.158) for different values of δ and τ.

References

1. Akutowicz, E.: On an explicit formula in least squares prediction. Math. Scand. **5**, 261–266 (1957)
2. Anderson, T.: The Statistical Analysis of Time Series. Wiley, New York (1994)
3. Anderson, T.: An Introduction to Multivariate Statistical Analysis. Wiley, Hoboken (2003)
4. Box, G., Jenkins, G., Reinsel, G.: Time Series Analysis: Forecasting and Control. Wiley-Blackwell, New York (2008)
5. Byers, J., Peel, D.: Bilinear quadratic arch and volatility spill-overs in inter-war exchange rate. Appl. Econom. Lett. **2**, 215–219 (1995)
6. Durbin, J.: Efficient estimation of parameters in moving-average models. Biometrika **46**, 306–316 (1959)
7. Fan, J., Yao, Q.: Nonlinear Time Series: Nonparametric and Parametric methods. Springer, New York (2003)
8. Granger, C., Andersen, A.: An Introduction to Bilinear Time Series Models. Vandenhoeck and Ruprecht, Gottingen (1978)
9. Hamilton, J.: Time Series Analysis. Princeton University Press, Princeton (1994)
10. Jammalamadaka, S., Wu, C., Wang, W.: The influence of numerical and observational errors on the likelihood of an ARMA series. J. Time Ser. Anal. **20**(2), 223–235 (1999)
11. Kharin, Yu.: Robustness of signal prediction under distortions. In: Signal Analysis and Prediction. Proceedings of the I European Conference on Signal Analysis and Prediction, vol. 1, pp. 219–221. ICT Press, Prague (1997)
12. Kharin, Yu.: Robustness of autoregressive forecasting under bilinear distortions. In: Computer Data Analysis and Modeling, vol. 1, pp. 124–128. BSU, Minsk (2007)
13. Kharin, Yu.: Optimality and Robustness in Statistical Forecasting (in Russian). BSU, Minsk (2008)
14. Kharin, Yu.: Optimality and robustness in statistical forecasting. In: Lovric, M. (ed.) International Encyclopedia of Statistical Science, vol. 1, pp. 1034–1037. Springer, New York (2011)
15. Kharin, Yu., Zenevich, D.: Robustness of statistical forecasting by autoregressive model under distortions. Theory Stoch. Process. **5**, 84–91 (1999)
16. Kharin, Yu., Zenevich, D.: Robustness of forecasting of autoregressive time series for additive distortions. J. Math. Sci. **127**(4), 2066–2072 (2005)
17. Kolmogorov, A.: On the use of statistically estimated forecasting formulae (in Russian). Geophys. J. **3**, 78–82 (1933)
18. Kolmogorov, A.: Interpolation and extrapolation of stationary stochastic series (in Russian). Izv. Akad. Nauk SSSR Ser. Mat **5**(1), 3–14 (1941)
19. Koroljuk, V.: Handbook on Probability Theory and Mathematical Statistics (in Russian). Nauka, Moscow (1985)
20. Launer, R., Wilkinson, G.: Robustness in Statistics: Proceedings of a Workshop. Academic, New York (1979)
21. Lutkepohl, H.: Introduction to Multiple Time Series Analysis. Springer, Berlin (1993)
22. Marcus, M., Minc, H.: A Survey of Matrix Theory and Matrix Inequalities. Allyn and Bacon, Boston (1964)
23. Martin, R.: Robust autoregressive analysis of time series. In: Launer, R., Wilkinson, G. (eds.) Robustness in Statistics, pp. 121–147. AP, New York (1979)
24. Priestley, M.: Spectral Analysis and Time Series. Academic, New York (1999)
25. Serfling, R.: Approximation Theorems of Mathematical Statistics. Wiley, New York (1980)
26. Terdik, G.: Bilinear Stochastic Models and Related Problems of Nonlinear Time Series Analysis. Springer, New York (1999)

Chapter 8
Optimality and Robustness of Vector Autoregression Forecasting Under Missing Values

Abstract In this chapter, robustness of vector autoregression time series forecasting is studied under the influence of missing values—a common type of distortion which is especially characteristic of large datasets. Assuming a non-stochastic missing values template, a mean square optimal forecasting statistic is constructed in the case of prior knowledge of the VAR model parameters, and its risk instability coefficient is evaluated under missing values and model specification errors. In the case of parametric prior uncertainty, a consistent forecasting statistic and an asymptotic expansion of the corresponding forecast risk are obtained. The chapter is concluded by considering plug-in forecasting under simultaneous influence of outliers and missing values.

8.1 VAR Time Series Models Under Missing Values

Missing values are a very common distortion of experimental data [4, 14, 31]. Vector time series in econometrics [22], biometrics, engineering, and numerous other applications often contain missing components, which are usually caused by the following reasons [6, 26]:

(1) the data doesn't exist (is undefined) at certain time points (for instance, economic analysts observing only monthly, quarterly, or yearly data);
(2) recording errors (due to malfunctions or other interruptions of the measurement process);
(3) some of the data points have been determined to be outliers and removed (for example, incorrectly registered or anomalous experimental data).

Under this type of distortion, many statistical software packages simply fill in the missing values based on the observed data and return to a standard forecasting problem. However, this approach leads to an uncontrolled increase of the forecast risk and cannot be justified from a scientific viewpoint. A fully mathematically

Y. Kharin, *Robustness in Statistical Forecasting*, DOI 10.1007/978-3-319-00840-0_8,

justified approach to forecasting under missing values requires, first of all, a definite mathematical model.

Assume that the observed d-variate time series Y_t satisfies the VAR(1) autoregression model (as mentioned in Sect. 3.5.4, this model can be used to approximate the more complex VARMA(p,q) model as accurately as required):

$$Y_t = BY_{t-1} + U_t, \quad t \in \mathbb{Z}, \tag{8.1}$$

where $Y_t = (Y_{t1}, \ldots, Y_{td})' \in \mathbb{R}^d$, $B = (B_{kl}) \in \mathbb{R}^{d\times d}$ is a matrix of (generally unknown) autoregression coefficients such that all of its eigenvalues lie inside the unit circle; $U_t = (U_{t1}, \ldots, U_{td}) \in \mathbb{R}^d$ is the d-variate innovation process; $\{U_t\}$ are jointly independent random vectors with zero expectations, $\mathbb{E}\{U_t\} = \mathbf{0}_d$, and the covariance matrix

$$\mathbb{E}\{U_t U_t'\} = \Sigma = (\sigma_{kl}) \in \mathbb{R}^{d\times d}$$

is positive-definite and independent of time t. Observations of the process $\{Y_t\}$ contain missing values. As in [10, 11], for each time point $t \in \mathbb{Z}$, we define the missing values template—a binary vector $O_t = (O_{t1}, \ldots, O_{td})' \in \{0,1\}^d$, where

$$O_{ti} = \begin{cases} 1, & \text{if the component } Y_{ti} \text{ is observed,} \\ 0, & \text{if the component } Y_{ti} \text{ is missing.} \end{cases}$$

Denote the minimal and maximal time points (bounds of the observation time period) as, respectively,

$$t_- = \min\left\{t \in \mathbb{Z} : \sum_{i=1}^{d} O_{ti} > 0\right\}, \quad t_+ = \max\left\{t \in \mathbb{Z} : \sum_{i=1}^{d} O_{ti} > 0\right\}.$$

Since, under the assumed conditions, the time series (8.1) is stationary, without loss of generality we are going to assume $t_- = 1$, $t_+ = T$, where T is the observation length. Absence of distortion, i.e., the case of the hypothetical model VAR(1) without missing values, corresponds to the template

$$O_{ti} \equiv 1, \quad t = 1, \ldots, T; \quad i = 1, \ldots, d.$$

The stationary time series (8.1) is characterized by a zero expectation, a covariance matrix G_0 and a covariance function G_τ [1]:

$$\begin{gathered} \mathbb{E}\{Y_t\} = \mathbf{0}_d, \qquad \mathbf{Cov}\{Y_t, Y_t\} = G_0 = \sum_{i=0}^{+\infty} B^i \Sigma (B')^i, \\ \mathbf{Cov}\{Y_t, Y_{t+\tau}\} = G_\tau = B^\tau G_0 \mathbf{1}(\tau) + G_0 (B')^{|\tau|} \mathbf{1}(-\tau), \quad \tau \neq 0. \end{gathered} \tag{8.2}$$

In the existing literature on statistical data analysis under missing values (see, e.g., [14]), two classes of assumptions are made about the missing values templates: missing at random and missing at nonrandom. In the better studied first class (values missing at random), it is assumed that $\{O_t\}$ are jointly independent binary random vectors independent of the observation vectors $\{Y_t\}$, and the probability that a value is missing $\mathbb{P}\{O_{ti} = 0\} = \varepsilon$ is sufficiently small. Optimality of statistical algorithms is evaluated not for a specific realization of the template $\{O_t\}$, but jointly over all possible realizations of the random process $\{O_t\}$. In the second class, the missing values template $\{O_t\}$ is assumed to be deterministic, and the optimal statistics are chosen for a fixed realization of the missing values template determined by the experimental results. We are going to follow the latter approach, assuming from now on that $\{O_1, \dots, O_T\}$ is a deterministic sequence of binary vectors serving as a parameter of the observation model.

To conclude the first section, let us formulate a family of additional assumptions on the innovation process ($A_1 - A_5$) and the missing values template ($A_6 - A_7$) that will be used in the remaining part of the chapter.

A1. The fourth moment of the innovation process is bounded:

$$\exists C \in (0, +\infty) : \mathbb{E}\{U_{ti}^4\} \leq C \quad \forall t \in \mathbb{Z}, \quad \forall i \in \{1, \dots, d\}.$$

A2. The covariance matrix of the innovation process is nonsingular:

$$|\mathbf{Cov}\{U_t, U_t\}| = |\Sigma| \neq 0, \quad t \in \mathbb{Z}.$$

A3. The eighth moment of the innovation process is bounded:

$$\exists C \in (0, +\infty) : \mathbb{E}\{U_{ti}^8\} \leq C \quad \forall t \in \mathbb{Z}, \ \forall i \in \{1, \dots, d\}.$$

A4. The fourth moment of the innovation process is independent of time:

$$\mathbb{E}\{U_{ti_1} \dots U_{ti_4}\} = \Sigma^{(4)}_{i_1,\dots,i_4}, \quad t \in \mathbb{Z}, \ i_1, i_2, i_3, i_4 \in \{1, \dots, d\}.$$

A5. The innovation process is normally distributed:

$$\mathcal{L}\{U_t\} = \mathcal{N}(\mathbf{0}_d, \Sigma), \quad t \in \mathbb{Z}, \ |\Sigma| \neq 0.$$

A6. For any pair of vector components, as the observation length tends to infinity, this pair is simultaneously observed infinitely often at single time points and at pairs of consecutive time points:

$$\sum_{t=1}^{T-k} O_{t+k,i} O_{tj} \underset{T \to +\infty}{\longrightarrow} +\infty, \quad k \in \{0, 1\}, \ i, j \in \{1, \dots, d\}.$$

A7. The following asymptotic relations are satisfied:

$$\frac{\sum\limits_{t=1}^{T-k} O_{t+k,i} O_{tj}}{T-k} \underset{T\to+\infty}{\longrightarrow} \vartheta_{k,i,j} \in (0,1],$$

$$\frac{\sum\limits_{t,t'=1}^{T-1} O_{t+k,i} O_{tj} O_{t'+k',i'} O_{t'j'} \delta_{t-t',\tau}}{T-|\tau|-1} \underset{T\to+\infty}{\longrightarrow} \widetilde{\vartheta}_{\tau,k,k',i,j,i',j'} \in [0,1],$$

where $\vartheta_{k,i,j}$ is the limit frequency of observing a vector component pair (i, j) at time points separated by k time units; $\widetilde{\vartheta}_{\tau,k,k',i,j,i',j'}$ is the limit frequency of observing a pair (i, j) jointly with a pair (i', j') at time points separated by $\tau+k-k'$ and τ time units w.r.t. the first pair for $\tau \in \mathbb{Z}$, $k, k' \in \{0, 1\}$, and the indices i, j, i', j' lie in $\{1, \ldots, d\}$. Obviously, the following relation holds for the limit frequencies: $\widetilde{\vartheta}_{0,k,k,i,j,i,j} = \vartheta_{k,i,j}$, where $k \in \{0, 1\}$, $i, j \in \{1, \ldots, d\}$.

Note some simple relations between these assumptions:

(1) A1 follows from A3.
(2) A1, A2, A3, A4 follow from A5.
(3) A6 follows from A7.

8.2 The Optimal Forecasting Statistic and Its Risk

Let us consider the problem of statistical forecasting in the autoregression model of time series (8.1) given a fixed missing values template $\{O_t\}$ for three levels of prior uncertainty in the model parameters B, Σ:

(1) true values B, Σ are a priori known;
(2) a specification error is present in the definition of the matrices B, Σ;
(3) the parameters B, Σ are unknown.

Let us begin by considering the case of complete prior knowledge. Assume that a τ-step-ahead, $\tau \in \mathbb{N}$, forecast of the future value $Y_{T+\tau} \in \mathbb{R}^d$ is constructed for a time series satisfying the vector autoregression model (8.1) based on a time series $Y_1, \ldots, Y_T$ of length $T \in \mathbb{N}$ with a corresponding missing values template $O_1, \ldots, O_T$ under prior knowledge of B, Σ.

Let us define a finite set

$$M = \{(t,i) : t \in \{1, \ldots, T\},\ i \in \{1, \ldots, d\},\ O_{ti} = 1\}$$

which contains pairs of time points and indices of the observed components (t, i) ordered lexicographically; let $K = |M|$ be the total number of the observed components. We are going to define a bijection $\chi(t, i) : M \leftrightarrow \{1, \ldots, K\}$ and its inverse function $\bar{\chi}(k) : \{1, \ldots, K\} \leftrightarrow M$. Let us construct a K-vector of all observed components:

$$X = (X_1, \ldots, X_K)' \in \mathbb{R}^K, \quad X_k = Y_{\bar{\chi}(k)}, \quad k \in \{1, \ldots, K\}.$$

For $O_t = \mathbf{1}_d, t \in \{1, \ldots, T\}$, the observations contain no missing values, and

$$K = Td, \quad X = \left(Y_1', \ldots, Y_T'\right)' \in \mathbb{R}^{Td},$$
$$\chi(t, i) = i + (t-1)d, \quad t \in \{1, \ldots, T\}, \ i \in \{1, \ldots, d\},$$
$$\bar{\chi}(k) = ((k-1) \text{ Div } d + 1, (k-1) \text{ Mod } d + 1), \quad k \in \{1, \ldots, K\},$$

where n Div m stands for the integer quotient of n divided by m, and n Mod m is the residue of the division. Let

$$\hat{Y}_{T+\tau} = \hat{Y}_{T+\tau}(X) : \mathbb{R}^K \to \mathbb{R}^d$$

be a τ-step-ahead forecasting statistic based on an observation of length T. As in Chap. 4, the performance of the forecast will be characterized by the matrix risk of forecasting,

$$R = \mathbb{E}\left\{\left(\hat{Y}_{T+\tau} - Y_{T+\tau}\right)\left(\hat{Y}_{T+\tau} - Y_{T+\tau}\right)'\right\} \in \mathbb{R}^{d \times d},$$

the scalar risk of forecasting: $r = \operatorname{tr}(R) \geq 0$, and (under assumption $A2$) the forecast risk instability coefficient (see Sect. 4.4):

$$\kappa = \frac{r - r_0}{r_0} \geq 0, \quad r_0 = \operatorname{tr}\left(\sum_{i=0}^{\tau-1} B^i \Sigma (B')^i\right) > 0,$$

where r_0 is the minimum risk. This risk value r_0 is attained for a τ-step-ahead forecast $\hat{Y}_{T+\tau} = B^\tau Y_T$ based on an observation of length T under prior knowledge of the parameters B, Σ in the absence of missing values. One can think of the forecast risk instability coefficient as the relative risk increment due to missing values w.r.t. the minimum risk.

Let us denote covariance matrices as follows:

$$F = (F_{ij}) = \mathbf{Cov}\{X, X\} \in \mathbb{R}^{K \times K},$$
$$H = (H_{ij}) = \mathbf{Cov}\{X, Y_{T+\tau}\} \in \mathbb{R}^{K \times d}, \tag{8.3}$$
$$A_0 = A_0(B, \Sigma) = H' F^{-1} \in \mathbb{R}^{d \times K}, \quad T, \tau \in \mathbb{N}.$$

Lemma 8.1. *The following equalities hold for the covariance matrices F and H:*

$$F_{ij} = \left(B^{\bar{\chi}_1(i)-\bar{\chi}_1(j)}G_0\right)_{\bar{\chi}_2(i),\bar{\chi}_2(j)}, \qquad i,j \in \{1,\dots,K\},\ i \geq j;$$

$$H_{ij} = \left(B^{T+\tau-\bar{\chi}_1(i)}G_0\right)_{j,\bar{\chi}_2(i)}, \qquad i \in \{1,\dots,K\},\ j \in \{1,\dots,d\},\ T,\tau \in \mathbb{N}.$$

Proof. To compute the matrix F, it is sufficient to observe that

$$F_{ij} = \mathbf{Cov}\left\{X_i, X_j\right\} = \mathbf{Cov}\left\{Y_{\bar{\chi}(i)}, Y_{\bar{\chi}(j)}\right\} = \left(\mathbf{Cov}\{Y_{\bar{\chi}_1(i)}, Y_{\bar{\chi}_1(j)}\}\right)_{\bar{\chi}_2(i),\bar{\chi}_2(j)} =$$
$$= \left(B^{\bar{\chi}_1(i)-\bar{\chi}_1(j)}G\right)_{\bar{\chi}_2(i),\bar{\chi}_2(j)}, \quad i,j \in \{1,\dots,K\},\ i \geq j.$$

The expression for the matrix H can be obtained similarly. □

Assuming prior knowledge of model parameters, the following theorem constructs an optimal (in the maximum likelihood sense) forecast for a vector autoregression model of time series under missing values and gives a formula for its risk.

Theorem 8.1. *In the model* (8.1), *if assumption A5 is satisfied, the parameters B, Σ are a priori known, and the covariance matrix F is nonsingular, $|F| \neq 0$, the optimal (in the sense of maximum likelihood) τ-step-ahead forecasting statistic based on an observation of length T is the following conditional expectation:*

$$\hat{Y}_{T+\tau} = \mathbb{E}\{Y_{T+\tau} \mid X\} = A_0 X, \tag{8.4}$$

and the respective matrix risk equals

$$R = G_0 - H' F^{-1} H. \tag{8.5}$$

Proof. Let $Y_+ = (X', Y'_{T+\tau})' \in \mathbb{R}^{K+d}$ denote the composite vector of all observed values and the unknown future value $Y_{T+\tau}$. Under the conditions of the theorem, the vector Y_+ is normally distributed. The likelihood function can be written as

$$L\left(Y_{T+\tau}; B, \Sigma\right) = n_K\left(X \mid 0_K, F\right) n_d\left(Y_{T+\tau} \mid H'F^{-1}X, G_0 - H'F^{-1}H\right), \tag{8.6}$$

where $n_K\left(X \mid \mu, \Sigma\right)$ is the K-variate normal probability density with parameters μ, Σ. An optimal forecast in the maximum likelihood sense is the solution of the following optimization problem:

$$L\left(Y_{T+\tau}; B, \Sigma\right) \to \max_{Y_{T+\tau} \in \mathbb{R}^d}.$$

The first multiplier in the right-hand side of (8.6) doesn't depend on $Y_{T+\tau}$. Thus, from well-known properties of joint normal distributions and (8.3), we obtain the following unique solution:

$$\hat{Y}_{T+\tau} = H' F^{-1} X = A_0 X,$$

which coincides with (8.4).

From the total expectation formula, we have

$$\begin{aligned} R &= \mathbb{E}\left\{\left(\hat{Y}_{T+\tau} - Y_{T+\tau}\right)(\hat{Y}_{T+\tau} - Y_{T+\tau})'\right\} = \\ &= \mathbb{E}\left\{\mathbb{E}\left\{(\mathbb{E}\{Y_{T+\tau} \mid X\} - Y_{T+\tau})(\mathbb{E}\{Y_{T+\tau} \mid X\} - Y_{T+\tau})' \mid X\right\}\right\} = \\ &= \mathbb{E}\left\{\mathbf{Cov}\left\{Y_{T+\tau}, Y_{T+\tau} \mid X\right\}\right\} = \mathbb{E}\left\{G_0 - H' F^{-1} H\right\} = G_0 - H' F^{-1} H, \end{aligned}$$

leading to the expression (8.5) for the matrix risk. □

Since the forecasting statistic (8.4) is defined as a conditional expectation, by Theorem 4.1 it delivers the minimum risk (8.5).

Corollary 8.1. *Under the conditions of Theorem 8.1, the scalar risk and the risk instability coefficient for the statistic* (8.4) *can be written as*

$$r = \operatorname{tr}(G_0) - \operatorname{tr}(H' F^{-1} H), \quad \kappa = \frac{\operatorname{tr}(G_0) - \operatorname{tr}(H' F^{-1} H)}{\operatorname{tr}\left(\sum_{i=0}^{\tau-1} B^i \Sigma (B')^i\right)} - 1. \tag{8.7}$$

8.3 Robustness of the Optimal Forecasting Statistic Under Specification Errors

As we can see from (8.3), (8.4), the matrix coefficient A_0 of the optimal forecasting statistic depends on model parameters—$(d \times d)$-matrices B, Σ. The expressions (8.5), (8.7) define the risk under prior knowledge of B and Σ. Let us evaluate the forecast risk under specification errors in the model (8.1).

Assume that the parameters of a vector autoregression model under missing values have been defined incorrectly, i.e., that the forecast

$$\hat{Y}_{T+\tau} = \hat{A}_0 X \in \mathbb{R}^d \tag{8.8}$$

is based on a deterministic matrix

$$\hat{A}_0 \neq A_0 \in \mathbb{R}^{d \times K}. \tag{8.9}$$

Theorem 8.2. *In the model* (8.1) *under assumption A5, let the forecast* (8.8) *be constructed from the misspecified parameters* (8.9)*, and let* $|F| \neq 0$*. Then the matrix forecast risk equals*

$$R = G_0 - H'F^{-1}H + \left(A_0 - \hat{A}_0\right) F \left(A_0 - \hat{A}_0\right)'.$$

Proof. It is sufficient to perform the following equivalent transformations:

$$\begin{aligned} R &= \mathbb{E}\left\{\left(Y_{T+\tau} - \hat{Y}_{T+\tau}\right)\left(Y_{T+\tau} - \hat{Y}_{T+\tau}\right)'\right\} = \\ &= \mathbb{E}\left\{\left(Y_{T+\tau} - A_0X + A_0X - \hat{A}_0X\right)\left(Y_{T+\tau} - A_0X + A_0X - \hat{A}_0X\right)'\right\} = \\ &= \mathbb{E}\left\{(Y_{T+\tau} - A_0X)(Y_{T+\tau} - A_0X)'\right\} + \mathbb{E}\left\{\left(A_0X - \hat{A}_0X\right)\left(A_0X - \hat{A}_0X\right)'\right\} = \\ &= G_0 - H'F^{-1}H + \left(A_0 - \hat{A}_0\right) F \left(A_0 - \hat{A}_0\right)'. \end{aligned}$$

□

Corollary 8.2. *Under the conditions of Theorem 8.2, the scalar risk and the risk instability coefficient for the forecasting statistic* (8.8) *can be written as*

$$r = \operatorname{tr}(G_0) - \operatorname{tr}(H'F^{-1}H) + \operatorname{tr}\left(\left(A_0 - \hat{A}_0\right) F \left(A_0 - \hat{A}_0\right)'\right), \tag{8.10}$$

$$\kappa = \frac{\operatorname{tr}(G_0) - \operatorname{tr}(H'F^{-1}H) + \operatorname{tr}\left(\left(A_0 - \hat{A}_0\right) F \left(A_0 - \hat{A}_0\right)'\right)}{\operatorname{tr}\left(\sum_{i=0}^{\tau-1} B^i \Sigma (B')^i\right)} - 1.$$

Looking at (8.10), it is easy to see that, under missing values, presence of a specification error (8.9) yields a positive increment of the forecast risk.

8.4 Modified Least Squares Estimators Under Missing Values

Now consider a situation where the parameters B, Σ defining the optimal forecasting statistic obtained in Sect. 8.2 are unknown.

Theorem 8.3. *In the model* (8.1) *under missing values, let assumption A5 be satisfied. If the model parameters* B, Σ *are a priori unknown, and the matrix* F *defined by Lemma 8.1 is nonsingular,* $|F| \neq 0$*, then the optimal forecasting statistic (w.r.t. the maximum likelihood) is the following plug-in statistic:*

$$\widetilde{Y}_{T+\tau} = A_0(\widetilde{B}, \widetilde{\Sigma})X, \tag{8.11}$$

where joint MLEs $\widetilde{B}$, $\widetilde{\Sigma}$ for the model parameters are the solutions of the following minimization problem:

$$l_1(B, \Sigma) = X'F^{-1}X + \ln|F| + \ln|G - H'F^{-1}H| \to \min_{B,\Sigma}. \tag{8.12}$$

Proof. By (8.6), the joint ML estimators for $Y_{T+\tau}$, B, Σ are the solutions of the maximization problem

$$L(Y_{T+\tau}; B, \Sigma) \to \max_{Y_{T+\tau}, B, \Sigma}.$$

Theorem 8.1 yields (8.11), where due to (8.6) the ML estimators $\widetilde{B}$, $\widetilde{\Sigma}$ are solutions of the maximization problem

$$n_k(X \mid \mathbf{0}_k, F) n_k(H'F^{-1}X \mid H'F^{-1}X, G - H'F^{-1}H) \to \max_{B,\Sigma}.$$

Taking a logarithm of the objective function of this maximization problem yields (8.12). □

Unfortunately, solving the minimization problem (8.12) is extremely computationally intensive due to significantly nonlinear dependence of the objective functions on the parameters B, Σ (see Lemma 8.1). Although ML estimators are known to be asymptotically optimal, a practical realization of the optimal forecasting statistic (8.11), (8.12) proves difficult. In applications, this difficulty is usually solved by applying the well-known numerical EM algorithm [14, 19]. However, multiextremality of the likelihood function implies that this algorithm doesn't necessarily yield the global minimum in (8.12).

Due to the above difficulties in computing ML and EM estimators, a special modification of the least squares estimator for the parameters B, Σ is proposed in the case of missing values. Let us explain the construction of this modified least squares estimator.

Assuming no missing values, i.e., $O_t = \mathbf{1}_d$, $t \in \{1, \ldots, T\}$, the least squares estimator of the matrix coefficient B is found from the following relation [1]:

$$\frac{1}{T-1}\sum_{t=1}^{T-1} Y_{t+1}Y_t' = \hat{B}\frac{1}{T-1}\sum_{t=1}^{T-1} Y_t Y_t'.$$

Obviously, the matrices

$$\frac{1}{T-1}\sum_{t=1}^{T-1} Y_{t+1}Y_t', \quad \frac{1}{T-1}\sum_{t=1}^{T-1} Y_t Y_t' \tag{8.13}$$

are consistent estimators for, respectively, the covariance matrices G_1 and G_0 defined by (8.2). Thus, knowing the equation that links the matrix coefficient B and the covariances (8.2) allows us to construct matrix coefficient estimators by first estimating the covariances (8.13). Let us use this idea to construct statistical estimators for the matrix parameters B, Σ of a vector autoregression model under missing values.

Note that the model (8.1) includes the following equations:

$$G_1 = BG_0, \quad G_0 = BG_1' + \Sigma.$$

Given a fixed observation length T and assuming that the missing values template $O_t, t \in \{1, \dots, T\}$, satisfies

$$\sum_{t=1}^{T-k} O_{t+k,i} O_{tj} > 0, \quad k \in \{0, 1\}, \ i, j \in \{1, \dots, d\},$$

which follows from $A6$ for a sufficiently large T, let us evaluate sample covariances $\hat{G}_k \in \mathbb{R}^{d \times d}$, $k \in \{0, 1\}$:

$$(\hat{G}_k)_{ij} = \frac{\sum_{t=1}^{T-k} Y_{t+k,i} Y_{tj} O_{t+k,i} O_{tj}}{\sum_{t=1}^{T-k} O_{t+k,i} O_{tj}}, \quad k \in \{0, 1\}, \ i, j \in \{1, \dots, d\}. \tag{8.14}$$

Assuming $|\hat{G}_0| \neq 0$, we can use the expressions (8.14) to obtain estimators for the parameters of a VAR(1) model under missing values:

$$\hat{B} = \hat{G}_1 \hat{G}_0^{-1}, \quad \hat{\Sigma} = \hat{G}_0 - \hat{G}_1 \hat{G}_0^{-1} \hat{G}_1' \in \mathbb{R}^{d \times d}. \tag{8.15}$$

To study asymptotic properties of these estimators, we are going to require the following auxiliary results [10, 11].

Lemma 8.2. *In the model* (8.1), *assuming $A1$, the following boundedness condition is satisfied for the covariances:* $\forall \lambda \in (\lambda_{\max}(B), 1)$ $\exists C \in (0, +\infty)$ *such that* $\forall t_1, t_2, t_3, t_4 \in \mathbb{Z}$, $\forall i_1, i_2, i_3, i_4 \in \{1, \dots, d\}$ *we have*

$$|\mathbf{Cov}\{Y_{t_1 i_1} Y_{t_2 i_2}, Y_{t_3 i_3} Y_{t_4 i_4}\}| \leq C \left(\lambda^{|t_1 - t_3| + |t_2 - t_4|} + \lambda^{|t_1 - t_4| + |t_2 - t_3|} \right).$$

If, in addition, we assume $A4$, then for all $i_1, i_2, i_3, i_4 \in \{1, \dots, d\}$ the functional dependence between the covariances $\mathbf{Cov}\{Y_{t_1 i_1} Y_{t_2 i_2}, Y_{t_3 i_3} Y_{t_4 i_4}\}$ *and the time points $t_1, t_2, t_3, t_4 \in \mathbb{Z}$ can be rewritten to depend only on the differences $t_2 - t_1$, $t_3 - t_1$, $t_4 - t_1$. If, in addition, we assume $A5$, the covariances can be written explicitly:*

$$\mathbf{Cov}\{Y_{t_1i_1}Y_{t_2i_2}, Y_{t_3i_3}Y_{t_4i_4}\} = (G_{t_1-t_3})_{i_1i_3}(G_{t_2-t_4})_{i_2i_4} + (G_{t_1-t_4})_{i_1i_4}(G_{t_2-t_3})_{i_2i_3},$$
$$t_1, t_2, t_3, t_4 \in \mathbb{Z}, \quad i_1, i_2, i_3, i_4 \in \{1, \ldots, d\},$$

and the following boundedness condition is satisfied for the eighth order moments: $\forall \lambda \in (\lambda_{\max}(B), 1)$ $\exists C \in (0, +\infty)$ *such that* $\forall t_1, \ldots, t_8 \in \mathbb{Z}$, $\forall i_1, \ldots, i_8 \in \{1, \ldots, d\}$ *we have*

$$|\mathbb{E}\{(Y_{t_1i_1}Y_{t_2i_2} - \mathbb{E}\{Y_{t_1i_1}Y_{t_2i_2}\}) \cdots (Y_{t_7i_7}Y_{t_8i_8} - \mathbb{E}\{Y_{t_7i_7}Y_{t_8i_8}\})\}| \leq$$
$$\leq C \sum_{(l_1,\ldots,l_8)\in C_8(2,2,2,2)} \lambda^{|t_{l_1}-t_{l_2}|+|t_{l_3}-t_{l_4}|+|t_{l_5}-t_{l_6}|+|t_{l_7}-t_{l_8}|},$$

where the set $C_8(2,2,2,2)$ *of cardinality* $|C_8(2,2,2,2)| = 60$ *is defined as follows:*

$$(l_1, \ldots, l_8) \in C_8(2,2,2,2) \Leftrightarrow \begin{cases} l_n \in \{1, \ldots, 8\}, \quad n = 1, 2, \ldots, 8; \\ l_n \neq l_m, \quad 1 \leq n, m \leq 8, \ n \neq m; \\ l_1 < l_2, \ l_3 < l_4, \ l_5 < l_6, \\ \qquad l_7 < l_8, \ l_1 < l_3 < l_5 < l_7; \\ (l_1, l_2), (l_3, l_4), (l_5, l_6), (l_7, l_8) \notin \\ \qquad \notin \{(1,2), (3,4), (5,6), (7,8)\}. \end{cases}$$

Corollary 8.3. *Under the model (8.1), let us assume A1 and consider a time series* $Y_t^{(m)} = \sum_{i=0}^{m} B^i U_{t-i} \in \mathbb{R}^d$, $t \in \mathbb{Z}$, *obtained from the time series (8.1) by truncating to the first* $m + 1$ *terms. Then the following two statements hold:*

1. *Covariances of the time series* $Y_t^{(m)}$, $t \in \mathbb{Z}$, *are bounded:* $\forall \lambda \in (\lambda_{\max}(B), 1)$ $\exists C \in (0, +\infty)$ *such that* $\forall m \in \mathbb{N} \cup \{0\}$, $\forall t_1, t_2, t_3, t_4 \in \mathbb{Z}$, $\forall i_1, i_2, i_3, i_4 \in \{1, \ldots, d\}$ *we have*

$$\left|\mathbf{Cov}\left\{Y_{t_1i_1}^{(m)}Y_{t_2i_2}^{(m)}, Y_{t_3i_3}^{(m)}Y_{t_4i_4}^{(m)}\right\}\right| \leq C\left(\lambda^{|t_1-t_3|+|t_2-t_4|} + \lambda^{|t_1-t_4|+|t_2-t_3|}\right).$$

2. *Covariances of the time series* $Y_t^{(m)}$, $t \in \mathbb{Z}$, *converge to the covariances of the time series* Y_t, $t \in \mathbb{Z}$:

$$\mathbf{Cov}\left\{Y_{t_1i_1}^{(m)}Y_{t_2i_2}^{(m)}, Y_{t_3i_3}^{(m)}Y_{t_4i_4}^{(m)}\right\} \xrightarrow[m\to+\infty]{} \mathbf{Cov}\{Y_{t_1i_1}Y_{t_2i_2}, Y_{t_3i_3}Y_{t_4i_4}\},$$
$$t_1, t_2, t_3, t_4 \in \mathbb{Z}, \quad i_1, i_2, i_3, i_4 \in \{1, \ldots, d\}.$$

If, in addition, we assume A4, then for all $m \in \mathbb{N} \cup \{0\}$, $i_1, i_2, i_3, i_4 \in \{1, \ldots, d\}$ *the functional dependence of the covariance* $\mathbf{Cov}\left\{Y_{t_1i_1}^{(m)}Y_{t_2i_2}^{(m)}, Y_{t_3i_3}^{(m)}Y_{t_4i_4}^{(m)}\right\}$ *on the time* *points $t_1, t_2, t_3, t_4 \in \mathbb{Z}$ can be rewritten to depend only on their differences $t_2 - t_1$, $t_3 - t_1$, $t_4 - t_1$.*

The next theorem establishes unbiasedness and mean square consistency of the sample covariances (8.14), as well as consistency in probability of the constructed statistical estimators (8.15) of vector autoregression model parameters under missing values.

Theorem 8.4. *In the vector autoregression model* (8.1), *let the missing values template O_t, $t \in \{1, \ldots, T\}$, where T is the observation length, satisfy the condition*

$$\sum_{t=1}^{T-k} O_{t+k,i} O_{tj} > 0, \quad k \in \{0, 1\}, \ i, j \in \{1, \ldots, d\}. \tag{8.16}$$

Then the sample covariances (8.14) *are unbiased covariance estimators:*

$$\mathbb{E}\{\hat{G}_k\} = G_k, \quad k \in \{0, 1\}.$$

If, in addition, we assume A1 and A6, then the sample covariances (8.14) *are mean square consistent estimators:*

$$\hat{G}_k \xrightarrow[T \to +\infty]{\text{m.s.}} G_k, \quad k \in \{0, 1\}.$$

If we also have $|G_0| \neq 0$, then the estimators for the parameters $\widehat{B}$, $\widehat{\Sigma}$ of the vector autoregression model under missing values (8.15) *are consistent in probability:*

$$\hat{B} \xrightarrow[T \to +\infty]{\mathbf{P}} B, \quad \hat{\Sigma} \xrightarrow[T \to +\infty]{\mathbf{P}} \Sigma.$$

Proof. It is easy to see that, under the model (8.1), the condition (8.16) implies the unbiasedness of the sample covariances (8.14):

$$\mathbb{E}\{\hat{G}_k\} = G_k, \quad k \in \{0, 1\}.$$

Under additional assumptions $A1$ and $A6$, the sample covariances (8.14) are also mean square consistent estimators:

$$\hat{G}_k \xrightarrow[T \to +\infty]{\text{m.s.}} G_k, \quad k \in \{0, 1\},$$

and from $A6$, it follows that there exists a minimum observation length such that for any larger observation interval, any pair of vector components is observed simultaneously at some time point, and there exists a pair of consecutive time points such that the first and the second vector components are observed, respectively, at the first and the second time points:

$$\exists T_0 \in \mathbb{N}: \quad \forall T \geq T_0, \ \forall k \in \{0,1\}, \ \forall i,j \in \{1,\ldots,d\} \ \sum_{t=1}^{T-k} O_{t+k,i} O_{tj} > 0.$$

Let us assume that $T \geq T_0$ and write the variance of sample covariances as

$$\mathbb{D}\left\{\left(\hat{G}_k\right)_{ij}\right\} = \mathbb{E}\left\{\left(\frac{\sum_{t=1}^{T-k} Y_{t+k,i} Y_{tj} O_{t+k,i} O_{tj}}{\sum_{t=1}^{T-k} O_{t+k,i} O_{tj}} - (G_k)_{ij}\right)^2\right\} =$$

$$= \mathbb{E}\left\{\left(\frac{\sum_{t=1}^{T-k} \left(Y_{t+k,i} Y_{tj} - \mathbb{E}\{Y_{t+k,i} Y_{tj}\}\right) O_{t+k,i} O_{tj}}{\sum_{t=1}^{T-k} O_{t+k,i} O_{tj}}\right)^2\right\} =$$

$$= \frac{\sum_{t,t'=1}^{T-k} \mathbf{Cov}\left\{Y_{t+k,i} Y_{tj}, Y_{t'+k,i} Y_{t'j}\right\} O_{t+k,i} O_{tj} O_{t'+k,i} O_{t'j}}{\left(\sum_{t=1}^{T-k} O_{t+k,i} O_{tj}\right)^2},$$

$$T \geq T_0, \quad k \in \{0,1\}, \quad i,j \in \{1,\ldots,d\}.$$

Applying Lemma 8.2 yields that $\forall \lambda \in (\lambda_{\max}(B), 1) \ \exists C \in (0, +\infty)$ such that

$$\forall T \in \mathbb{N}, \ T \geq T_0, \quad \forall k \in \{0,1\}, \quad \forall i,j \in \{1,\ldots,d\}$$

we have

$$\mathbb{D}\left\{\left(\hat{G}_k\right)_{ij}\right\} \leq C \, \frac{\sum_{t,t'=1}^{T-k} \lambda^{|t-t'|} O_{t+k,i} O_{tj} O_{t'+k,i} O_{t'j}}{\left(\sum_{t=1}^{T-k} O_{t+k,i} O_{tj}\right)^2} \leq$$

$$\leq C \, \frac{\left(1 + 2\lambda + \cdots + 2\lambda^{T-k-1}\right) \sum_{t=1}^{T-k} O_{t+k,i} O_{tj}}{\left(\sum_{t=1}^{T-k} O_{t+k,i} O_{tj}\right)^2} = C \, \frac{1 + 2\lambda + \cdots + 2\lambda^{T-k-1}}{\sum_{t=1}^{T-k} O_{t+k,i} O_{tj}}.$$

Mean square consistency follows from the convergence

$$\frac{1 + 2\lambda + \cdots + 2\lambda^{T-k-1}}{\sum_{t=1}^{T-k} O_{t+k,i} O_{tj}} \xrightarrow[T\to+\infty]{} 0$$

for $\lambda \in (\lambda_{\max}(B), 1), k \in \{0, 1\}, i, j \in \{1, \ldots, d\}$.

If, in addition, the covariance matrix G_0 of the time series (8.1) is positive-definite, then the relations

$$G_1 = BG_0, \quad G_0 = BG_1' + \Sigma$$

imply that the estimators (8.15) are consistent in probability. □

Corollary 8.4. *In the model* (8.1) *under missing values, assuming A*1*, A*2*, and A*6*, the estimators* (8.15) *for the autoregression model parameters are consistent in probability:*

$$\hat{B} \xrightarrow[T\to+\infty]{\mathbf{P}} B, \quad \hat{\Sigma} \xrightarrow[T\to+\infty]{\mathbf{P}} \Sigma.$$

Proof. It is sufficient to observe that assumption $A2$ implies that the covariance G_0 of the time series (8.1) is positive-definite. □

Note that if we have $O_t = \mathbf{1}_d$, $t \in \{1, \ldots, T\}$, i.e., the time series doesn't contain missing values, then the results of Theorem 8.4 and Corollary 8.4 coincide with earlier results [1].

Let us establish the conditions for asymptotic normality of the estimator (8.14). Assume $A4$ and denote the covariances as follows:

$$g_{t-t',k,k',i,j,i',j'} = \mathbf{Cov}\left\{Y_{t+k,i}Y_{tj}, Y_{t'+k',i'}Y_{t'j'}\right\},$$
$$t, t' \in \mathbb{Z}, \quad k, k' \in \{0, 1\}, \quad i, j, i', j' \in \{1, \ldots, d\}.$$

By Lemma 8.2, assumption $A4$ implies that for all $k, k' \in \{0, 1\}$, $i, j, i', j' \in \{1, \ldots, d\}$ the dependence of the covariance

$$\mathbf{Cov}\left\{Y_{t+k,i}Y_{tj}, Y_{t'+k',i'}Y_{t'j'}\right\}$$

on time points $t, t' \in \mathbb{Z}$ can be expressed in terms of their differences $t - t'$. Thus, the above notation is valid.

Assuming $A7$, let us define some numerical characteristics of the limit behavior of the missing values template:

$$C_{\tau,k,k',i,j,i',j'} = \frac{\tilde{\vartheta}_{\tau,k,k',i,j,i',j'}}{\vartheta_{k,i,j}\vartheta_{k',i',j'}},$$

where $\tau \in \mathbb{Z}$, $k, k' \in \{0,1\}$, $i, j, i', j' \in \{1, \ldots, d\}$. Assuming that the missing values template O_t, $t \in \{1, \ldots, T\}$, where T is the observation length, satisfies the condition

$$\sum_{t=1}^{T-k} O_{t+k,i} O_{tj} > 0, \quad k \in \{0,1\}, \ i, j \in \{1, \ldots, d\},$$

let us define the following functions depending on the missing values template:

$$c_{\tau,k,k',i,j,i',j'}(T) = \frac{T \sum\limits_{t,t'=1}^{T-1} O_{t+k,i} O_{tj} O_{t'+k',i'} O_{t'j'} \delta_{t-t',\tau}}{\sum\limits_{t=1}^{T-k} O_{t+k,i} O_{tj} \sum\limits_{t=1}^{T-k'} O_{t+k',i'} O_{tj'}}$$

for $\tau \in \mathbb{Z}$, $k, k' \in \{0,1\}$, $i, j, i', j' \in \{1, \ldots, d\}$.

Lemma 8.3. *Assuming A7, we have*

$$c_{\tau,k,k',i,j,i',j'}(T) \xrightarrow[T \to +\infty]{} C_{\tau,k,k',i,j,i',j'},$$

where $\tau \in \mathbb{Z}$, $k, k' \in \{0,1\}$, $i, j, i', j' \in \{1, \ldots, d\}$. The limit properties of the missing values template satisfy the following symmetry condition:

$$C_{\tau,k,k',i,j,i',j'} = C_{-\tau,k',k,i',j',i,j}, \quad \tau \in \mathbb{Z}, \ k, k' \in \{0,1\}, \ i, j, i', j' \in \{1, \ldots, d\}.$$

Proof. The convergence is obvious: under assumption $A7$, there exists a minimum observation length such that for any larger observation interval, any pair of vector components is observed simultaneously at some time point, and there exists a pair of consecutive time points such that the first and the second vector components are observed respectively at the first and the second time points:

$$\exists T_0 \in \mathbb{N}: \quad \forall T \geq T_0, \ \forall k \in \{0,1\}, \ \forall i, j \in \{1, \ldots, d\} \quad \sum_{t=1}^{T-k} O_{t+k,i} O_{tj} > 0.$$

Functions of the missing values template satisfy the following symmetry condition:

$$c_{\tau,k,k',i,j,i',j'}(T) = c_{-\tau,k',k,i',j',i,j}(T),$$

$$T \in \mathbb{N}, \quad T \geq T_0, \quad \tau \in \mathbb{Z}, \quad k, k' \in \{0,1\}, \quad i, j, i', j' \in \{1, \ldots, d\},$$

which implies the relations between the limit properties of the template. □

Asymptotic normality of the statistical estimator (8.15) under missing values will be proved by applying the following central limit theorem [16,28] for m_T-dependent random vectors.

Theorem 8.5. *Let*

$$\left\{Z_t^{(T)},\ t \in \{1,2,\dots,k_T\}\right\}$$

be a sequence of $\tilde{m}_T$-dependent random variables such that $\mathbb{E}\left\{Z_t^{(T)}\right\} = 0$ holds for all t and T, and let

$$k_T \to \infty \ \text{as}\ T \to \infty, \qquad S_T = \sum_{t=1}^{k_T} Z_t^{(T)}, \qquad D_T^2 = \mathbb{E}\left\{S_T^2\right\},$$

$$\bar{D}_T^2 = \sum_{t=1}^{k_T} \mathbb{E}\left\{\left(Z_t^{(T)}\right)^2\right\}, \quad F_T(x) = \mathbb{P}\{S_T \le D_T x\}, \quad \Delta_T(x) = |F_T(x) - \Phi(x)|,$$

where $\Phi(x)$ is the standard normal distribution function. Also define

$$\gamma_\delta = \frac{\delta(\delta+2)}{2(\delta^2+4\delta+2)}, \quad 0 < \delta \le 1, \qquad \varepsilon_T = D_T^{-2}\tilde{m}_T^{\frac{3\delta+2}{\delta}}.$$

Assume that

$$\mathbb{E}\left\{\left|Z_t^{(T)}\right|^{2+\delta}\right\} < +\infty,$$

and that for a sufficiently large T_0, the following conditions are satisfied:

$$1)\ D_T^2 \longrightarrow \infty, \qquad 2)\ \bar{D}_T^2 = O\left(D_T^2\right), \qquad 3)\ \sum_{t=1}^{k_T} \mathbb{E}\left\{\left|Z_t^{(T)}\right|^{2+\delta}\right\} = O\left(D_T^2\right),$$

$$4)\ k_T = O\left(D_T^2\right), \quad 5)\ D_T^8 \tilde{m}_T^{-6} \le k_T^7,\ T > T_0, \quad 6)\ \varepsilon_T \longrightarrow 0.$$

Then for all x we have

$$\Delta_T(x) \le \frac{C}{(1+|x|)^{2+\delta}} \varepsilon_T^{\gamma_\delta},$$

where C is a constant independent of T and x.

Theorem 8.6. *In the model (8.1), assuming A2, A3, A4, A7, the d^2-vectorization of the $d \times d$-matrix $\sqrt{T}\left(\hat{B} - B\right)$ is asymptotically normally distributed as $T \to +\infty$ with a zero expectation and the covariance*

$$\mathbf{Cov}\left\{\sqrt{T}\left(\hat{B}-B\right)_{ij}, \sqrt{T}\left(\hat{B}-B\right)_{i'j'}\right\} =$$

$$= \sum_{k,k'=0}^{1} \sum_{n,m,n',m'=1}^{d} \sum_{\tau=-\infty}^{+\infty} (-1)^{k+k'} \left(B^{1-k}\right)_{in} \left(B^{1-k'}\right)_{i'n'} \times$$

$$\times g_{\tau,k,k',n,m,n',m'} C_{\tau,k,k',n,m,n',m'} \left(G_0^{-1}\right)_{mj} \left(G_0^{-1}\right)_{m'j'}$$

for $i, j, i', j' \in \{1, \dots, d\}$.

Proof. The proof is based on three facts:

(1) Theorem 8.5;
(2) Shiryaev's theorem [30]: a random vector is asymptotically normally distributed if and only if any linear combination of the components of this vector is asymptotically normally distributed;
(3) theorem on continuous functional transformations of asymptotically normal random vectors [27].

For a detailed proof, refer to [10]. □

Let us consider some special cases of Theorem 8.6.

Corollary 8.5. *Under the conditions of Theorem 8.6, assume that the number of missing values is bounded:*

$$\exists T_{\text{obs}} : \ \forall T \geq T_{\text{obs}} \quad O_t = \mathbf{1}_d, \ t \in \{T_{\text{obs}}, \dots, T\},$$

and assumption A5 is satisfied. Then for $T \to +\infty$ *the vectorization of the matrix* $\sqrt{T}\left(\hat{B}-B\right)$ *is asymptotically normally distributed with a zero expectation and the covariances*

$$\mathbf{Cov}\left\{\sqrt{T}\left(\hat{B}-B\right)_{ij}, \sqrt{T}\left(\hat{B}-B\right)_{i'j'}\right\} = \Sigma_{ii'}\left(G_0^{-1}\right)_{jj'}, \quad i, j, i', j' \in \{1, \dots, d\}.$$

Corollary 8.6. *Under the conditions of Theorem 8.6, assume that the limit properties of the missing values template do not depend on the time lag:*

$$C_{\tau,k,k',i,j,i',j'} = C_{0,k,k',i,j,i',j'}, \quad \tau \in \mathbb{Z}, \ k,k' \in \{0,1\}, \ i,j,i',j' \in \{1,\dots,d\},$$

assumption A5 holds, and the time series Y_t *is univariate,* $d = 1$. *Then for* $T \to +\infty$ *the normalized random deviation* $\sqrt{T}\left(\hat{B}-B\right) \in \mathbb{R}$ *of the estimator* $\hat{B}$ *is asymptotically normally distributed with a zero expectation and the variance*

$$\mathfrak{D} = \frac{1}{\vartheta_{0,1,1}}(1 - B^2).$$

Corollary 8.7. *Under the conditions of Theorem 8.6, assume that the time series* Y_t *is univariate,* $d = 1$*, the number of missing values is bounded:*

$$\exists T_{\text{obs}} : \ \forall T \geq T_{\text{obs}} \quad O_t = \mathbf{1}_d, \ t \in \{T_{\text{obs}}, \ldots, T\},$$

and assumption A5 is satisfied. Then for $T \to +\infty$*, the normalized random deviation* $\sqrt{T}\left(\hat{B} - B\right) \in \mathbb{R}$ *of the estimator* $\hat{B}$ *is asymptotically normally distributed with a zero expectation and the variance*

$$\mathcal{D} = 1 - B^2.$$

Note that if $O_t = \mathbf{1}_d, t \in \{1, \ldots, T\}$, i.e., in the absence of missing values, the results of Corollaries 8.5 and 8.7 coincide with earlier results [1].

8.5 Least Squares Forecasting and Its Risk Under Missing Values

As established in Sect. 8.4, least squares forecasting

$$\widehat{Y}_{T+\tau} = A_0(\widehat{B}, \widehat{\Sigma})X \tag{8.17}$$

can be used to construct τ-step-ahead forecasts of vector autoregression time series (8.1) under missing values. Here $\widehat{B}, \widehat{\Sigma}$ are modified least squares estimators defined by (8.14), (8.15). They have the required asymptotic properties, as shown by Theorems 8.4, 8.6. Let us give a detailed description of this forecasting algorithm.

Assume that a time series $Y_1, \ldots, Y_T$ of length T with a missing values template $O_1, \ldots, O_T$ is observed. Let us construct a τ-step ahead, $\tau \in \mathbb{N}$, statistical forecast $\hat{Y}_{T+\tau} \in \mathbb{R}^d$ of the future value $Y_{T+\tau} \in \mathbb{R}^d$ if the parameters $B, \Sigma \in \mathbb{R}^{d \times d}$ are a priori unknown. This statistical forecasting algorithm consists of ten steps.

1. Verify the data sufficiency condition:

$$\sum_{t=1}^{T-k} O_{t+k,i} O_{tj} > 0, \quad k \in \{0, 1\}, \ i, j \in \{1, \ldots, d\}.$$

   If the condition isn't satisfied, increase observation length T and repeat step 1.
2. Calculate sample covariances $\hat{G}_k \in \mathbb{R}^{d \times d}, k \in \{0, 1\}$:

$$(\hat{G}_k)_{ij} = \frac{\sum_{t=1}^{T-k} Y_{t+k,i} Y_{tj} O_{t+k,i} O_{tj}}{\sum_{t=1}^{T-k} O_{t+k,i} O_{tj}}, \quad k \in \{0, 1\}, \ i, j \in \{1, \ldots, d\}.$$

3. Verify nonsingularity of the sample covariance matrix:

$$\left|\hat{G}_0\right| \neq 0.$$

In the case of singularity, increase observation length T and go to step 2.

4. Construct statistical estimators

$$\hat{B} = \hat{G}_1 \hat{G}_0^{-1}, \quad \hat{\Sigma} = \hat{G}_0 - \hat{G}_1 \hat{G}_0^{-1} \hat{G}_1'.$$

5. Construct a discrete set

$$M = \{(t, i),\ t \in \{1, \ldots, T\},\ i \in \{1, \ldots, d\} : O_{ti} = 1\},$$

which contains pairs of time points and indices of the observed vector components. Order the set M lexicographically.

6. Compute the total number of observed components $K = |M|$.
7. Define a bijection

$$\chi(t, i) : M \leftrightarrow \{1, \ldots, K\}$$

and its inverse function

$$\bar{\chi}(k) : \{1, \ldots, K\} \leftrightarrow M.$$

8. Construct a K-vector of all observed components:

$$X = (X_1, \ldots, X_K)' \in \mathbb{R}^K, \quad X_k = Y_{\bar{\chi}(k)}, \quad k \in \{1, \ldots, K\}.$$

9. Substitute the estimators $\hat{B}, \hat{\Sigma} \in \mathbb{R}^{d \times d}$ defined by (8.15) in place of the true values $B, \Sigma \in \mathbb{R}^{d \times d}$ in the expressions for the covariance matrices $F \in \mathbb{R}^{K \times K}$, $H \in \mathbb{R}^{K \times d}$ obtained in Lemma 8.1. This yields covariance matrix estimators $\hat{F} \in \mathbb{R}^{K \times K}$, $\hat{H} \in \mathbb{R}^{K \times d}$, and for $|\hat{F}| \neq 0$ we obtain

$$\hat{A}_0 = \hat{H}' \hat{F}^{-1} \in \mathbb{R}^{d \times K}.$$

10. Apply (8.17) to construct the forecast $\hat{Y}_{T+\tau} = \hat{A}_0 X$.

Let us study the asymptotic behavior of the matrix risk for this forecasting algorithm. For brevity, let us assume that $O_T = \mathbf{1}_d$, i.e., the vector Y_T has been observed in full, and only the vectors $Y_1, \ldots, Y_{T-1}$ may contain missing values. The one-step-ahead ($\tau = 1$) forecasting statistic (8.17) has the simple form

$$\hat{Y}_{T+1} = \hat{B} Y_T. \tag{8.18}$$

The next theorem [10, 11] proves an asymptotic expansion of the matrix risk of plug-in vector autoregression forecasting under missing values.

Theorem 8.7. *In the model* (8.1)*, assuming A5, A7, and* $O_T = \mathbf{1}_d$*, consider a one-step-ahead forecast* ($\tau = 1$) *based on the plug-in forecasting statistic* (8.18) *and the estimators* (8.15)*. Further, assume that* $\forall n \in \mathbb{N}\ \exists C \in (0, +\infty),\ \exists T_0 \in \mathbb{N}$*:*

$$\forall T \geq T_0,\ \forall \theta \in (0,1) \quad \mathbb{E}\left\{ \frac{1}{\left| \mathbf{I}_d + \theta G_0^{-1} \left(\hat{G}_0 - G_0 \right) \right|^{2n}} \right\} \leq C. \tag{8.19}$$

Then the following asymptotic expansion holds for the matrix risk as $T \to +\infty$*:*

$$R = \Sigma + T^{-1} A + o\left(T^{-1}\right) 1_{d \times d}, \tag{8.20}$$

where

$$\begin{aligned} A = (A_{kl}) = & \sum_{i,j,i',j'=1}^{d} \sum_{\tau=-\infty}^{+\infty} BI(i,j) \left(BI(i',j') G_0^{-1} \right)' \times \\ & \times \left((G_\tau)_{ii'} (G_\tau)_{jj'} + (G_\tau)_{ij'} (G_\tau)_{ji'} \right) C_{\tau,0,0,i,j,i',j'} - \\ & - \sum_{i,j,i',j'=1}^{d} \sum_{\tau=-\infty}^{+\infty} \left(BI(i,j) \left(I(i',j') G_0^{-1} \right)' + I(i',j') \left(BI(i,j) G_0^{-1} \right)' \right) \times \\ & \times \left((G_{\tau-1})_{ii'} (G_\tau)_{jj'} + (G_\tau)_{ij'} (G_{\tau-1})_{ji'} \right) C_{\tau,0,1,i,j,i',j'} + \\ & + \sum_{i,j,i',j'=1}^{d} \sum_{\tau=-\infty}^{+\infty} I(i,j) \left(I(i',j') G_0^{-1} \right)' \times \\ & \times \left((G_\tau)_{ii'} (G_\tau)_{jj'} + (G_{\tau+1})_{ij'} (G_{\tau-1})_{ji'} \right) C_{\tau,1,1,i,j,i',j'} \in \mathbb{R}^{d \times d}, \end{aligned}$$

$$I(i,j) \in \mathbb{R}^{d \times d}, \quad (I(i,j))_{kl} = \begin{cases} 1, & \text{if } (k,l) = (i,j) \\ 0, & \text{if } (k,l) \neq (i,j) \end{cases}, \quad i,j,k,l \in \{1, \ldots, d\}.$$

Presented below are some special cases of Theorem 8.7 [10].

Corollary 8.8. *Under the conditions of Theorem 8.7, let the number of missing values be bounded:*

$$\exists T_{\text{obs}} : \ \forall T \geq T_{\text{obs}} \quad O_t = \mathbf{1}_d, \ t \in \{T_{\text{obs}}, \ldots, T\}.$$

Then the following asymptotic expansion holds for the matrix forecast risk:

$$R = \Sigma + d T^{-1} \Sigma + o\left(1 T^{-1}\right) \mathbf{1}_{d \times d} \quad \text{as } T \to +\infty.$$

Corollary 8.9. *In the univariate case ($d = 1$) of the model* (8.1)*, assuming A5, A7, and that the last observed vector contains no missing values ($O_T = \mathbf{1}_d$), consider the plug-in forecasting statistic* (8.18) *based on the estimators* (8.15) *for a one-step-ahead forecast ($\tau = 1$). Then the condition* (8.19) *is satisfied, and the following asymptotic expansion for the mean square risk holds as $T \to +\infty$:*

$$R = \Sigma + cAT^{-1} + o\left(T^{-1}\right),$$

where

$$A = \frac{\Sigma}{1 - B^2}\left(2B^2 \sum_{\tau=-\infty}^{+\infty} B^{2|\tau|} C_{\tau,0,0,1,1,1,1} - 4B \sum_{\tau=-\infty}^{+\infty} B^{|\tau-1|+|\tau|} C_{\tau,0,1,1,1,1,1} + \right.$$
$$\left. + \sum_{\tau=-\infty}^{+\infty} \left(B^{2|\tau|} + B^{|\tau+1|+|\tau-1|}\right) C_{\tau,1,1,1,1,1,1}\right).$$

Corollary 8.10. *Under the conditions of Corollary 8.9, let the number of missing values be bounded:*

$$\exists T_{\text{obs}} : \forall T \geq T_{\text{obs}} \quad O_t = 1, \; t \in \{T_{\text{obs}}, \ldots, T\}.$$

Then the following asymptotic expansion of the mean square risk holds as $T \to +\infty$:

$$R = \Sigma + \Sigma T^{-1} + o\left(T^{-1}\right).$$

In the absence of missing values, a result similar to Corollary 8.8 was stated by Lutkepohl [15].

Theorem 8.7 and its corollaries imply that the risk of least squares forecasting tends to the risk of the optimal forecasting statistic as $T \to +\infty$.

8.6 Results of Computer Experiments

Performance of the modified least squares estimators and the forecasting statistics based on these estimators was evaluated by performing a series of computer simulations. The simulations were based on three well-known sets of real-world statistical data, which are adequately described by autoregression models. Then, using different missing values templates, $N = 1{,}000$ realizations of time series were generated, and the quality of the respective statistical inferences was evaluated by comparing the experimental data with the theoretical values.

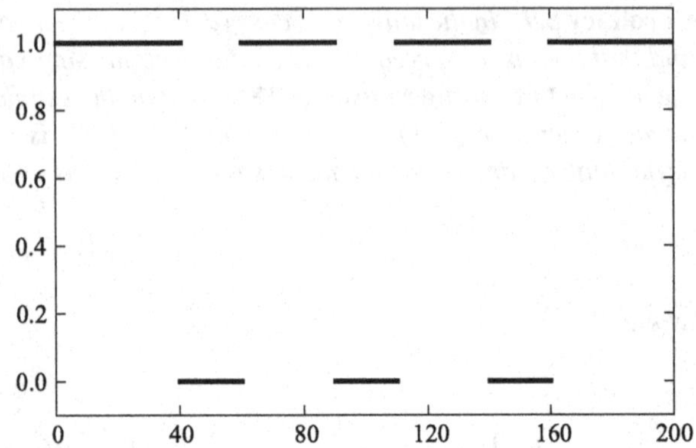

Fig. 8.1 Missing values template

8.6.1 Performance of the Estimator $\widehat{B}$

Performance of the statistical parameter estimation algorithm (8.1) for vector autoregression time series models under missing values was evaluated by a simulation experiment based on a real-world dataset—Wolf numbers of solar activity, which are a sequence of yearly measurements of solar activity collected between 1610 and 1960 [1]. Yule's autoregression model [36] was fitted to that data, yielding the following parameters:

$$p = 2, \quad (b_1, b_2)' = (1.34254, -0.65504)', \quad \sigma = 15.41,$$

and the following missing values template was used in the simulations:

$$o_t = \begin{cases} 1, & t \in D_{mv} \\ 0, & \text{otherwise}, \end{cases} \tag{8.21}$$

where the domain of missing values D_{mv} was chosen as

$$D_{mv} = \{t \in \mathbb{Z} : \quad |t \text{ Mod } 200 - 50| \leq 10 \vee \\ \vee\, |t \text{ Mod } 200 - 100| \leq 10 \vee |t \text{ Mod } 200 - 150| \leq 10\}.$$

A plot of o_t is shown in Fig. 8.1.

This model was reduced to the respective vector autoregression model defined by the parameters

$$d = 2, \quad B = \begin{pmatrix} 1.34254 & -0.65504 \\ 1 & 0 \end{pmatrix}, \quad \Sigma = \begin{pmatrix} 237.4681 & 0 \\ 0 & 0 \end{pmatrix},$$

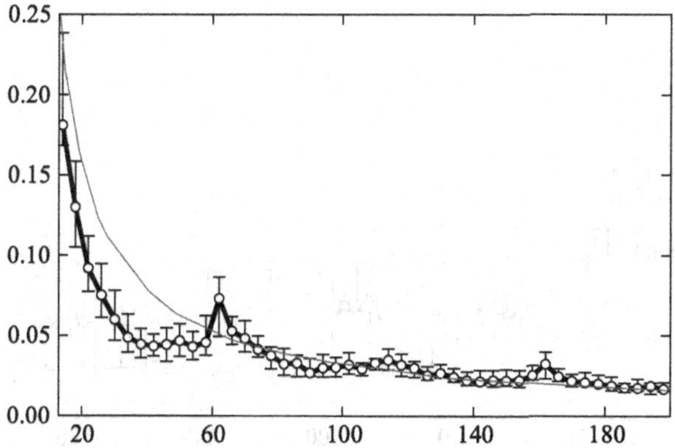

Fig. 8.2 Theoretical and experimental variances computed for Wolf data under missing values

and the vector missing values template

$$O_{ti} = o_{t-i+1}, \quad t \in \mathbb{Z}, \ i \in \{1, 2\}.$$

The results of estimating the matrix coefficient B of this vector autoregression model under missing values are shown in Fig. 8.2, presenting the main term of the trace of the theoretical covariance matrix (8.20) computed for the constructed estimator (8.15), and the trace $V\{\hat{B}\}$ of the sample covariance matrix of the estimator (8.15) obtained in the simulations:

$$V\{\hat{B}\} = \frac{\sum_{i=1}^{N} \sum_{n,m=1,2} \left(\hat{B}_{nm}^{(i)} - B_{nm}\right)^2}{N},$$

where $N = 1{,}000$. The error bars indicate 95 % confidence bounds for $V\{\hat{B}\}$, and the simulated observation length is plotted on the x-axis. Simulation results show a high level of agreement between the theoretical and the experimental results.

8.6.2 Experimental Evaluation of the Forecast Risk

Robustness of vector autoregression (8.1) forecasting algorithms under missing values was evaluated by simulations based on real-world data—the Beverage price index of wheat on European markets recorded from 1670 to 1869 [1]. Anderson's AR(2) autoregression model [1] with the following parameters:

$$p = 2, \quad (b_1, b_2)' = (0.7368, -0.311)', \quad \sigma^2 = 0.6179,$$

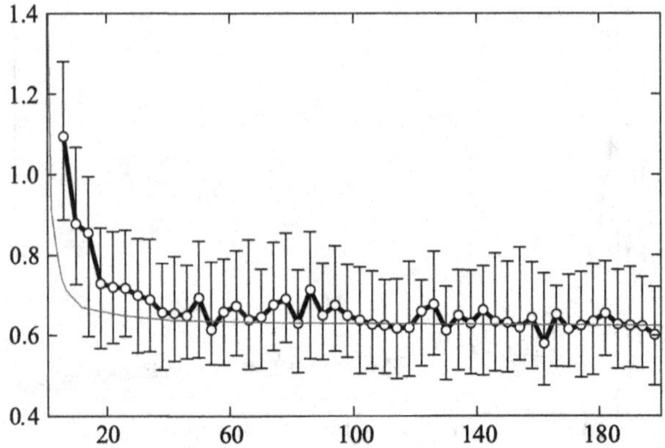

Fig. 8.3 Theoretical and experimental risks for the Beverage dataset under missing values

and the same missing values template (8.21) as in Sect. 8.6.1 was reduced to the corresponding vector autoregression model defined by the parameters

$$d = 2, \quad B = \begin{pmatrix} 0.7368 & -0.311 \\ 1 & 0 \end{pmatrix}, \quad \Sigma = \begin{pmatrix} 0.6179 & 0 \\ 0 & 0 \end{pmatrix},$$

and the missing values template

$$O_{ti} = o_{t-i+1}, \quad t \in \mathbb{Z}, \quad i \in \{1, 2\}.$$

Forecasts were based on the estimator (8.15) obtained in Theorem 8.7. The plot in Fig. 8.3 presents the simulation results for one-step-ahead forecasts ($\tau = 1$) in a setting where the last observation vector doesn't contain missing values ($O_T = \mathbf{1}_d$). It includes the main term of the theoretical asymptotic matrix risk defined by (8.20), the trace $\hat{r}$ of the sample matrix risk of forecasting,

$$\hat{r} = \frac{\sum_{i=1}^{N} \sum_{j=1}^{2} \left(\hat{Y}_{T+1,j}^{(i)} - Y_{T+1,j}^{(i)} \right)^2}{N},$$

where $N = 1{,}000$, and the respective 95 % confidence limits. As in the previous experiment, simulation data agrees with the theoretical results.

The third simulation-based experiment used another classical dataset—a centered time series of annual Canadian lynx trappings collected over 113 years [32]. The forecasts were based on an AR(11) autoregression model ($d = 11$):

$$y_t = 1.0938 y_{t-1} - 0.3571 y_{t-2} - 0.1265 y_{t-4} + 0.3244 y_{t-10} - 0.3622 y_{t-11} + \xi_t,$$

where $\mathbb{D}\{\xi_t\} = \sigma^2 = 0.04405$. Model parameters were estimated from observations at times $t = 1, 2, \ldots, T = 113$, and the value y_{T+1} was simulated under this model.

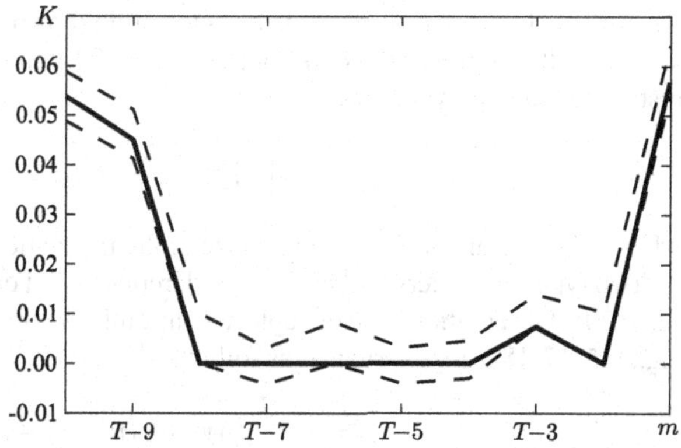

Fig. 8.4 Risk instability coefficient for the Canadian lynx dataset under missing values

A single missing value at time $m \in \{T-d+1, \ldots, T-1\}$ was introduced into the time series y_t, and forecasts of y_{T+1} based on $\{y_t : t = 1, \ldots, m-1, m+1, \ldots, T\}$ were made at two levels of prior information:

(a) prior knowledge of autoregression coefficients;
(b) parametric prior uncertainty—unknown autoregression coefficients.

The model AR(11) above was then transformed into a model (8.1) of a vector time series $Y_t = (y_t, y_{t-1}, \ldots, y_{t-10}) \in \mathbb{R}^{11}$.

In the case (a), Fig. 8.4 shows a dependence of the theoretical value of the risk instability coefficient $\kappa(m)$ computed from (8.5) and the Corollary 8.1 on the coordinate m of the missing value. The 95 % confidence limits for $\kappa(m)$ computed from the simulation results are also presented. The plots for the case (b) are omitted since the confidence limits only change by 1–2 % compared to the former case.

Figure 8.4 shows that the location of a missing value can have significant effect on the forecast risk. For example, missing values at time points $T-1, T-9, T-10$ have greater effect on forecasting accuracy compared to other time points.

8.7 Robust Plug-In Forecasting Under Simultaneous Influence of Outliers and Missing Values

8.7.1 A Mathematical Model of Simultaneous Distortion by Outliers and Missing Values

In applications, an observed time series is often simultaneously distorted by outliers and registered with missing values. We are going to study the effect of this simultaneous distortion on forecasting under $AR(p)$ autoregression models:

$$y_t + b_1 y_{t-1} + \cdots + b_p y_{t-p} = \xi_t, \quad t \in \mathbb{Z}, \tag{8.22}$$

where $b = (b_1, \dots, b_p)' \in \mathbb{R}^p$ is a vector of unknown autoregression coefficients satisfying the stationarity condition; $\{\xi_t : t \in \mathbb{Z}\}$ is a sequence of i.i.d. normally distributed random variables,

$$\mathcal{L}\{\xi_t\} = \mathcal{N}_1(0, \sigma^2),$$

where σ^2 is the unknown finite variance of the innovation process ξ_t.

Two types of outliers will be discussed, replacement outliers and additive outliers [2, 4, 20]. Observations under replacement outliers (the so-called RO-model: see, e.g., [15, 17, 18]) are distributed as follows:

$$z_t = (1 - \eta_t)y_t + \eta_t \upsilon_t, \quad t \in \mathbb{Z}, \tag{8.23}$$

where $\{\eta_t \in \{0, 1\}\}$ are i.i.d. Bernoulli random variables,

$$P\{\eta_t = 1\} = 1 - P\{\eta_t = 0\} = \varepsilon, \tag{8.24}$$

and $0 \le \varepsilon < 1$ is the (sufficiently small) outlier probability, $\{\upsilon_t \in \mathbb{R}\}$ are i.i.d. random variables independent of $\{y_t\}$ with some unknown symmetric probability distribution, zero means $\mathbb{E}\{\upsilon_t\} = 0$, and finite variances $\mathbb{D}\{\upsilon_t\}$. The magnitude of the outlier variance is determined by the value $\beta = \sqrt{D\{\upsilon_t\}/D\{y_t\}}$. Observations under additive outliers (the so-called AO-model: cf. [15, 17]) satisfy the equation

$$z_t = y_t + \eta_t \upsilon_t, t \in \mathbb{Z}. \tag{8.25}$$

As in Sect. 8.1, missing values will be described by a missing values template

$$o_t = \{1, \text{ if } z_t \text{ is observed}; \ 0, \text{ if } z_t \text{ is a missing value}\}, \quad t \in \mathbb{Z}. \tag{8.26}$$

Let $\{t_k\}$ be an increasing sequence of time moments, where $o_{t_k} = 1$. For $t_- = \min_k t_k$, $t_+ = \max_k t_k$, without loss of generality it can be assumed that $t_- = 1$, $t_+ = T$, where T is the length of the observation period. The observed time series is represented as

$$x_k = z_{t_k}, \quad k = 1, \dots, K, \tag{8.27}$$

where $K \le T$ is the total number of registered observations.

Consider a problem of constructing a robust statistical forecast for the future value y_{T+1} based on K registered observations $x_{t_1}, \dots, x_{t_K}$. We are going to use the plug-in forecasting statistic obtained by substituting robust estimators for b and σ^2 into the mean square optimal forecasting statistic determined in Sect. 8.2. The plug-in approach reduces the initial forecasting problem to statistical estimation of the vector autoregression coefficients of a stationary $AR(p)$ model with possible missing observations and additive or replacement outliers.

Let us give a short review of published results on robust estimation of $AR(p)$ parameters. Analytic reviews on this topic can be found in [4, 17], and [20]. One possible approach consists of robust estimation of the covariance function $\sigma_\tau = \mathrm{cov}\{y_t, y_{t+\tau}\} = E\{y_t y_{t+\tau}\}$, $\tau = 0, 1, \ldots, p$, and construction of an estimator for b by solving the Yule–Walker equations [1]. A number of robust estimators for covariances under outliers have been developed (see, e.g., [5, 17, 33, 35]); let us mention some of them: the Huber estimator [8], the median coefficient of correlation and its generalizations [29], estimators based on nonparametric measures [3], and elimination of detected outliers [7, 24]. M-estimators are another type of robust estimators for b under outliers [8, 15, 18]:

$$\hat{b} = \arg\min_{b} \sum_{t=p+1}^{T} \varrho(x_t - b_1 x_{t-1} - \cdots - b_p x_{t-p}),$$

where ϱ is a specially constructed function. Finally, let us mention the approach based on robust filters which was developed in [17].

A number of robust statistical methods have been developed for the case of missing data values without outliers:

- imputation of missing values [25, 26];
- special MLE-estimators and iterative EM-estimators [14];
- Monte-Carlo techniques [23];
- empirical Bayesian methods;
- procedures based on special properties of the ARIMA model [13, 21].

However, robust estimation of autoregression coefficients under simultaneous influence of outliers and missing values has, until recently, remained undeveloped. This section presents the results of the paper [12] which was the first attempt to

(a) investigate the setting where an autoregression model is distorted by both outliers and missing values;
(b) construct parametric robust estimators for correlations using special properties of the Cauchy probability distribution.

8.7.2 A Family of Robust Estimators for Correlations Based on the Cauchy Probability Distribution

Let θ_τ be the correlation function of the time series y_t defined by (8.22):

$$\theta_\tau = \mathbf{Corr}\{y_t, y_{t+\tau}\} = \sigma_\tau / \sigma_0, \quad \tau \in \mathbb{Z}. \tag{8.28}$$

First, let us modify the robust Huber estimator defined in [8] for the correlation coefficient θ_τ so that it can be applied in the case of missing values. The resulting algorithm to compute the modified Huber robust estimator consists of three steps.

1. Compute the values:

$$a = 1/\operatorname{Med}\left\{|z_t| : o_t = 1,\ t = 1,\dots,T-\tau\right\},$$
$$b = 1/\operatorname{Med}\left\{|z_t| : o_t = 1,\ t = \tau+1,\dots,T\right\}.$$

2. Compute the values:

$$S_+ = \operatorname{Med}\left\{|az_t + bz_{t+\tau}| : o_t = o_{t+\tau} = 1,\ t = 1,\dots,T-\tau\right\},$$
$$S_- = \operatorname{Med}\left\{|az_t - bz_{t+\tau}| : o_t = o_{t+\tau} = 1,\ t = 1,\dots,T-\tau\right\}.$$

3. Evaluate the modified Huber estimator:

$$\hat{\theta}_\tau = \frac{S_+^2 - S_-^2}{S_+^2 + S_-^2}. \tag{8.29}$$

In computer experiments described in Sect. 8.7.6, the robustness of the new estimators proposed by the author was evaluated against the estimator (8.29).

Construction of a new robust estimator for θ_τ will be based on the following special properties of the Cauchy probability distribution [12].

Lemma 8.4. *If y_t, $t \in \mathbb{Z}$, is a stationary normal time series with a zero mean and a correlation function* (8.28), *then for any $\tau \neq 0$ the ratio*

$$\zeta_\tau = \frac{y_t}{y_{t+\tau}}$$

follows the Cauchy probability distribution:

$$\mathcal{L}\{\zeta_\tau\} = \mathcal{C}\left(\theta_\tau, \sqrt{1-\theta_\tau^2}\right).$$

Lemma 8.5. *If $\phi : \mathbb{R}^2 \to \mathbb{R}$ is a bounded function which is odd w.r.t. both arguments:*

$$\phi(-u_1, u_2) = \phi(u_1, -u_2) = -\phi(u_1, u_2) \quad \forall\, u_1, u_2 \in \mathbb{R},$$

and z_t is a time series following the RO-model defined by (8.22)–(8.24), *then*

$$\mathbb{E}\{\phi(z_t, z_{t+\tau})\} = (1-\varepsilon)^2 \mathbb{E}\{\phi(y_t, y_{t+\tau})\}. \tag{8.30}$$

Note that under the AO-model (8.22), (8.25), (8.24), we have

$$\begin{aligned}\mathbb{E}\{\phi(z_t, z_{t+\tau})\} &= (1-\varepsilon)^2\mathbb{E}\{\phi(y_t, y_{t+\tau})\} + \varepsilon^2\mathbb{E}\{\phi(y_t + \upsilon_t, y_{t+\tau} + \upsilon_{t+\tau})\} + \\ &+ \varepsilon(1-\varepsilon)\big(\mathbb{E}\{\phi(y_t, y_{t+\tau} + \upsilon_{t+\tau})\} + \mathbb{E}\{\phi(y_t + \upsilon_t, y_{t+\tau})\}\big),\end{aligned} \tag{8.31}$$

as in Lemma 8.5. The expression (8.31) is different from (8.30) in the last two terms. These last terms of (8.31) are negligible w.r.t. the main term (8.30) for two reasons:

(1) the main term is of the order $O(1)$ and the last terms—of the order $O(\varepsilon)$;
(2) the random variables y_t and $y_{t+\tau}+\upsilon_{t+\tau}$, as well as $y_t+\upsilon_t$ and $y_{t+\tau}$, are weakly correlated under large outliers ($\beta \gg 1$):

$$\mathbf{Corr}\{y_t, y_{t+\tau}+\upsilon_{t+\tau}\} = \frac{\mathbb{E}\{y_t y_{t+\tau}\}}{\sqrt{\mathbb{D}\{y_t\}\left(\mathbb{D}\{y_{t+\tau}\}+\mathbb{D}\{\upsilon_{t+\tau}\}\right)}} = \frac{\theta_\tau}{\sqrt{1+\beta^2}}.$$

Let us choose the function $\phi(\cdot)$ in the following special form:

$$\phi(z_t, z_{t+\tau}) = \psi(z_t/z_{t+\tau}), \quad |\psi(u)| \le c_0 < +\infty, \quad \psi(-u) = -\psi(u), \tag{8.32}$$

where $\psi(\cdot)$ is a bounded odd function. Let us introduce the following bounded odd function $[-1,+1] \to \mathbb{R}$:

$$f_\psi(\theta_\tau) ::= \mathbb{E}\{\psi(\zeta_\tau)\} = \frac{\sqrt{1-\theta_\tau^2}}{\pi} \int_{-\infty}^{+\infty} \frac{\psi(z)}{1-\theta_\tau^2+(z-\theta_\tau)^2} dz, \tag{8.33}$$

where $\Psi(\cdot)$ is defined by (8.32), and $\zeta_\tau = y_t/y_{t+\tau}$. For $0 < \tau < T$, denote:

$$s_{\psi,\tau,t} = \psi(z_t/z_{t+\tau}), \qquad N_{\tau,T} = \sum_{t=1}^{T-\tau} o_t o_{t+\tau},$$

$$S_{\psi,\tau,T} = \frac{1}{N_{\tau,T}} \sum_{t=1}^{T-\tau} s_{\psi,\tau,t} o_t o_{t+\tau}, \qquad \Sigma_\tau^\psi(\lambda) = \sum_{t\in\mathbb{Z}} \mathbf{Cov}\{s_{\psi,\tau,t}, s_{\psi,\tau,0}\} \cos(t\lambda).$$

Theorem 8.8. *In the RO-model with missing values (8.22)–(8.24), (8.26), (8.25), if the function $\psi(\cdot)$ is chosen so that the function $f_\psi(\cdot)$ defined by (8.33) has a continuous inverse function $f_\psi^{-1}(\cdot)$, then for $0 < \tau < T$, and $N_{\tau,T} \to \infty$ as the observation time increases, $T \to \infty$, the statistic*

$$\hat{\theta}_\tau = f_\psi^{-1}\left((1-\varepsilon)^{-2} S_{\psi,\tau,T}\right) \tag{8.34}$$

is a consistent estimator for the correlation coefficient θ_τ: $\hat{\theta}_\tau \xrightarrow{\mathbf{P}} \theta_\tau$.

Theorem 8.8 was proved in [12]. Note that under the AO-model, the estimator (8.34) is biased:

$$\hat{\theta}_\tau \xrightarrow{\mathbf{P}} \theta_\tau + \Delta\theta_\tau,$$

but the remark following (8.31) implies that this bias vanishes for large outliers:

$$\Delta\theta_\tau \to 0 \quad \text{as} \quad \beta \to \infty.$$

Choosing different functions $\psi(\cdot)$ in (8.33) and applying Theorem 8.8 yields a family (8.34) of robust ψ-estimators for θ_τ that are consistent under replacement outliers and missing values (8.23), (8.26). Let us present four examples of ψ-estimators.

$$\psi s\text{-}estimator: \quad \psi(x) = \mathrm{sign}(x), \quad f_\psi(\theta) = \frac{2}{\pi}\arcsin\theta, \quad f_\psi^{-1}(\theta) = \sin\frac{\pi u}{2};$$

$$\psi a\text{-}estimator: \quad \psi(x) = \arctan x, \quad f_\psi(\theta) = \frac{1}{2}\arcsin\theta, \quad f_\psi^{-1}(\theta) = \sin(2u);$$

$$\psi t\text{-}estimator: \quad \psi(x) = \frac{2x}{1+x^2}, \quad f_\psi(\theta) = \frac{\theta}{1+\sqrt{1-\theta^2}}, \quad f_\psi^{-1}(\theta) = \frac{2u}{1+u^2};$$

$$\psi x\text{-}estimator: \quad \psi(x) = \begin{cases} x, |x| \le 1, \\ 1/x, |x| > 1, \end{cases} \qquad f_\psi(\theta) = \theta + \frac{\sqrt{1-\theta^2}}{\pi}\ln\frac{1-\theta}{1+\theta}.$$

For example, the ψs-estimator of θ_τ is

$$\hat{\theta}_\tau = \sin\left((N_{\tau,T})^{-1}\frac{\pi}{2(1-\varepsilon)^2}\sum_{t=1}^{T-\tau}\mathrm{sign}(z_t z_{t+\tau})o_t o_{t+\tau}\right).$$

Theorem 8.9 below establishes asymptotic normality of ψ-estimators [12].

Theorem 8.9. *Under the conditions of Theorem 8.8, if* $\min_{\lambda\in[-\pi,\pi]}\Sigma_\tau^\psi(\lambda) > 0$, *then the* ψ*-estimator* (8.34) *is asymptotically normal:*

$$\frac{\hat{\theta}_\tau - \theta_\tau}{\sqrt{\mathbb{D}_T}} \xrightarrow[T\to\infty]{\mathbf{D}} \mathcal{N}_1(0,1),$$

with the asymptotic mean θ_τ *and the asymptotic variance*

$$\mathbb{D}_T = \frac{(1-\varepsilon)^{-4}}{N_{\tau,T}}\frac{\Sigma_\tau^\psi(\lambda_T^*)}{(f_\psi'(\theta_\tau))^2}, \tag{8.35}$$

where $\lambda_T^* \in [-\pi,\pi]$.

8.7.3 Minimizing Asymptotic Variance of ψ-Estimators

To find the best estimator in the family of ψ-estimators (8.34), let us consider the problem of minimizing the asymptotic variance (8.35) in $\psi(\cdot)$. Unfortunately, the sequence λ_T^* depends on the index sequence of missing values (8.26) in an extremely complicated way, so let us minimize the integral asymptotic variance for λ_T^* in the interval $[-\pi, \pi]$:

$$\frac{(1-\varepsilon)^{-4}}{N_{\tau,T}(f_\psi'(\theta_\tau))^2}\int_{-\pi}^{\pi}\Sigma_\tau^\psi(\lambda)d\lambda = \frac{2\pi(1-\varepsilon)^{-4}\mathbb{D}\{s_{\psi,\tau,t}\}}{N_{\tau,T}(f_\psi'(\theta_\tau))^2} \to \min_{\psi(\cdot)}.$$

Because of (8.30) in the RO-model and (8.31) in the AO-model, we have:

$$\mathbb{D}\{s_{\psi,\tau,t}\} = \mathbb{D}\{\psi(\zeta_\tau)\} + O(\varepsilon),$$

where $\zeta_\tau = y_t/y_{t+\tau}$ has the Cauchy probability distribution $\mathcal{C}(\theta_\tau, \sqrt{1-\theta_\tau^2})$, as established by Lemma 8.4. Taking the main term of this expression yields the following minimization problem:

$$V(\psi,\theta_\tau) = \frac{\mathbb{D}\{\psi(\zeta_\tau)\}}{(f_\psi'(\theta_\tau))^2} \to \min_{\psi(\cdot)}, \quad f_\psi(\theta_\tau) = \mathbb{E}\{\psi(\zeta_\tau)\}. \tag{8.36}$$

Let us denote the Cauchy probability distribution function and its even and odd parts as, respectively,

$$p(x;\theta) = \frac{1}{\pi\sqrt{1-\theta^2}}\left(1+\frac{(x-\theta)^2}{1-\theta^2}\right)^{-1}, \quad x \in \mathbb{R};$$

$$C(x,\theta) = \frac{p(x;\theta)+p(-x;\theta)}{2}, \quad S(x,\theta) = \frac{p(x;\theta)-p(-x;\theta)}{2};$$

and define

$$I_1\{f\} = \int_{-\infty}^{+\infty} f(x)p(x;\theta)dx, \quad I_2\{f\} = \int_{-\infty}^{+\infty} f(x)p_\theta'(x;\theta)dx;$$

$$\omega = \left(1 - \int_{-\infty}^{+\infty}\frac{S^2(x,\theta)}{C(x,\theta)}dx\right)\left(\int_{-\infty}^{+\infty}\frac{S(x,\theta)S_\theta'(x,\theta)}{C(x,\theta)}dx\right)^{-1}.$$

Theorem 8.10. *Assume that ζ_τ is a random variable with a probability distribution function $p(x;\theta)$ which is differentiable in θ. Then the solution of the minimization problem* (8.36) *can be written as*

$$\psi_*(x) = \frac{S(x,\theta)+\omega S_\theta'(x,\theta)}{C(x,\theta)}, \quad V(\psi_*,\theta) = \omega\left(\frac{f_{\psi_*}(\tau)}{f_{\psi_*}'(\tau)}\right)_{\tau=\theta}.$$

Note that changing a parameter of the probability distribution function $p(x;\theta)$ by taking $\alpha = \alpha(\theta)$ leads to $V(\psi,\alpha) = V(\psi,\theta)(\alpha'(\theta))^2$, and the solution of the problem (8.36) remains unchanged. Based on this fact, let us make a suitable parameter change in the Cauchy distribution $\mathcal{C}(\theta, \sqrt{1-\theta^2})$: $\alpha(\theta) = \arcsin\theta$; the Cauchy distribution then assumes the form $\mathcal{C}(\sin\alpha, \cos\alpha)$.

Theorem 8.11. *The solution of the minimization problem* (8.36) *within the class of odd functions* ψ *for the Cauchy probability distribution* $\mathcal{C}(\sin\alpha, \cos\alpha)$ *is*

$$\psi_*(x) = \frac{2x}{1+x^2}\left(\frac{2(1+x^2)^2\cos\alpha}{1+x^4+2x^2\cos 2\alpha} - 1\right),$$

$$f_{\psi_*}(y) = y\left(\frac{2}{\cos\alpha + \sqrt{1-y^2}} - \frac{1}{1+\sqrt{1-y^2}}\right),$$

$$V(\psi_*,\alpha) = \frac{2\cos^2\alpha(1+\cos\alpha)^2}{1+2\cos\alpha+\cos^4\alpha}. \tag{8.37}$$

Theorems 8.10 and 8.11 were proved in [12].

Note that the Cramer–Rao bound for the Cauchy distribution $\mathcal{C}(\sin\alpha, \cos\alpha)$ equals $I^{-1}(\alpha) = 2\cos^2\alpha$, where $I(\alpha)$ is the Fisher information. Therefore, the efficiency loss of the asymptotic variance (8.37) compared to this bound is less than 10 %:

$$\frac{V(\psi_*,\alpha)}{I^{-1}(\alpha)} = \frac{1+2\cos\alpha+\cos^2\alpha}{1+2\cos\alpha+\cos^4\alpha} \le \frac{13+5\sqrt{5}}{22} \approx 1.099, \quad \alpha \in \left[-\frac{\pi}{2}, \frac{\pi}{2}\right].$$

Corollary 8.11. *The optimal function* $\psi_*(\cdot)$ *for the estimation of* θ_τ *is*

$$\psi_*(x;\theta_\tau) = \frac{2x}{1+x^2}\left(\frac{2(1+x^2)^2\sqrt{1-\theta_\tau^2}}{1+x^4+2x^2(1-2\theta_\tau^2)} - 1\right). \tag{8.38}$$

Figure 8.5 presents a parametric family (8.38) of curves $\psi_*(x;\theta)$ in the first quadrant for $\theta = 0; 0.7; 0.9; 0.9999$ (from top to bottom).

Note that the optimal function $\psi_* = \psi_*(x;\theta_\tau)$ defined by (8.38) depends on the value of θ_τ. Constructing the optimal ψ-estimator $\hat{\theta}_\tau^*$, which is going to be called the ψo-estimator, requires solving the following nonlinear equation

$$\hat{\theta}_\tau^* = G(\hat{\theta}_\tau^*), \quad G(\theta) = f_{\psi_*}^{-1}\left(\frac{N_T^{-1}}{(1-\varepsilon)^2}\sum_{t=1}^{T-\tau}\psi_*\left(\frac{z_t}{z_{t+\tau}};\theta\right)o_t o_{t+\tau}\right). \tag{8.39}$$

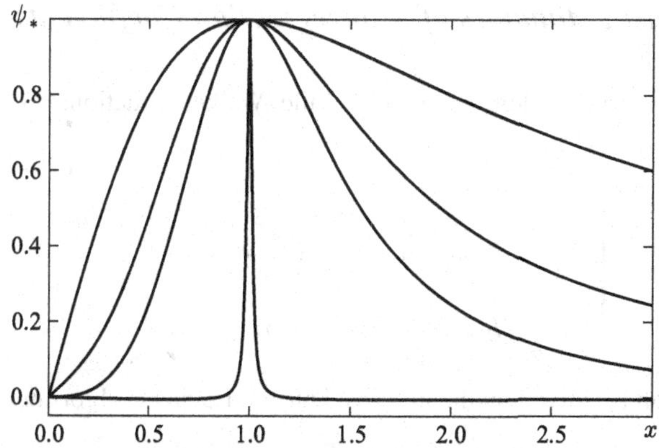

Fig. 8.5 The function $\psi_*(x;\theta)$ plotted for $\theta = 0; 0.7; 0.9; 0.9999$ (from *top* to *bottom*). Modified from [12]. Published with kind permission © Elsevier 2011. All Rights Reserved

By inversion, (8.39) can be represented in the form

$$f_{\psi_*(\cdot;\theta)}(\theta) = \frac{\theta}{\sqrt{1-\theta^2}(1+\sqrt{1-\theta^2})} = f_{\psi_*(\cdot;\theta)}(G(\theta)) =$$

$$= \frac{N_T^{-1}}{(1-\varepsilon)^2} \sum_{t=1}^{T-\tau} \psi_* \left(\frac{z_t}{z_{t+\tau}}; \theta \right) o_t o_{t+\tau}.$$

Since the right-hand side of this equation is bounded, and the left-hand side can assume any real value, this equation has a solution

$$\theta = \hat{\theta}_\tau^* \in [-1, 1]$$

which can be found by applying one of the classical numerical algorithms.

It can easily be shown that the Fisher information on the parameter θ_τ contained in the ratio $y_t/y_{t+\tau}$ is 50 % smaller compared to the Fisher information contained in the vector $(y_t, y_{t+\tau})$:

$$I_{y_t/y_{t+\tau}}(\theta_\tau) = \frac{1}{2(1-\theta_\tau^2)^2}, \quad I_{(y_t, y_{t+\tau})}(\theta_\tau) = \frac{1}{(1-\theta_\tau^2)^2}.$$

This reduction of the Fisher information obviously leads to reduced efficiency of the ψo-estimator compared to the ML-estimator in the absence of outliers, which is due to the invariance of the ψo-estimator w.r.t. the scale parameter σ. We require this invariance since it is currently impossible to construct consistent estimators $\hat{\sigma}$ in *AR* models under simultaneous influence of outliers and missing values [34].

8.7.4 Robust Estimators of Autoregression Coefficients

Let us consider the system of $p + 1$ Yule–Walker equations in $\{b_i\}$ for the autoregression model (8.22):

$$\begin{cases} \theta_0 + b_1\theta_1 + \cdots + b_p\theta_p = \sigma^2/\sigma_0, \\ \theta_{-1} + b_1\theta_0 + \cdots + b_p\theta_{p-1} = 0, \\ \quad \cdots \\ \theta_{-p} + b_1\theta_{1-p} + \cdots + b_p\theta_0 = 0. \end{cases}$$

This system can be solved by the iterative Durbin–Levinson algorithm [17]:

$$\begin{aligned} \hat{\phi}_{m,m} &= \frac{\hat{\theta}_m - \sum_{i=1}^{m-1} \hat{\theta}_{m-i}\hat{\phi}_{m-1,i}}{1 - \sum_{i=1}^{m-1} \hat{\theta}_i\hat{\phi}_{m-1,i}}, \\ \hat{\phi}_{m,i} &= \hat{\phi}_{m-1,i} - \hat{\phi}_{m,m}\hat{\phi}_{m-1,m-i}, \quad 1 \le i \le m-1, \\ \hat{b}_i &= -\hat{\phi}_{p,i}, \quad i = 1, 2, \ldots, p. \end{aligned} \tag{8.40}$$

Thus, autoregression coefficients $b = (b_i)$ can be estimated as follows:

1. If the outlier probability ε is unknown, then construct its estimator $\hat{\varepsilon}$ (see Sect. 8.7.5 below).
2. Pick a suitable ψ-estimator from the estimators defined in Sect. 8.7.2 and the optimal ψ_O-estimator defined by (8.39). Apply Theorem 8.8 to estimate the correlations $\{\theta_\tau : \tau = 1, 2, \ldots, p\}$ by using either the a priori known ε or its estimator $\hat{\varepsilon}$.
3. Use the iterative procedure (8.40) to compute the estimators $\{\hat{b}_i\}$ for the autoregression coefficients.

Let us introduce the following notation:

$$\forall n \in \mathbb{N}, \quad \Xi(n) = \left(\Xi_{ij}(n)\right) \in \mathbb{R}^{(n+1)\times(n+1)}, \quad \Xi_{ij}(n) = \theta_{i-j}.$$

Theorem 8.12. *Under the conditions of Theorem 8.9, the following convergence is satisfied:*

$$\mathbb{P}\left(||\hat{b} - b|| \le K\,||\hat{\theta} - \theta||\right) \underset{T\to\infty}{\longrightarrow} 1,$$

$$K = \frac{1}{\lambda_{min}(\Xi(p-1))}\left(1 + \sqrt{\frac{2p}{\pi}\int_{-\pi}^{\pi} |B(e^{ix})|^2 dx}\right),$$

where $\hat{\theta} = (\hat{\theta}_i)$ *and* $B(\lambda) = \lambda^p + b_1\lambda^{p-1} + \cdots + b_p$ *is the characteristic polynomial of the difference equation* (8.22)*;* $\lambda_{min}(Q)$ *is the minimal eigenvalue of a matrix* Q*.*

This theorem was proved in [12].

8.7.5 Estimation of the Contamination Level ε

The proposed ψ-estimators (8.34) rely on prior knowledge of the contamination level ε. Let us propose an approach for statistical estimation of ε. Assume, in addition, that the outliers are normally distributed: $\mathcal{L}\{\upsilon_t\} = \mathcal{N}_1(0, \beta_1\sigma^2)$, where $\beta_1 > 0$ determines the outlier magnitude. Under this assumption, the probability distribution of $\{z_t\}$ in the RO-model (8.23) is a mixture of two normal distributions with the mean values equal to zero and the variances $\sigma_1^2 = \mathbb{D}\{\upsilon_t\} = \sigma^2\beta_1$, $\sigma_2^2 = \mathbb{D}\{y_t\} = \sigma^2\beta_2$, where $\beta_2 = (2\pi)^{-1}\int_{-\pi}^{\pi} |B(e^{ix})|^{-2}dx$. Therefore, the corresponding characteristic function is the following mixture:

$$f_z(\lambda) = \varepsilon \exp\left\{-\sigma_1^2\lambda^2/2\right\} + (1-\varepsilon)\exp\left\{-\sigma_2^2\lambda^2/2\right\}, \quad \lambda \in [-\pi, \pi].$$

For large outliers ($\beta \gg 1$) we have $\beta_1 = \beta^2\beta_2 \gg \beta_2$, and the approximation

$$f_z(\lambda) \approx (1-\varepsilon)\exp\left\{-\sigma_2^2\lambda^2/2\right\} \tag{8.41}$$

holds for sufficiently large values of $|\lambda|$. The larger the outlier magnitude β, the higher the accuracy of (8.41). Note that in the AO-model (8.25), the characteristic function can be written as

$$f_z(\lambda) = (1-\varepsilon)\exp\left\{-\sigma_2^2\lambda^2/2\right\}\left(1 + \varepsilon\exp\left\{-\sigma_1^2\lambda^2/2\right\}\right),$$

and thus, the approximation (8.41) remains valid.

Further, using (8.41), we have:

$$f_z(\lambda_2)/f_z(\lambda_1) \approx \exp\left\{-\sigma_2^2\left(\lambda_2^2 - \lambda_1^2\right)/2\right\}. \tag{8.42}$$

Let us fix some values of λ_1 and λ_2, $\lambda_1 \neq \lambda_2$. The numerator and the denominator in the left-hand side of (8.42) can be estimated separately by using the sample characteristic function, thus leading to an estimator for the parameter σ_2 present in the right-hand side. Substituting this estimator into (8.41) in place of σ_2 yields an equation that can be used to obtain an estimator for ε. Thus, estimation of ε consists of four steps [12]:

1. Choose λ_1 and λ_2, $\lambda_1 \neq \lambda_2$.
2. Evaluate the sample characteristic function at the chosen points:

$$\hat{f}_z(\lambda_i) = \frac{\sum_{t=1}^{T} o_t \cos(\lambda_i z_t)}{\sum_{t=1}^{T} o_t}, \quad i = 1, 2.$$

3. Construct an estimator for σ_2^2:

$$\hat{\sigma}_2^2 = 2\left(\lambda_1^2 - \lambda_2^2\right)^{-1} \ln\left(\hat{f}_z(\lambda_2) \,/\, \hat{f}_z(\lambda_1)\right).$$

4. Substitute it into (8.41):

$$\sigma_2 ::= \hat{\sigma}_2, \qquad \lambda ::= \hat{\sigma}_2^{-1},$$

and compute

$$\hat{\varepsilon} = 1 - \sqrt{e}\, \hat{f}_z\left(\hat{\sigma}_2^{-1}\right).$$

The constructed algorithm includes two tuning parameters, λ_1 and λ_2.

We propose the following experimentally obtained optimal (in the sense of minimal mean square error of $\hat{\varepsilon}$) values of these parameters [12]:

$$\lambda_1^2 = \frac{\sum_{t=1}^{T} o_t}{\sum_{t=1}^{T} o_t z_t^2}, \qquad \lambda_2 = 2\lambda_1.$$

8.7.6 *Simulation-Based Performance Evaluation of the Constructed Estimators and Forecasting Algorithms*

Performance of the constructed estimators was evaluated by two series of computer simulations. The first series was based on $AR(5)$ models, $p = 5$, under RO-outliers (8.23) or AO-outliers (8.25) and missing values (8.26) [12].

The contamination distribution was normal: $\mathcal{L}\{v_t\} = \mathcal{N}_1\left(0, \beta^2\mathbb{D}\{y_t\}\right)$ with different values of the outlier magnitude β. The missing values (8.26) were generated by periodically repeating the pattern $\tilde{o} = \{1011101\}$ in the index sequence o_t (thus, roughly 30 % of the elements were missing in each of the observed time series). Each simulation experiment consisted of 10^4 Monte-Carlo simulation rounds.

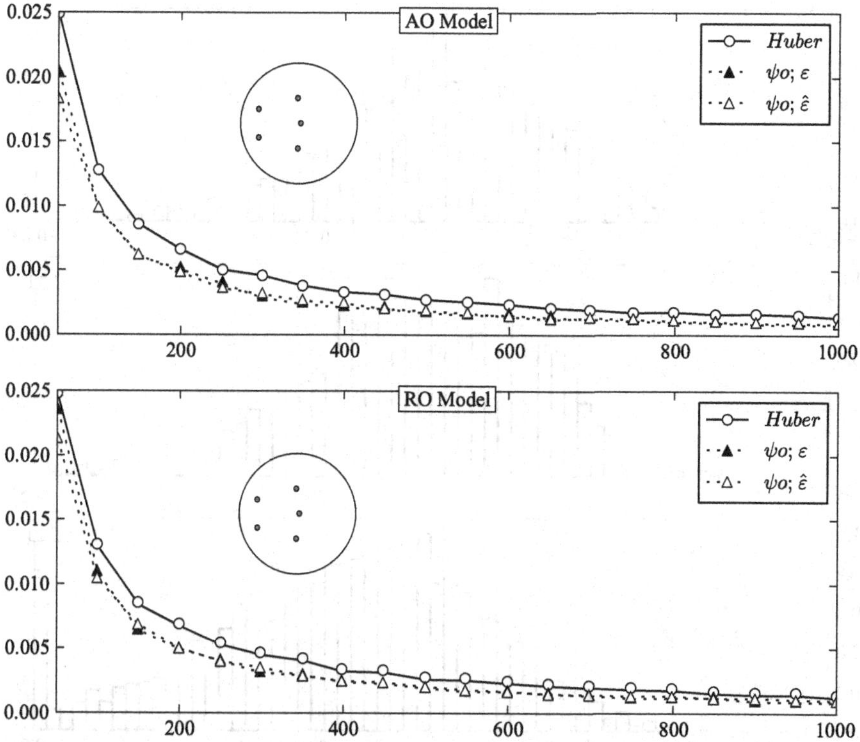

Fig. 8.6 Mean square error $\mathbb{E}\{||\hat{b}-b||^2\}$ for $\beta = 5$, $\varepsilon = 0.1$. Modified from [12]. Published with kind permission © Elsevier 2011. All Rights Reserved

Three robust estimates for autoregression coefficients $\{b_i\}$ were computed for each simulated time series: the Huber estimate based on (8.29), the ψo-estimate based on (8.38) under prior knowledge of ε, and the ψo-estimate for an unknown ε.

Results of the first series of simulations are presented in Figs. 8.6–8.10. Histograms obtained by estimating θ_1 under the RO-model are presented in Fig. 8.7. The true value $\theta_1 = -0.9$ is shown by thin vertical lines. In Figs. 8.6, 8.9, 8.10, the mean square error $\mathbb{E}\{||\hat{b} - b||^2\}$ is plotted against, respectively, T, ε, and β. Figure 8.8 illustrates the dependence between the MSE $\mathbb{E}\{||\hat{b} - b||^2\}$ and T under missing values only ($\varepsilon = 0$, no outliers were present in the simulated time series); the traditional LS-estimator is used for comparison.

AR coefficients $\{b_i\}$ are presented in Figs. 8.6–8.10 as roots of the characteristic polynomial $B(\lambda) = \lambda^p + b_1\lambda^{p-1} + \cdots + b_p$ in relation to the unit circle.

Figure 8.7 shows that applying the ψo-estimator $\hat{\theta}_1$ results in a smaller bias compared to the Huber estimator. The plot in Fig. 8.8 indicates that in a setting without outliers, the efficiency of the ψo-estimator is only slightly lower when evaluated against the optimal LS-estimator. In Fig. 8.9, the performance of the ψo-estimator for an unknown ε is compared to the case where ε is a priori known.

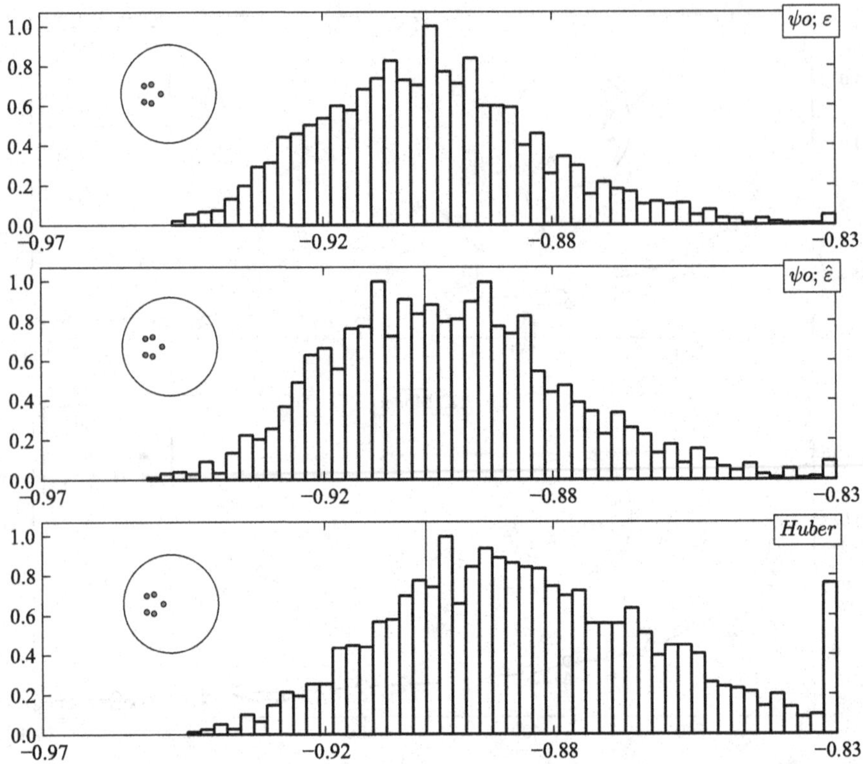

Fig. 8.7 Histograms of $\hat{\theta}_1$. Modified from [12]. Published with kind permission © Elsevier 2011. All Rights Reserved

The efficiency loss resulting from estimation of ε is shown to be small for $\varepsilon \leq 0.2$. Figure 8.6 illustrates sufficiently good performance of the ψo-estimators under both AO- and RO-models. Figure 8.10 shows that the performance of the ψo-estimator under unknown ε is stable w.r.t. the outlier magnitude β.

The second series of simulations [12] was based on a well-known real-world dataset—"Numbers of weekly births in Quebec from May, 15, 1981 till February, 13, 1983" from the Time Series Data Library [9]. This data was centered w.r.t. the sample mean, and then the set of 100 observations was split into two subsets. First $T = 90$ values $\{y_t\}$ were artificially contaminated following the distortion model (8.23)–(8.25) with $\varepsilon = 0.1$, $\beta = 2$. The contaminated data was used to estimate $p = 10$ coefficients of an $AR(10)$ model by applying the ψt-estimator (assuming a priori unknown ε) and the Huber estimator.

Based on the calculated estimates of autoregression coefficients, plug-in forecasts (constructed similarly to Sect. 7.3.2) were computed for the remaining time points $t = 91, 92, \ldots, 100$ and compared to the actual values from the second subset of the data.

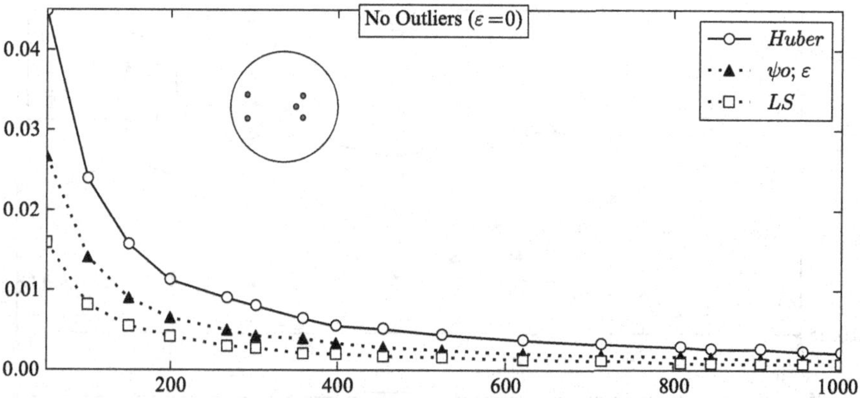

Fig. 8.8 Mean square error $\mathbb{E}\{||\hat{b} - b||^2\}$ for $\varepsilon = 0$. Modified from [12]. Published with kind permission © Elsevier 2011. All Rights Reserved

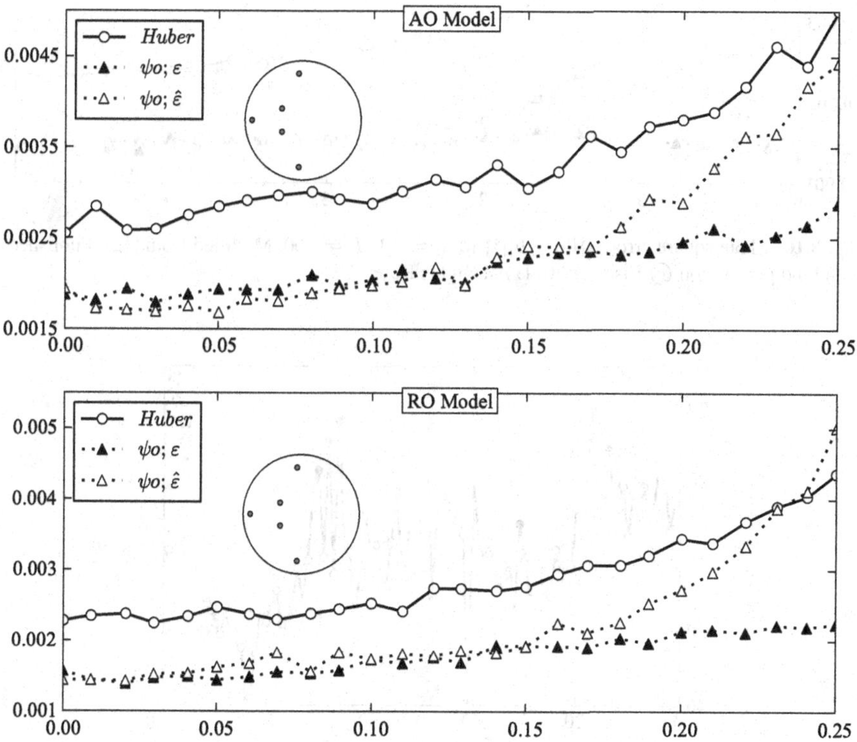

Fig. 8.9 Mean square error $\mathbb{E}\{||\hat{b} - b||^2\}$ for $\beta = 5$, $T = 500$. Modified from [12]. Published with kind permission © Elsevier 2011. All Rights Reserved

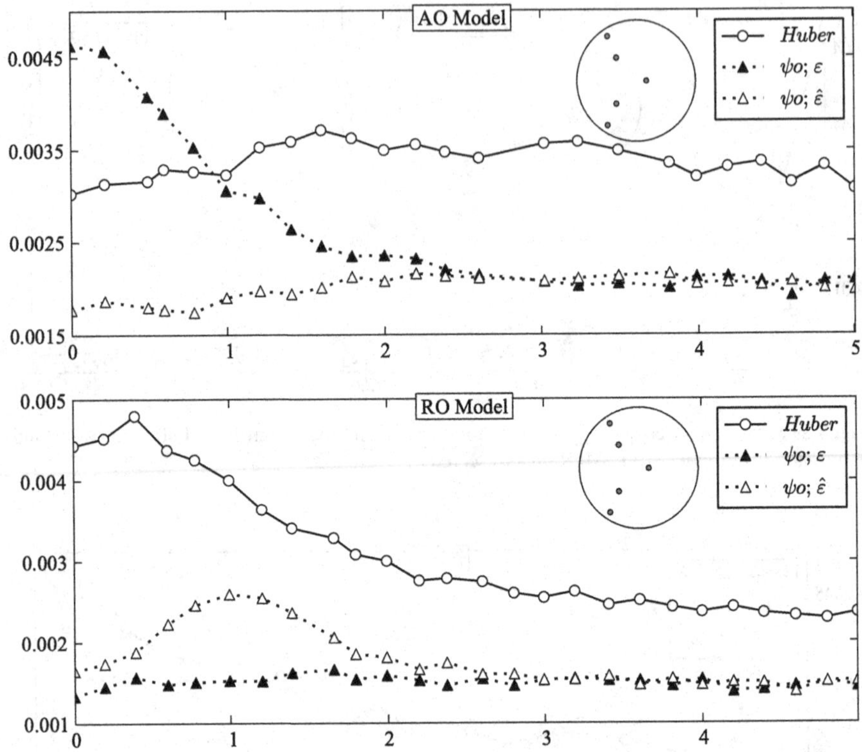

Fig. 8.10 Mean square error $\mathbb{E}\{||\hat{b} - b||^2\}$ for $\varepsilon = 0.1$, $T = 500$. Modified from [12]. Published with kind permission © Elsevier 2011. All Rights Reserved

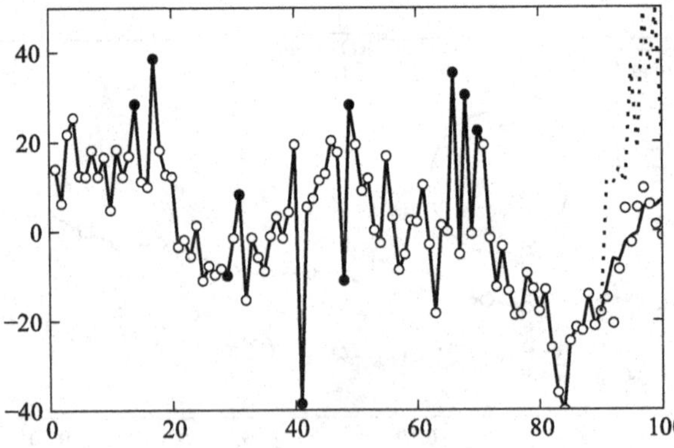

Fig. 8.11 The ten-step linear forecasts by the ψt-estimator under unknown ε (*solid line*) and by the Huber estimator (*dashed line*). Modified from [12]. Published with kind permission © Elsevier 2011. All Rights Reserved

In Fig. 8.11, the uncontaminated data is shown as empty circles; the outliers—as filled circles; the observed time series for $t \in \{1, 2, \ldots, 90\}$ and the forecasts based on the ψt-estimator—as a solid line; the forecasts based on the Huber estimator—as a dashed line. It is easy to see that, for this dataset, the forecasting procedure based on the ψt-estimator is more accurate.

References

1. Anderson, T.: The Statistical Analysis of Time Series. Wiley, New York (1994)
2. Atkinson, A., Riani, M.: Robust Diagnostic Regression Analysis. Springer, New York (2000)
3. Blomqvist, N.: On a measure of dependence between two random variables. Ann. Math. Stat. **21**, 593–600 (1950)
4. Davies, P., Gather, U.: Robust statistics. In: Handbook of Computational Statistics, pp. 655–695. Springer, Berlin (2004)
5. Genton, M.: Highly robust estimation of the autocovariance function. J. Time Anal. **21**(6), 663–684 (2000)
6. Greene, W.: Econometric Analysis. Macmillan, New York (2000)
7. Hampel, F., Ronchetti, E., Rousseeuw, P., Stahel, W.: Robust Statistics: The Approach Based on Influence Functions. Wiley, New York (1986)
8. Huber, P.: Robust Statistics. Wiley, New York (1981)
9. Hyndman, R.: Time series data library. http://www.robhyndman.info/TSDL (2008)
10. Kharin, Y., Huryn, A.: Plug-in statistical forecasting of vector autoregressive time series with missing values. Austrian J. Stat. **34**(2), 163–174 (2005)
11. Kharin, Y., Huryn, A.: Sensitivity analysis of the risk of forecasting for autoregressive time series with missing values. Pliska Studia Math. Bulg. **17**, 137–146 (2005)
12. Kharin, Y., Voloshko, V.: Robust estimation of AR coefficients under simultaneously influencing outliers and missing values. J. Stat. Plan. Inference **141**(9), 3276–3288 (2011)
13. Kohn, R., Ansley, C.: Estimation, prediction and interpolation for ARIMA models with missing data. J. Acoust. Soc. Am. **81**(395), 751–761 (1986)
14. Little, R., Rubin, D.: Statistical Analysis with Missing Data. Wiley, New York (1987)
15. Lutkepohl, H.: Introduction to Multiple Time Series Analysis. Springer, Berlin (1993)
16. Maejima, M.: A non-uniform estimate in the central limit theorem for m-dependent random variables. KEIO Eng. Rep. **31**(2), 15–20 (1978)
17. Maronna, R., Martin, R., Yohai, V.: Robust Statistics: Theory and Methods. Wiley, New York (2006)
18. Maronna, R., Zamar, R.: Robust multivariate estimates for high-dimensional datasets. Technometrics **44**, 307–317 (2002)
19. McLachlan, G., Krishnan, T.: The EM Algorithm and Extensions. Wiley, New York (1997)
20. Morgenthaler, S.: A survey of robust statistics. Stat. Meth. Appl. **15**, 271–293 (2007)
21. Nijman, T., Palm, F.: Parameter identification in ARMA processes in the presence of regular but incomplete sampling. J. Time Anal. **11**(3), 219–238 (1990)
22. Pantula, S., Shin, D.: Testing for a unit root in ar processes with systematic but incomplete sampling. Stat. Probab. Lett. **18**, 183–190 (1993)
23. Roth, P., Switzer, F., Switzer, D.: Missing data in multiple item scales: a Monte Carlo analysis of missing data techniques. Organ. Res. Meth. **2**(3), 211–232 (1999)
24. Rousseeuw, P., Leroy, A.: Robust Regression and Outlier Detection. Chapman & Hall, London (1987)
25. Rubin, D.: Multiple Imputation for Nonresponse in Surveys. Wiley, New York (1997)
26. Schafer, J.: Analysis of Incomplete Multivariate Data. Chapman & Hall, New York (1997)
27. Serfling, R.: Approximation Theorems of Mathematical Statistics. Wiley, New York (1980)

28. Shergin, V.: Estimate of the error term in the central limit theorem for m-dependent random variables. Lietuvos Matematikos Rinkinys **16**(4), 245–250 (1976)
29. Shevlyakov, G., Vilchevski, N.: Robustness Data Analysis: Criteria and Methods. VSP, Vilnius (1997)
30. Shiryaev, A.: Probability. Springer, New York (1996)
31. Stockinger, N., Dutter, R.: Robust time series analysis: a survey. Kybernetika **23**, 1–90 (1987)
32. Tong, H.: Non-linear Time Series. Clarendon Press, Oxford (1993)
33. Van Aelst, S., Yohai, V., Zamar, R.: Propagation of outliers in multivariate data. Ann. Stat. **37**(1), 311–331 (2009)
34. Varmuza, K., Filzmoser, P.: Introduction to Multivariate Statistical Analysis in Chemometrics. Taylor & Francis, Boca Raton (2009)
35. Visuri, S., Koivunen, V., Oja, H.: Sign and rank covariance matrices. J. Stat. Plann. Infer. **91**, 557–575 (2000)
36. Yule, G.: On a method for investigating periodicities in disturbed series, with special references to Wolfer's sunspot numbers. Phil. Trans. Roy. Soc. Lond. **226**, 267–298 (1927)

Chapter 9
Robustness of Multivariate Time Series Forecasting Based on Systems of Simultaneous Equations

Abstract Systems of simultaneous equations (SSE) are a well-studied class of multivariate time series models with applications to macroeconometrics. In this chapter, we analyze robustness of forecasting statistics based on the least squares method and its modifications under model specification errors and ε-drift of model coefficients.

9.1 Systems of Simultaneous Equations

9.1.1 SSE Model

As mentioned in Sect. 3.5.5, systems of simultaneous equations (SSE) are usually applied in econometrics when modeling and forecasting such complex economic systems as national economies [3, 4], major economic unions such as the EU, and other large economic conglomerates. SSEs are also applied in engineering, microeconomics, and social studies [2, 4, 8, 10, 15].

Following econometric conventions, the dependent (interior) variables will be called *endogenous*, and the independent (exterior) variables—*exogenous*. Time lagged endogenous variables, as well as current (corresponding to the current time) and lagged exogenous variables, are called *predetermined* variables.

A linear *system of simultaneous equations* SSE(N, K) is defined as the following system of N stochastic equations [3]:

$$y_{jt} = \sum_{l \in A_j} \alpha^0_{lj} y_{lt} + \sum_{k \in B_j} \beta^0_{kj} x_{kt} + \xi_{jt}, \quad j = 1, \ldots, N, \ t = 1, \ldots, n. \tag{9.1}$$

Here $y_{lt} \in \mathbb{R}^1$ is the value of the lth endogenous variable at time t; $x_{kt} \in \mathbb{R}^1$ is the value of the kth predetermined variable at time t; $\{\alpha^0_{lj}\}, \{\beta^0_{kj}\}$ are (generally unknown) true values of the model parameters; $\xi_{jt} \in \mathbb{R}^1$ is a random variable which

Y. Kharin, *Robustness in Statistical Forecasting*, DOI 10.1007/978-3-319-00840-0_9,

is going to be characterized later; $A_j = \{l_1, \dots, l_{N_j}\} \subseteq \{1, \dots, N\}$ is the set of N_j endogenous variables (excepting y_{jt}) which are present in the jth equation,

$$|A_j| = N_j, \quad 0 \leq |A_1 \cup A_2 \cup \dots \cup A_N| \leq N;$$

N is the number of equations (and also endogenous variables) in the SSE; $B_j = \{k_1, \dots, k_{K_j}\} \subseteq \{1, \dots, K\}$ is the set of K_j predetermined variables included in the jth equation:

$$|B_j| = K_j, \quad 0 \leq |B_1 \cup B_2 \cup \dots \cup B_N| = K;$$

K is the total number of predetermined variables included in the SSE; n is the length of the observation period.

The system of equations (9.1) is referred to as *the structural form of the SSE*. Presented below are several other equivalent representations of SSE(N, K).

Let us introduce the following matrix notation:

$$y^j = \begin{pmatrix} y_{j1} \\ \vdots \\ y_{jn} \end{pmatrix}, \quad x^k = \begin{pmatrix} x_{k1} \\ \vdots \\ x_{kn} \end{pmatrix}, \quad \xi^j = \begin{pmatrix} \xi_{j1} \\ \vdots \\ \xi_{jn} \end{pmatrix} \in \mathbb{R}^n,$$

$$Y^j = (y^{l_1} \vdots \dots \vdots y^{l_{N_j}}) \in \mathbb{R}^{n \times N_j}$$

is the $(n \times N_j)$-matrix of endogenous variables in the right-hand side of the jth equation of (9.1);

$$X^j = (x^{k_1} \vdots \dots \vdots x^{k_{K_j}}) \in \mathbb{R}^{n \times K_j}$$

is the $(n \times K_j)$-matrix of predetermined variables in the jth equation (9.1);

$$\alpha_j^0 = (\alpha_{l_1 j}^0 \ \dots \ \alpha_{l_{N_j} j}^0)' \in \mathbb{R}^{N_j}, \quad \beta_j^0 = (\beta_{k_1 j}^0 \ \dots \ \beta_{k_{K_j} j}^0)' \in \mathbb{R}^{K_j}$$

are, respectively, N_j- and K_j-vectors of the unknown true parameter values. Then the jth equation of the system (9.1) can be rewritten in a matrix form:

$$y^j = Y^j \alpha_j^0 + X^j \beta_j^0 + \xi^j, \quad j = 1, \dots, N. \tag{9.2}$$

Let us introduce additional matrix notation: $A^0 = (\alpha_{lj}^0)$ and $B^0 = (\beta_{kj}^0)$ are, respectively, $(N \times N)$- and $(K \times N)$-matrices, where

$$\alpha_{lj}^0 = 0 \ \text{ if } \ l \notin A_j, \quad \beta_{kj}^0 = 0 \ \text{ if } \ k \notin B_j, \quad j = 1, \dots, N;$$

$Y = (y^1 \vdots \dots \vdots y^N)$ is the $(n \times N)$-matrix of all endogenous variables in the SSE; $X = (x^1 \vdots \dots \vdots x^K)$ is the $(n \times K)$-matrix of all predetermined variables in the SSE;

$\Xi = (\xi^1 \vdots \ldots \vdots \xi^N)$ is the $(n \times N)$-matrix of random variables in the SSE. Then the structural matrix form of the SSE (9.2) is defined as follows [3]:

$$Y = YA^0 + XB^0 + \Xi. \tag{9.3}$$

Introducing the following notation for, respectively, the vectors of endogenous, predetermined, and the random variables at time t,

$$y_t = (y_{1t} \ldots y_{Nt})' \in \mathbb{R}^N, \quad x_t = (x_{1t} \ldots x_{kt})' \in \mathbb{R}^K, \quad \xi_t = (\xi_{1t} \ldots \xi_{Nt})' \in \mathbb{R}^N,$$

the system (9.3) can be rewritten in another equivalent form:

$$y_t = A^{0'} y_t + B^{0'} x_t + \xi_t, \quad t = 1, \ldots, n. \tag{9.4}$$

The model SSE(N, K) is called *complete* if $|\mathbf{I}_N - A^0| \neq 0$. In this case, denoting

$$\theta^0 = B^0(\mathbf{I}_N - A^0)^{-1} \in \mathbb{R}^{K \times N}, \quad U = \Xi(\mathbf{I}_N - A^0)^{-1} = (u_1 \vdots \ldots \vdots u_n)' \in \mathbb{R}^{n \times N},$$

and solving (9.3), (9.4) in Y and $\{y_t\}$ yields the *reduced form of an SSE* [3]:

$$Y = X\theta^0 + U, \tag{9.5}$$

$$y_t = \theta^{0'} x_t + u_t, \quad t = 1, \ldots, n. \tag{9.6}$$

Observe that this form coincides with the definition of a multivariate linear regression (cf. Chap. 5).

The random variables $\{\xi_{j1}, \ldots, \xi_{jn}\}$ are assumed to be jointly independent and normally distributed:

$$\begin{aligned} \mathcal{L}\{\xi^j\} = \mathcal{N}_n(\mathbf{0}_n, \sigma_{jj}\mathbf{I}_n), \quad \mathbb{E}\{\xi^i {\xi^j}'\} = \sigma_{ij} I_n, \quad i, j = 1, \ldots, N, \\ \mathbb{E}\{\xi_t \xi_\tau'\} = \delta_{t\tau}\Sigma, \quad t, \tau = 1, \ldots, n, \end{aligned} \tag{9.7}$$

where $\Sigma = (\sigma_{ij}) \succeq \mathbf{0}$ is the $(N \times N)$ covariance matrix. Note that Σ can be singular since the initial form of the SSE may contain identities.

Recall that in Chap. 6 we have considered optimality and robustness of forecasting based on multivariate linear regression models. Although we have formally reduced the SSE(N, K) to a linear regression model (9.5), (9.6), a separate treatment of the model (9.1)–(9.4) is necessary because of the following reasons:

1. In econometrics, it is often important to estimate the matrices A^0 and B^0, and estimates of θ^0 (which is a function of A^0, B^0) are insufficient;
2. An SSE can contain identities, leading to singularity of the matrix Σ, in which case the established methods of regression analysis are inapplicable.

9.1.2 Example of an SSE: Klein's Model I

As an example of an SSE, let us consider the econometric model proposed by Lawrence R. Klein [3] to analyze the impact of government policies on the crucial indicators of the American economy. This dynamic model defines a dependence of six endogenous variables ($N = 6$) on three exogenous variables ($K = 3$). Note that this is an extremely simple model, which doesn't quite agree with the real-world data. In actuality, the endogenous variables of this model depend on many other quantities. Practically applicable macroeconomic models developed in the more recent years, such as the Wharton Econometric Forecasting Model or the Brukings model [3], contain hundreds of equations.

Let us describe the exogenous and endogenous variables (at time t), the equations and the identities of Klein's Model I.

The exogenous variables:

- $W2_t$ is the wage bill of the government sector;
- G_t is the government expenditure plus net exports;
- TX_t is the amount of business taxes.

The endogenous variables:

- C_t is the private consumption expenditure;
- $W1_t$ is the wage bill of the private sector;
- Π_t is the profits net of business taxes;
- I_t is the (net) private investment;
- K_t is the stock of (private) capital goods;
- Y_t is the gross national product.

The functional equations:

- Consumption function:

$$C_t = a_1 + a_2(W1_t + W2_t) + a_3\Pi_{t-1} + \xi_{1t};$$

- Investment function:

$$I_t = b_1 + b_2\Pi_t + b_3\Pi_{t-1} + b_4K_{t-1} + \xi_{2t};$$

- Labor resources demand function:

$$W1_t = c_1 + c_2(Y_t + TX_t - W2_t) + c_3(Y_{t-1} + TX_{t-1} - W2_{t-1}) + c_4t + \xi_{3t}.$$

The identities:

- Gross national product equation:

$$Y_t = C_t + I_t - G_t - TX_t;$$

- Profits equation:

$$\Pi_t = Y_t - (W1_t + W2_t);$$

- Capital stock equation:

$$K_t = K_{t-1} + I_t.$$

In [3], a three-step least squares method (3LS method) was used to estimate the parameters of Klein's model I, and the following representation of the model in the form (9.4) was obtained:

$$y_t = A^{0'} y_t + B^{0'} x_t + \xi_t,$$

where $y_t' = (C_t, I_t, W1_t, Y_t, \Pi_t, K_t)$ is the vector of endogenous variables and

$$x_t' = \big(1, W2_t, W2_{t-1}, \Pi_{t-1}, K_{t-1}, TX_t, TX_{t-1}, Y_{t-1}, G_t, t\big)$$

is the vector of predetermined variables; the matrices are defined by

$$A^0 = \begin{pmatrix} 0 & 0 & 0 & -1 & 0 & 0 \\ 0 & 0 & 0 & 1 & 0 & 1 \\ 0.7901 & 0 & 0 & 0 & -1 & 0 \\ 0 & 0 & 0.4005 & 0 & 1 & 0 \\ 0.1249 & -0.01308 & 0 & 0 & 0 & 0 \\ 0 & 0 & 0 & 0 & 0 & 0 \end{pmatrix},$$

$$B^0 = \begin{pmatrix} 16.44 & 28.18 & 15.08 & 0 & 0 & 0 \\ 0.7901 & 0 & -0.4005 & 0 & 1 & 0 \\ 0 & 0 & 0.1813 & 0 & 0 & 0 \\ 0.1631 & 0.7557 & 0 & 0 & 0 & 0 \\ 0 & -0.1948 & 0 & 0 & 0 & 1 \\ 0 & 0 & 0.1813 & -1 & 0 & 0 \\ 0 & 0 & 0.1813 & 0 & 0 & 0 \\ 0 & 0 & 0.1813 & 0 & 0 & 0 \\ 0 & 0 & 0 & 1 & 0 & 0 \\ 0 & 0 & 0.1497 & 0 & 0 & 0 \end{pmatrix}.$$

The sequence $\{\xi_t\}$ is composed of independent normal random vectors with zero mean values and a singular covariance matrix Σ:

$$\Sigma = \begin{pmatrix} 3.6 & 0.8 & 2.1 & 0 & 0 & 0 \\ & 2.0 & 1.4 & 0 & 0 & 0 \\ & & 6.0 & 0 & 0 & 0 \\ & & & 0 & 0 & 0 \\ & & & & 0 & 0 \\ & & & & & 0 \end{pmatrix}.$$

9.1.3 The Optimal Forecasting Statistic Under the SSE Model

First, let us consider a setting where the parameters $\{\alpha^0_{lj}\}$, $\{\beta^0_{kj}\}$ defining the matrices A^0, B^0 are a priori known. Let us construct an optimal forecasting statistic for $y_{n+1} \in \mathbb{R}^N$ (for simplicity, assume one-step-ahead forecasting, $\tau = 1$).

Theorem 9.1. *In the complete SSE(N, K) model defined by* (9.1), *given a vector of predetermined variables x_{n+1} and a priori known parameters $\{\alpha^0_{lj}\}$, $\{\beta^0_{kj}\}$, the mean square optimal forecasting statistic is linear:*

$$\hat{y}_{n+1} = \theta^{0'} x_{n+1}, \tag{9.8}$$

where the $(K \times N)$-matrix θ^0 is defined as $\theta^0 = B^0(\mathbf{I}_N - A^0)^{-1}$; the minimum mean square forecast risk equals

$$r_0 = \mathbb{E}\{|\hat{y}_{n+1} - y_{n+1}|^2\} = \operatorname{tr}\left((\mathbf{I}_N - A^0)^{-1}\Sigma(\mathbf{I}_N - A^{0'})^{-1}\right). \tag{9.9}$$

Proof. Since the model is complete, it can be rewritten in the reduced form (9.6):

$$y_{n+1} = \theta^{0'} x_{n+1} + u_{n+1}. \tag{9.10}$$

It is known (see Chap. 7) that the optimal forecasting statistic can be defined as the conditional expectation

$$\hat{y}_{n+1} = \mathbb{E}\{y_{n+1} \mid x_{n+1}, \ldots, x_1\}. \tag{9.11}$$

By (9.4), (9.6), (9.7), we have

$$\begin{aligned} u_t = (\mathbf{I}_N - A^0)^{-1}\xi_t, \qquad & \mathbb{E}\{u_t\} = \mathbf{0}_N, \\ & \mathbb{E}\{u_t u'_\tau\} = \delta_{t\tau}(\mathbf{I}_N - A^0)^{-1}\Sigma(\mathbf{I}_N - A^{0'})^{-1}. \end{aligned} \tag{9.12}$$

The equalities (9.10)–(9.12) imply (9.8), as well as the expression for the risk:

$$r_0 = \mathbb{E}\{u'_{n+1}u_{n+1}\} = \mathbb{E}\{\operatorname{tr}(u_{n+1}u'_{n+1})\} = \operatorname{tr}\left((\mathbf{I}_N - A^0)^{-1}\Sigma(\mathbf{I}_N - A^{0'})^{-1}\right),$$

which coincides with (9.9). □

9.2 Robustness of SSE-Based Forecasting Under Specification Errors

Let us consider the situation where the parameters of the SSE(N, K) model are distorted by specification errors, and instead of A^0, B^0, the matrices

$$\hat{A} = A^0 + \delta A, \quad \hat{B} = B^0 + \delta B, \tag{9.13}$$

are known, where δA and δB are, respectively $(N \times N)$ and $(K \times N)$ deterministic matrices characterizing the deviations from the true parameter values. A forecast is made using a statistic of the form (9.8):

$$\hat{y}_{n+1} = \hat{\theta} x_{n+1}, \quad \hat{\theta} = \hat{B}(\mathbf{I}_N - \hat{A})^{-1}; \tag{9.14}$$

we assume that under the specification errors present in the model, the SSE remains complete:

$$|\mathbf{I}_N - \hat{A}| \neq 0.$$

Theorem 9.2. *In the SSE model defined by* (9.1) *under specification errors* (9.13), *the bias and the mean square forecast risk of the statistic* (9.14) *used to forecast* $y_{n+1} \in \mathbb{R}^N$ *can be written as*

$$b = \mathbb{E}\{\hat{y}_{n+1} - y_{n+1}\} = D x_{n+1}, \tag{9.15}$$

$$r = \mathbb{E}\{|\hat{y}_{n+1} - y_{n+1}|^2\} = r_0 + x'_{n+1} D' D x_{n+1}, \tag{9.16}$$

where the $(N \times K)$*-matrix* D *is equal to*

$$D = \left(\mathbf{I}_N - A^{0T} - \delta A^T\right)^{-1}\left(B^{0T}\delta A^T\left(\mathbf{I}_N - A^{0T}\right)^{-1} + \delta B^T\right). \tag{9.17}$$

Proof. Let us find $\hat{y}_{n+1} - y_{n+1}$:

$$\hat{y}_{n+1} - y_{n+1} = \left(\left(\mathbf{I}_N - \hat{A}'\right)^{-1}\hat{B}' - \left(\mathbf{I}_N - A^{0'}\right)^{-1}B^{0'}\right)x_{n+1} - \left(\mathbf{I}_N - A^{0'}\right)^{-1}\xi_{n+1}.$$

Consider the following expression:

$$\hat{B}\left(\mathbf{I}_N-\hat{A}\right)^{-1}-B^0\left(\mathbf{I}_N-A^0\right)^{-1}=B^0\left(\left(\mathbf{I}_N-\hat{A}\right)^{-1}-\left(\mathbf{I}_N-A^0\right)^{-1}\right)+\delta B(\mathbf{I}_N-\hat{A})^{-1}.$$

Let us determine $\left(\mathbf{I}_N-\hat{A}\right)^{-1}$ by using the following matrix identity [5]:

$$(\mathbf{I}_m-F)^{-1}=\mathbf{I}_m+F(\mathbf{I}_m-F)^{-1},$$

where F is a nonsingular $(m\times m)$-matrix. Then

$$\begin{aligned}\left(\mathbf{I}_N-\hat{A}\right)^{-1}&=(\mathbf{I}_N-A^0-\delta A)=(\mathbf{I}_N-(\mathbf{I}_N-A^0)^{-1}\delta A)(\mathbf{I}_N-A^0)^{-1}=\\&=(\mathbf{I}_N+(\mathbf{I}_N-A^0)^{-1}\delta A(\mathbf{I}_N-(\mathbf{I}_N-A^0)^{-1}\delta A)^{-1})(\mathbf{I}_N-A^0)^{-1}=\\&=(\mathbf{I}_N-A^0)^{-1}+(\mathbf{I}_N-A^0)^{-1}\delta A(\mathbf{I}_N-A^0-\delta A)^{-1}.\end{aligned}$$

Thus, we have

$$\left(\mathbf{I}_N-\hat{A}\right)^{-1}-(\mathbf{I}_N-A^0)^{-1}=(\mathbf{I}_N-A^0)^{-1}\delta A(\mathbf{I}_N-A^0-\delta A)^{-1}$$

and

$$\hat{B}\left(\mathbf{I}_N-\hat{A}\right)^{-1}-B^0\left(\mathbf{I}_N-A^0\right)^{-1}=D',$$

where

$$D'=\left((\mathbf{I}_N-A^0)^{-1}\delta AB^0+\delta B\right)(\mathbf{I}_N-A^0-\delta A)^{-1},$$

which coincides with (9.17).

Therefore, we can write

$$\hat{y}_{n+1}-y_{n+1}=Dx_{n+1}+\left(\mathbf{I}_N-A^{0'}\right)^{-1}\xi_{n+1}.$$

This, in turn, leads to

$$\begin{aligned}b&=\mathbb{E}\{\hat{y}_{n+1}-y_{n+1}\}=Dx_{n+1}+\left(\mathbf{I}_N-A^{0'}\right)^{-1}\mathbb{E}\{\xi_{n+1}\}=Dx_{n+1},\\r&=\mathbb{E}\{(\hat{y}_{n+1}-y_{n+1})'(\hat{y}_{n+1}-y_{n+1})\}=\\&=x'_{n+1}D'Dx_{n+1}+\mathbb{E}\left\{\xi'_{n+1}\left(\mathbf{I}_N-A^0\right)^{-1}\left(\mathbf{I}_N-A^{0'}\right)^{-1}\xi_{n+1}\right\}=\\&=r_0+x'_{n+1}D'Dx_{n+1}.\end{aligned}$$

The last two expressions coincide, respectively, with (9.15) and (9.16). □

Note that if $\delta A = \mathbf{0}, \delta B = \mathbf{0}$, then (9.15), (9.16) are consistent with the statement of Theorem 9.1.

Corollary 9.1. *Under the conditions of Theorem 9.2, if $|x_{n+1}| \leq c$, then the risk instability coefficient equals*

$$\kappa_+ = \frac{c\lambda_{max}(D'D)}{r_0}.$$

Assume that δA and δB are matrices of the form

$$\delta A = \varepsilon_A \left(\mathbf{I}_N - A^0\right), \quad \delta B = \varepsilon_B B^0, \tag{9.18}$$

where $\varepsilon_A, \varepsilon_B \geq 0$ are quantities characterizing the level of specification error, and $\varepsilon_A \neq 1$ (this holds since the system is complete).

Corollary 9.2. *Under the conditions of Theorem 9.2, assuming* (9.18), *the bias and the mean square forecast risk can be written as*

$$b = D_\varepsilon x_{n+1}, \quad r = r_0 + x'_{n+1} D'_\varepsilon D_\varepsilon x_{n+1},$$

where

$$D_\varepsilon = \frac{\varepsilon_A + \varepsilon_B}{1 - \varepsilon_A} \left(\mathbf{I}_N - A^{0'}\right)^{-1} B^{0'}.$$

To conclude the subsection, let us note that for a given maximum forecast risk r_{max} we can estimate the admissible levels $(\varepsilon_A^*, \varepsilon_B^*)$ of specification errors (9.13), (9.18) corresponding to this maximum risk:

$$r_{max} - r_0 = \frac{(\varepsilon_A^* + \varepsilon_B^*)^2}{(1 - \varepsilon_A^*)^2} g,$$

where $g = x'_{n+1} B^0 \left(\mathbf{I}_N - A^0\right)^{-1} \left(\mathbf{I}_N - A^{0'}\right)^{-1} B^{0'} x_{n+1}$.

9.3 Plug-In Forecasting Statistics in the SSE Model

In applications, the parameters $\{\alpha_{ij}^0\}$ (or A^0), $\{\beta_{kj}^0\}$ (or B^0), and Σ of the SSE(N, K) model defined by (9.1) are usually unknown, making it impossible to construct the optimal forecasting statistic (9.8). Since the dimensions N, K of the models used to solve modern applied problems are usually quite high, and joint ML estimation and forecasting is highly computationally intensive, construction of the forecasting statistics is usually split into two stages [3]:

1. Identification of the SSE to construct statistical estimators $\hat{A}$, $\hat{B}$, $\hat{\Sigma}$ based on an observed realization $y_1, \ldots, y_n$;
2. Computing the statistical estimator $\hat{\theta} = \hat{B}(\mathbf{I}_N - \hat{A})^{-1}$ for the matrix θ^0 and constructing a plug-in forecasting statistic

$$\hat{y}_{n+1} = \hat{\theta}' x_{n+1}. \tag{9.19}$$

If $\hat{\theta}$ is a consistent estimator (for $n \to +\infty$), then the forecasting statistic (9.19) is also consistent:

$$r_n = \mathbb{E}\{|\hat{y}_{n+1} - y_{n+1}|^2\} \to r_0.$$

Known methods of SSE identification can be divided into two classes [3]. The methods in the first class rely on statistical analysis of every equation in the SSE independently from the other equations. This class includes such methods as the two-step least squares method (2LS method) and the ML method with limited information.

1. *2LS method.* The least squares method for multiple linear regression is applied to each of the equations, where the endogenous variables in the right-hand sides are replaced by their least squares forecasts. Under the hypothetical assumptions, this yields consistent asymptotically unbiased estimators [3].
2. *ML method with limited information.* For the jth equation in the system, where $j = 1, \ldots, N$, under a normality assumption on the random observation error a log likelihood function is constructed for every time point $t \in \{1, 2, \ldots, n\}$. This function depends on the corresponding blocks of the matrices A^0, B^0, Σ; it is maximized numerically. The resulting estimators are consistent, asymptotically unbiased, and asymptotically equivalent to the 2LS estimators [3].

The second class contains three so-called systemic estimation methods: the 3LS method, the ML method with complete information, and the approximation method.

1. *3LS method.* The estimators $\hat{A}$, $\hat{B}$, $\hat{\Sigma}$ are constructed by applying the 2LS method. Then all of the identities are removed from the system, and generalized LS method is used to estimate all of the parameters of the system, substituting an estimator $\hat{\Sigma}$ for the unknown covariance error matrix Σ. The resulting estimators are consistent and asymptotically unbiased [3].
2. *ML method with complete information.* Assuming normality of the random errors $\{\xi_t\}$, a log likelihood function for the sample $\{y_t\}$ is constructed and maximized. The resulting estimators are consistent, asymptotically unbiased, and asymptotically efficient. It can be shown that they are asymptotically equivalent to the 3LS estimators [3].

3. *The approximation method for SSE identification* [2]. The method lies in finding such estimators $\{\tilde{\alpha}_j, \tilde{\beta}_j\}$ that the matrix of "adjusted values"

$$\tilde{Y} = (\tilde{y}_{jt}) = X\tilde{B}(\mathbf{I}_N - \tilde{A})^{-1}$$

is the l_2-optimal approximation for the matrix of recorded experimental data $Y = (y_{jt})$:

$$\sum_{j=1}^{N}\sum_{t=1}^{n}\left(\tilde{y}_{jt} - y_{jt}\right)^2 \to \min_{\{\tilde{\alpha}_j, \tilde{\beta}_j\}}. \tag{9.20}$$

An exact solution of the optimization problem (9.20) is hard to obtain due to nonlinear dependence of the objective function on the SSE coefficients. In [2], it was proposed to approximate the objective function in (9.20) by a different objective function, which can be easily minimized:

$$\sum_{j=1}^{N}\sum_{k=1}^{K}\left(\sum_{t=1}^{n} x_{kt}\left(y_{jt} - \left(Y^j\tilde{\alpha}_j - X^j\tilde{\beta}_j\right)_t\right)\right)^2 \to \min_{\{\tilde{\alpha}_j, \tilde{\beta}_j\}}. \tag{9.21}$$

In the paper [9], it was shown that the estimators $\{\tilde{\alpha}_j, \tilde{\beta}_j\}$ found as explicit solutions of the minimization problem (9.21) are consistent, and asymptotic behavior of the bias and the mean square error of these estimators, as well as the forecast risk of the statistic (9.19), were investigated under Tukey–Huber distortions of $\{\xi_t\}$.

9.4 Asymptotic Properties of the Least Squares Estimator Under Drifting Coefficients

9.4.1 Drifting Coefficient Models for SSEs

As mentioned in Sect. 9.1, SSEs (9.1) are widely used in econometric modeling due to their ability to account for the relations between endogenous and exogenous variables not only within the same equation but also across different equations. Presently, some researchers are voicing their concern about the inadequacy of the classical SSE models, where the coefficients $\{\alpha_{lj}, \beta_{kj}\}$ are constant and independent of time, for modeling real-world economic data which is often subject to *parameter drift* (changes in model parameters as time passes). Influence of parameter drift on several statistical techniques was analyzed in [6, 7, 11, 13, 14]; however, the author was the first to evaluate the effect of drifting parameters on statistical forecast risk under SSE models.

Let us emphasize the p lagged endogenous variables in the reduced SSE form (9.4) by introducing the notation

$$(\mathbf{I}_N - A^{0'})y_t - B^0 x_t ::= -A_0 y_t - A_1 y_{t-1} - \cdots - A_p y_{t-p} - B x_t, \quad \xi_t ::= u_t.$$

We obtain

$$\sum_{i=0}^{p} A_i y_{t-i} + Bx_t = u_t, \quad t \in \mathbb{Z}, \tag{9.22}$$

where $y_t = (y_{ti}) \in \mathbb{R}^N$ and $x_t = (x_{tj}) \in \mathbb{R}^K$ are column vectors of, respectively, endogenous and deterministic exogenous variables at time t; the $(N \times N)$-matrix A_i represents the coefficients in front of the endogenous variables at time $(t - i)$; the $(N \times K)$-matrix B represents the coefficients in front of the exogenous variables at time t; $\{u_t\}$ are jointly independent identically distributed random vectors (the innovation process):

$$\mathbb{E}\{u_t\} = \mathbf{0}_N, \quad \mathbf{Cov}\{u_t, u_\tau\} = \mathbb{E}\{u_t u_\tau'\} = \delta_{t\tau}\Sigma. \tag{9.23}$$

In the theory of statistical time series analysis, the model (9.22), (9.23) is known as the VARX$(p, 0)$ model of order p with exogenous variables (see, e.g., [12]).

Let us consider the following model of drifting coefficients in the SSE (9.22):

$$B = B(t) = B_0 + \varepsilon B_1 \psi(t), \qquad t \in \mathbb{Z}, \tag{9.24}$$

$$A_i = A_i(t) = A_{i0} + \varepsilon_i A_{i1} \psi_i(t), \quad i = 0, 1, \ldots, p, \tag{9.25}$$

where B_0, $\{A_{i0}\}$ are the hypothetical values of matrix coefficients (in the absence of drift); $\varepsilon, \varepsilon_i \geq 0$ are distortion levels (or drift levels); $\psi(t), \psi_i(t) \in \mathbb{R}^1$ are nonrandom real-valued functions defining the model of parameter variation; the matrices B_1, $\{A_{i1}\}$ are fixed (independent of time).

Let us present three special cases of the general drift model (9.24), (9.25) which are often used to model real-world statistical data [7].

1. *The exponential model:*

$$A_i(t) = A_{i0} + \varepsilon_i A_{i1} \exp(-\lambda_i t),$$
$$B(t) = B_0 + \varepsilon B_1 \exp(-\mu t),$$

where $\lambda_i > 0$ and $\mu > 0$ are a priori known, and for $d, f \in \{0, 1\}$ the matrices A_{id}, B_f no longer depend on time. It is recommended to use this model if some of the coefficients are monotonous increasing (or decreasing) in time and tend to some limit (equilibrium) value.

2. *The harmonic model:*

$$A_i(t) = A_{i0} + \varepsilon_i A_{i1} \sin(t\omega_i + \phi_i),$$
$$B(t) = B_0 + \varepsilon B_1 \sin(t\omega + \phi),$$

where ω_i, ϕ_i, ϕ are given, and $\{A_{i0} > 0\}$, B_0 are independent of time. This model is used if the coefficients are periodic, e.g., due to seasonal factors. The parameters $\{\omega_i\}$ are the oscillation frequencies.

3. *The change point model:*

$$A_i(t) = \begin{cases} A_{i0}, & t \le \tau_0, \\ A_{i0} + \varepsilon_i A_{i1}, & t > \tau_0, \end{cases}$$
$$B(t) = \begin{cases} B_0, & t \le \tau_0, \\ B_0 + \varepsilon B_1, & t > \tau_0, \end{cases} \tag{9.26}$$

where $\{A_{i1}\}$, B_1 are fixed nonzero matrices, and τ_0 is a change point. This dependence is used to model processes characterized by jumps in model parameters. For instance, the change point model can be used to simulate introduction of new laws or regulations, as well as large-scale political events (e.g., elections).

Note that sometimes the cases where only $\{A_l(t)\}$, only $B(t)$, or both $\{A_l(t)\}$ and $B(t)$ depend on time are considered separately.

9.4.2 LS Parameter Estimators Under Parameter Drift

Let us evaluate the properties of LS estimators for the coefficients of the SSE models (9.22) under drifting model parameters.

Define the following block matrices:

$$Y_t = \begin{pmatrix} y_t \\ y_{t-1} \\ \vdots \\ y_{t-p+1} \end{pmatrix} \in \mathbb{R}^{Np}, U_t = \begin{pmatrix} u_t \\ 0 \\ \vdots \\ 0 \end{pmatrix} \in \mathbb{R}^{Np}, X_t = \begin{pmatrix} x_t \\ x_{t-1} \\ \vdots \\ x_{t-p+1} \end{pmatrix} \in \mathbb{R}^{Kp}, \tag{9.27}$$

$$A = \begin{pmatrix} A_1 & A_2 & \cdots & A_{p-1} & A_p \\ -A_0 & 0 & \cdots & 0 & 0 \\ 0 & -A_0 & \cdots & 0 & 0 \\ \vdots & \vdots & \vdots & \vdots & \vdots \\ 0 & 0 & \cdots & -A_0 & 0 \end{pmatrix} \in \mathbb{R}^{Np \times Np}, \tag{9.28}$$

$$\alpha = \begin{pmatrix} A_0 & 0 & \cdots & 0 \\ 0 & A_0 & \cdots & 0 \\ \vdots & \vdots & \vdots & \vdots \\ 0 & 0 & \cdots & A_0 \end{pmatrix}, \quad \beta = \begin{pmatrix} B & 0 & \cdots & 0 \\ 0 & 0 & \cdots & 0 \\ \vdots & \vdots & \vdots & \vdots \\ 0 & 0 & \cdots & 0 \end{pmatrix} \in \mathbb{R}^{Np \times Kp}. \tag{9.29}$$

Lemma 9.1. *If $|A_0| \neq 0$, then the SSE (9.22) with N endogenous and K exogenous variables (including the lagged variables) can be represented in the following equivalent forms: the structural form*

$$\alpha Y_t + AY_{t-1} + \beta X_t = U_t, \tag{9.30}$$

and the reduced form

$$Y_t + \tilde{A} Y_{t-1} + \tilde{\beta} X_t = \tilde{U}_t, \tag{9.31}$$

where we have

$$\mathbb{E}\{U_t\} = \mathbf{0}_{Np}, \qquad \mathbb{E}\{\tilde{U}_t\} = \mathbf{0}_{Np},$$
$$\mathbf{Cov}\{U_t, U_t\} = S, \qquad \mathbf{Cov}\{\tilde{U}_t, \tilde{U}_t\} = \alpha^{-1} S \alpha'^{-1},$$

with $\tilde{A} = \alpha^{-1} A$, $\tilde{\beta} = \alpha^{-1}\beta$, $\tilde{U}_t = \alpha^{-1} U_t$,

$$S = \begin{pmatrix} \Sigma & \mathbf{0}_{N\times N} & \cdots & \mathbf{0}_{N\times N} \\ \mathbf{0}_{N\times N} & \mathbf{0}_{N\times N} & \cdots & \mathbf{0}_{N\times N} \\ \vdots & \vdots & \vdots & \vdots \\ \mathbf{0}_{N\times N} & \mathbf{0}_{N\times N} & \cdots & \mathbf{0}_{N\times N} \end{pmatrix} \in \mathbb{R}^{Np \times Np}.$$

Proof. The relation (9.30) is obtained from (9.22) by introducing the matrix notation (9.27)–(9.29). To obtain (9.31), it suffices to multiply (9.30) from the left by α^{-1}, which exists by the nonsingularity of A_0. □

Corollary 9.3. *The Np-dimensional time series $\{Y_t\}$ defined by (9.31) can be represented as a linear combination of $\{\tilde{U}_t\}$ and $\{X_t\}$:*

$$Y_t = \sum_{\tau=0}^{\infty} \left(-\tilde{A}\right)^{\tau} \tilde{U}_{t-\tau} - \sum_{\tau=0}^{\infty} \left(-\tilde{A}\right)^{\tau} \tilde{\beta} X_{t-\tau}, \tag{9.32}$$

$$\tilde{A} = \begin{pmatrix} A_0^{-1}A_1 & A_0^{-1}A_2 & \cdots & A_0^{-1}A_{p-1} & A_0^{-1}A_p \\ -\mathbf{I}_N & \mathbf{0}_{N\times N} & \cdots & \mathbf{0}_{N\times N} & \mathbf{0}_{N\times N} \\ \mathbf{0}_{N\times N} & -\mathbf{I}_N & \cdots & \mathbf{0}_{N\times N} & \mathbf{0}_{N\times N} \\ \vdots & \vdots & \vdots & \vdots & \vdots \\ \mathbf{0}_{N\times N} & \mathbf{0}_{N\times N} & \cdots & -\mathbf{I}_N & \mathbf{0}_{N\times N} \end{pmatrix} \in \mathbb{R}^{Np\times Np},$$

$$\tilde{\beta} = \begin{pmatrix} A_0^{-1}B & \mathbf{0}_{N\times K} & \cdots & \mathbf{0}_{N\times K} \\ \mathbf{0}_{N\times K} & \mathbf{0}_{N\times K} & \cdots & \mathbf{0}_{N\times K} \\ \vdots & \vdots & \vdots & \vdots \\ \mathbf{0}_{N\times K} & \mathbf{0}_{N\times K} & \cdots & \mathbf{0}_{N\times K} \end{pmatrix} \in \mathbb{R}^{Np\times Kp}.$$

Proof. To obtain (9.32), it suffices to use (9.31) as a recurrence relation. □

Let us obtain statistical parameter estimators for an SSE model in the reduced form. Assume that the investigated stochastic process is observed at T time points $t \in \{1, 2, \ldots, T\}$, and $\gamma = \begin{pmatrix} \tilde{A}' \\ \tilde{\beta}' \end{pmatrix}$ is the composite $((N + K)p \times Np)$-matrix of parameters.

Define the following block matrices:

$$Q = \begin{pmatrix} Q_{11} & Q_{12} \\ Q_{12}' & Q_{22} \end{pmatrix}, \quad s = \begin{pmatrix} s_1 \\ s_2 \end{pmatrix},$$

$$Q_{11} = \frac{1}{T}\sum_{t=1}^{T} Y_{t-1}Y_{t-1}', \quad Q_{12} = \frac{1}{T}\sum_{t=1}^{T} Y_{t-1}X_t', \quad Q_{22} = \frac{1}{T}\sum_{t=1}^{T} X_t X_t',$$

$$s_1 = \frac{1}{T}\sum_{t=1}^{T} Y_{t-1}Y_t', \quad s_2 = \frac{1}{T}\sum_{t=1}^{T} X_t Y_t'. \tag{9.33}$$

Then, assuming $|Q| \neq 0$, we have the following expression for the least squares estimator:

$$\hat{\gamma} = -Q^{-1}s. \tag{9.34}$$

Note that the condition $|Q| \neq 0$ implies

$$(N + K)Np^2 < T.$$

Assume that the sequence of exogenous variables $\{X_t\}$ is such that for $T \to \infty$, the following limit is defined:

$$\lim_{T\to\infty} \frac{1}{T} \sum_{t=1}^{T} X_{t-\tau} X'_{t-\tau'} = M_{\tau-\tau'}, \tag{9.35}$$

where M_τ is a symmetric $(Kp \times Kp)$-matrix, and $\tau, \tau' \in \{0, 1, \dots\}$.
Taking into account (9.27)–(9.29), let us denote

$$F = \sum_{\tau=0}^{\infty} \tilde{A}^\tau \mathbb{E}\left\{\tilde{U}_{t-\tau}\tilde{U}'_{t-\tau}\right\} \tilde{A}'^\tau = \sum_{\tau=0}^{\infty} \tilde{A}^\tau \tilde{\Sigma} \tilde{A}'^\tau, \tag{9.36}$$

$$\tilde{\Sigma} = \left(\begin{array}{c|c|c|c} A_0^{-1}\Sigma A_0'^{-1} & 0 & \cdots & 0 \\ \hline 0 & 0 & \cdots & 0 \\ \hline \vdots & \vdots & \vdots & \vdots \\ \hline 0 & 0 & \cdots & 0 \end{array}\right) \in \mathbb{R}^{Np\times Np}, \tag{9.37}$$

$$w_t = -\sum_{\tau=0}^{\infty} \left(-\tilde{A}\right)^\tau \tilde{\beta} X_{t-\tau}. \tag{9.38}$$

Lemma 9.2. *Under the assumption* (9.35), *the following limit expressions hold:*

$$H = \lim_{T\to\infty} \frac{1}{T} \sum_{t=1}^{T} w_t w'_t = \sum_{\tau,\tau'=0}^{\infty} \left(-\tilde{A}\right)^\tau \tilde{\beta} M_{\tau-\tau'} \tilde{\beta}' \left(-\tilde{A}'\right)^{\tau'}, \tag{9.39}$$

$$L_{\tau'} = \lim_{T\to\infty} \frac{1}{T} \sum_{t=1}^{T} w_{t-1} X'_{t-\tau'} = -\sum_{\tau,\tau'=0}^{\infty} \left(-\tilde{A}\right)^\tau \tilde{\beta} M_{\tau+1-\tau'}. \tag{9.40}$$

Proof. By applying (9.35), (9.36)–(9.38), we obtain

$$\lim_{T\to\infty} \frac{1}{T} \sum_{t=1}^{T} w_t w'_t = \lim_{T\to\infty} \sum_{\tau,\tau'=0}^{\infty} \left(-\tilde{A}\right)^\tau \tilde{\beta} \frac{1}{T} \sum_{t=1}^{T} X_{t-\tau} X'_{t-\tau'} \tilde{\beta}' \left(-\tilde{A}'\right)^{\tau'} =$$

$$= \sum_{\tau,\tau'=0}^{\infty} \left(-\tilde{A}\right)^\tau \tilde{\beta} M_{\tau-\tau'} \tilde{\beta}' \left(-\tilde{A}'\right)^{\tau'};$$

$$\lim_{T\to\infty}\frac{1}{T}\sum_{t=1}^{T} w_{t-1}w'_{t-\tau'} = \lim_{T\to\infty}\left(-\sum_{\tau=0}^{\infty}(-\tilde{A})^{\tau}\tilde{\beta}\frac{1}{T}\sum_{t=1}^{T}X_{t-\tau-1}X'_{t-\tau'}\right) =$$

$$= -\sum_{\tau=0}^{\infty}(-\tilde{A})^{\tau}\tilde{\beta}M_{\tau+1-\tau'}. \qquad \square$$

Now let us assume that the matrix coefficients multiplying the variables in (9.22) are subject to drift, i.e., are changing with time. First, we are going to consider a general drift model (9.24) for the matrix coefficient multiplying the exogenous variables in (9.22):

$$B = B(t) = B_0 + \varepsilon B_1\psi(t),$$

where $\varepsilon \geq 0$ is the distortion level (drift level); B_0 and B_1 are fixed $(N \times K)$-matrices (now they are assumed to be independent of time); $\psi(t) : \mathbb{R}^1 \to \mathbb{R}^1$ is a fixed real-valued function defining the parameter drift model.

Then, similarly to (9.32), we can define

$$\tilde{\beta} = \tilde{\beta}(t) = \alpha^{-1}\beta(t) = \left(\begin{array}{c|c|c|c} A_0^{-1}B(t) & 0 & \cdots & 0 \\ \hline 0 & 0 & \cdots & 0 \\ \hline \vdots & \vdots & \vdots & \vdots \\ \hline 0 & 0 & \cdots & 0 \end{array}\right) = \tilde{\beta}_0 + \varepsilon\tilde{B}_1\psi(t), \tag{9.41}$$

where

$$\tilde{\beta}_0 = \left(\begin{array}{c|c|c|c} A_0^{-1}B_0 & 0 & \cdots & 0 \\ \hline 0 & 0 & \cdots & 0 \\ \hline \vdots & \vdots & \vdots & \vdots \\ \hline 0 & 0 & \cdots & 0 \end{array}\right), \quad \tilde{\beta}_1 = \left(\begin{array}{c|c|c|c} A_0^{-1}B_1 & 0 & \cdots & 0 \\ \hline 0 & 0 & \cdots & 0 \\ \hline \vdots & \vdots & \vdots & \vdots \\ \hline 0 & 0 & \cdots & 0 \end{array}\right). \tag{9.42}$$

Assuming that there is no parameter drift, i.e., $\varepsilon = 0$, and $B = B_0$, let $\hat{\gamma}$ be the traditional least squares estimator defined by (9.34). Let $\hat{\gamma}^{\varepsilon}$ be the least squares estimator defined by (9.34) and based on ε-distorted data under parameter drift: $B = B(t) = B_0 + \varepsilon B_1\psi(t)$. Let us evaluate the difference between these estimators, i.e., the deviation of the estimator in the distorted model from the undistorted case:

$$\Delta_{\gamma} = \hat{\gamma}^{\varepsilon} - \hat{\gamma}. \tag{9.43}$$

Theorem 9.3. *For an identifiable SSE (9.22), let $|Q| \neq 0$; $X_t = 0$ for $t = 0, -1, \ldots$. Assume that the matrix coefficient multiplying the vector of exogenous* *variables is drifting,* $B = B(t) = B_0 + \varepsilon B_1 \psi(t)$*, where* $\varepsilon \geq 0$ *is the distortion level;* $\psi(\cdot)$ *is a real function,* $\psi(t) : \mathbb{R}^1 \to \mathbb{R}^1$*; and* B_0*,* B_1 *are fixed* $(N \times K)$*-matrices. Then the deviation* (9.43) *satisfies the following asymptotic expansion:*

$$\Delta_\gamma = \varepsilon Q^{-1} \left(q_{(1)} Q^{-1} s - s_{(1)} \right) + O(\varepsilon^2) \mathbf{1}_{(N+K)p \times Np}, \tag{9.44}$$

where

$$s_{(1)} = \begin{pmatrix} \frac{1}{T} \sum_{t=1}^{T} (Y_{t-1} V_t' + V_{t-1} Y_t') \\ \hline \frac{1}{T} \sum_{t=1}^{T} X_t V_t' \end{pmatrix};$$

$$q_{(1)} = \left(\begin{array}{c|c} \frac{1}{T} \sum_{t=1}^{T} (Y_{t-1} V_{t-1}' + V_{t-1} Y_{t-1}') & \frac{1}{T} \sum_{t=1}^{T} V_{t-1} X_t' \\ \hline \frac{1}{T} \sum_{t=1}^{T} X_t V_{t-1}' & 0 \end{array} \right);$$

$$V_t = -\sum_{\tau=0}^{\infty} \psi(t-\tau) \left(-\tilde{A} \right)^\tau \tilde{\beta}_1 X_{t-\tau};$$

Q *and* s *are defined by the formulas* (9.33)*, and the remainder term is understood in the sense of almost sure convergence:*

$$\frac{1}{\varepsilon^2} O(\varepsilon^2) \xrightarrow{\text{a.s.}} c \quad \text{as } \varepsilon \to 0, \quad \text{where } |c| < \infty.$$

Proof. From (9.32), (9.41), (9.42), let us obtain an expression for the observed data under ε-distortion (ε-drift):

$$Y_t^\varepsilon = \sum_{\tau=0}^{\infty} \left(-\tilde{A} \right)^\tau \tilde{U}_{t-\tau} - \sum_{\tau=0}^{\infty} \left(-\tilde{A} \right)^\tau \tilde{\beta}_0 X_{t-\tau} - \varepsilon \sum_{\tau=0}^{\infty} \left(-\tilde{A} \right)^\tau \psi(t-\tau) \tilde{\beta}_1 X_{t-\tau} =$$

$$= Y_t - \varepsilon \sum_{\tau=0}^{\infty} \left(-\tilde{A} \right)^\tau \psi(t-\tau) \tilde{\beta}_1 X_{t-\tau} = Y_t + \varepsilon V_t, \quad \tau < t - \tau_0, \tag{9.45}$$

where

$$V_t = -\sum_{\tau=0}^{\infty} \left(-\tilde{A} \right)^\tau \psi(t-\tau) \tilde{\beta}_1 X_{t-\tau}.$$

From (9.45), (9.33) we have

$$
\begin{aligned}
Q_{11}^{\varepsilon} &= Q_{11} + \varepsilon \left(\frac{1}{T} \sum_{t=1}^{T} Y_{t-1} V_{t-1}' + \frac{1}{T} \sum_{t=1}^{T} V_{t-1} Y_{t-1}' \right) + O(\varepsilon^2) \mathbf{1}_{N_p \times N_p}, \\
Q_{12}^{\varepsilon} &= Q_{12} + \varepsilon \frac{1}{T} \sum_{t=1}^{T} V_{t-1} X_t', \quad Q_{22}^{\varepsilon} = Q_{22} = \frac{1}{T} \sum_{t=1}^{T} X_t X_t', \\
s_1^{\varepsilon} &= s_1 + \varepsilon \left(\frac{1}{T} \sum_{t=1}^{T} Y_{t-1} V_t' + \frac{1}{T} \sum_{t=1}^{T} V_{t-1} Y_t' \right) + O(\varepsilon^2) \mathbf{1}_{N_p \times N_p}, \qquad (9.46) \\
s_2^{\varepsilon} &= s_2 + \varepsilon \frac{1}{T} \sum_{t=1}^{T} X_t V_t'.
\end{aligned}
$$

Thus, we obtain

$$
\hat{\gamma}^{\varepsilon} = -\left(Q + \varepsilon q_{(1)} + \varepsilon^2 q_{(2)}\right)^{-1} \left(s + \varepsilon s_{(1)} + \varepsilon^2 s_{(2)}\right),
$$

where $q_{(1)}, q_{(2)}$ can be found from the expressions (9.46).

Now applying the well-known matrix expansion [5]

$$
(\mathbf{I}_N + X)^{-1} = \mathbf{I}_N - X + O(||X||^2)\, \mathbf{1}_{N \times N},
$$

where $||X||$ is the norm of the matrix X, proves the theorem. □

In applications, statisticians often encounter the special case described by (9.24), where the function $\psi(t)$ is the Heaviside step function:

$$
\psi(t) = \mathbf{1}(t - \tau_0) = \begin{cases} 1, & t > \tau_0, \\ 0, & t \leq \tau_0. \end{cases}
$$

This means that the dynamic behavior of the matrix coefficient B is defined by the jump function (9.26):

$$
B(t) = \begin{cases} B_0, & t \leq \tau_0, \\ B_0 + \varepsilon B_1, & t > \tau_0, \end{cases}
$$

where the parameter $\tau_0 \in \{1, 2, \ldots, T\}$ is known as the change point.

Theorem 9.4. *Under the conditions of Theorem 9.3, let $\psi(t) = \mathbf{1}(t - \tau_0)$, where τ_0 is a change point, and assume the asymptotics*

$$
T \to \infty, \quad \tau_0 \to \infty, \quad \lim_{T \to \infty} \frac{\tau_0}{T} = \lambda_0, \ 0 < \lambda_0 < 1.
$$

Then the deviation $\Delta_\gamma = \hat{\gamma}^\varepsilon - \hat{\gamma}$ *of the estimator* $\hat{\gamma}^\varepsilon$ *from* $\hat{\gamma}$ *has a limit in probability:*

$$\Delta_\gamma \xrightarrow{\mathbf{P}} a \neq \mathbf{0}_{(N+K)p\times Np}, \tag{9.47}$$

and the limit value can be written asymptotically as

$$a = \varepsilon a_0 + O(\varepsilon^2)\mathbf{1}_{(N+K)p\times Np}, \quad \text{where} \quad a_0 = \tilde{Q}^{-1}\left(\tilde{q}_{(1)}\tilde{Q}^{-1}\tilde{s} - \tilde{s}_{(1)}\right), \tag{9.48}$$

and the matrices $\tilde{Q}$, $\tilde{q}_{(1)}$, $\tilde{s}$, $\tilde{s}_{(1)}$ *are the following limits in probability:*

$$\tilde{Q} = p\lim_{T\to\infty} Q = \left(\begin{array}{c|c} F+H & L_0 \\ \hline L_0' & M_0 \end{array}\right), \tag{9.49}$$

$$\tilde{s} = p\lim_{T\to\infty} s = \left(\begin{array}{c} -(F+H)\tilde{A}' - L_0\tilde{\beta}' \\ \hline -L_0'\tilde{A}' - M_0\tilde{\beta}' \end{array}\right), \tag{9.50}$$

$$\tilde{s}_{(1)} = p\lim_{T\to\infty} s_{(1)} = -(1-\lambda_0)\times$$
$$\times \left(\begin{array}{c} \sum\limits_{\tau=0}^{\infty} L_1\tilde{\beta}_1'\left(-\tilde{A}'\right)^\tau + \sum\limits_{\tau=0}^{\infty}\left(\left(-\tilde{A}\right)^\tau\tilde{\beta}_1\left(L_{\tau+1}'\left(-\tilde{A}\right) - M_{-\tau-1}\tilde{\beta}'\right)\right)' \\ \hline \sum\limits_{\tau=0}^{\infty} M_{-\tau-1}\tilde{\beta}_1'\left(-\tilde{A}'\right)^\tau \end{array}\right), \tag{9.51}$$

$$\tilde{q}_{(1)} = p\lim_{T\to\infty} q_{(1)} = -(1-\lambda_0)\times$$
$$\times \left(\begin{array}{c|c} \sum\limits_{\tau=0}^{\infty} L_{\tau+1}\tilde{\beta}_1'\left(-\tilde{A}'\right)^\tau + \sum\limits_{\tau=0}^{\infty}\left(-\tilde{A}\right)^\tau\tilde{\beta}_1 L_{\tau+1}' & \sum\limits_{\tau=0}^{\infty}\left(-\tilde{A}\right)^\tau\tilde{\beta}_1 M_{-\tau-1}' \\ \hline \sum\limits_{\tau=0}^{\infty} M_{-\tau-1}\tilde{\beta}_1'\left(-\tilde{A}'\right)^\tau & \mathbf{0} \end{array}\right) \tag{9.52}$$

in the notation of (9.35)–(9.40).

Proof. If we prove (9.49)–(9.52), then (9.47)–(9.48) will follow from the expansion (9.44) and the well-known relation between the convergence almost surely and the convergence in probability.

The expression (9.49) is quite well known (see, for example, [1]).

To prove (9.50)–(9.52), we are going to use the result of Lemma 9.2 together with (9.35), (9.36)–(9.38). By (9.31), we obtain

$$\tilde{s} = p \lim_{T\to\infty} \begin{pmatrix} -\frac{1}{T}\sum_{t=1}^{T} Y_{t-1}Y_{t-1}'\tilde{A}' - \frac{1}{T}\sum_{t=1}^{T} Y_{t-1}X_t'\tilde{\beta}' + \frac{1}{T}\sum_{t=1}^{T} Y_{t-1}\tilde{U}_t' \\ \hdashline -\frac{1}{T}\sum_{t=1}^{T} X_tY_{t-1}'\tilde{A}' - \frac{1}{T}\sum_{t=1}^{T} X_tX_t'\tilde{\beta}' + \frac{1}{T}\sum_{t=1}^{T} X_t\tilde{U}_t' \end{pmatrix} =$$

$$= \begin{pmatrix} -(F+H)\tilde{A}' - L_0\tilde{\beta}' \\ \hdashline -L_0'\tilde{A}' - M_0\tilde{\beta}' \end{pmatrix}.$$

To study the convergence of $\tilde{s}_{(1)}$, let us consider the following expressions, taking into account that $X_t = 0$ for $t = 0, -1, -2, \ldots$:

$$p \lim_{T\to\infty} \frac{1}{T}\sum_{t=1}^{T} Y_{t-1}V_t' = p \lim_{T\to\infty} \left(\sum_{\tau=0}^{\infty} \left(-\frac{1}{T}\sum_{t=1}^{T} Y_{t-1}X_{t-\tau}'\tilde{\beta}_1' \left(-\tilde{A}'\right)^{\tau} \mathbf{1}(t-\tau-\tau_0) \right) \right) =$$

$$= p \lim_{T\to\infty} \left(\sum_{\tau=0}^{\infty} \left(-\frac{1}{T} \left(\frac{\tau+\tau_0}{\tau+\tau_0} \sum_{t=1}^{\tau+\tau_0} Y_{t-1}X_{t-\tau}'\tilde{\beta}_1' \left(-\tilde{A}'\right)^{\tau} \mathbf{1}(t-\tau-\tau_0) \right) - \right.\right.$$

$$\left.\left. -\frac{1}{T} \sum_{t=\tau+\tau_0+1}^{T} Y_{t-1}X_{t-\tau}'\tilde{\beta}_1' \left(-\tilde{A}'\right)^{\tau} \mathbf{1}(t-\tau-\tau_0) \right) \right) =$$

$$= -(1-\lambda_0) \sum_{\tau=0}^{\infty} L_{\tau}\tilde{\beta}_1' \left(-\tilde{A}'\right)^{\tau}.$$

Similarly, we have

$$p \lim_{T\to\infty} \frac{1}{T}\sum_{t=1}^{T} Y_tV_{t-1}' = -(1-\lambda_0) \sum_{\tau=0}^{\infty} \left(-\tilde{A}\right)^{\tau} \tilde{\beta}_1 \left(L_{\tau+1}' \left(-\tilde{A}\right) - M_{-\tau-1}\tilde{\beta}' \right),$$

$$p \lim_{T\to\infty} \frac{1}{T}\sum_{t=1}^{T} X_tV_t' = -(1-\lambda_0) \sum_{\tau=0}^{\infty} M_{-\tau}\tilde{\beta}_t' \left(-\tilde{A}'\right)^{\tau}.$$

Applying the same argument yields the following expressions for the components of $\tilde{q}_{(1)}$:

$$p \lim_{T\to\infty} \frac{1}{T}\sum_{t=1}^{T} Y_{t-1}V_{t-1}' = -(1-\lambda_0) \sum_{\tau=0}^{\infty} L_{\tau+1}\tilde{\beta}_t' \left(-\tilde{A}'\right)^{\tau},$$

$$p \lim_{T\to\infty} \frac{1}{T}\sum_{t=1}^{T} X_tV_{t-1}' = -(1-\lambda_0) \sum_{\tau=0}^{\infty} M_{-\tau-1}\tilde{\beta}_1' \left(-\tilde{A}'\right)^{\tau}.$$

The limit expressions (9.49)–(9.52) have been proved. □

Now let us consider the case (9.25), where parameter drift is affecting the matrix coefficients multiplying the dependent variables. As above, let us study the deviation (9.43). Similarly to Corollary 9.3, we can obtain an expansion of the Np-vector Y_t^ε, where the symbol ε indicates the presence of distortion (in this case, drift).

Lemma 9.3. *If in* (9.31) *the block* $(Np \times Np)$*-matrix of coefficients* $\tilde{A}$ *depends on time,*

$$\tilde{A} = \tilde{A}(t) = \tilde{A}^0 + \varepsilon \tilde{A}^1 \psi(t),$$

where $\varepsilon \geq 0$, $\tilde{A}^0$ *and* $\tilde{A}^1$ *are* $(Np \times Np)$*-matrices independent of time, and* $\psi(t)$ *is a scalar function, then the Np-vector* Y_t^ε, $t \in \mathbb{Z}$, *can be represented as follows:*

$$Y_t^\varepsilon = \sum_{\tau=0}^{\infty} \prod_{i=1}^{\tau} \left(-\tilde{A}(t-i+1)\right) \tilde{U}_{t-\tau} - \sum_{\tau=0}^{\infty} \prod_{i=1}^{\tau} \left(-\tilde{A}(t-i+1)\right) \tilde{\beta} X_{t-\tau}. \qquad (9.53)$$

Proof. Write

$$\begin{aligned}
Y_t^\varepsilon &= -\tilde{A}(t) Y_{t-1} - \tilde{\beta} X_t + \tilde{U}_t = \\
&= -\tilde{A}(t) \left(-\tilde{A}(t-1) Y_{t-2} - \tilde{\beta} X_{t-1} + \tilde{U}_{t-1}\right) - \tilde{\beta} X_t + \tilde{U}_t = \\
&= \tilde{A}(t)\tilde{A}(t-1) Y_{t-2} + \left(-\tilde{A}(t)\right)\left(-\tilde{\beta}\right) X_{t-1} - \tilde{\beta} X_t + \left(-\tilde{A}(t)\right) \tilde{U}_{t-1} + \tilde{U}_t = \\
&= \cdots = \sum_{\tau=0}^{\infty} \prod_{i=1}^{\tau} \left(-\tilde{A}(t-i+1)\right) \tilde{U}_{t-\tau} - \sum_{\tau=0}^{\infty} \prod_{i=1}^{\tau} \left(-\tilde{A}(t-i+1)\right) \tilde{\beta} X_{t-\tau}.
\end{aligned}$$

□

Lemma 9.4. *Under the conditions of Lemma 9.3, we have the following stochastic expansion of the Np-vector* Y_t^ε*:*

$$Y_t^\varepsilon = Y_t + \varepsilon V_t^\varepsilon + O(\varepsilon^2) \mathbf{1}_{Np \times 1}, \qquad (9.54)$$

where

$$V_t^\varepsilon = \sum_{\tau=0}^{\infty} \left(\sum_{s=0}^{\tau} \left(-\tilde{A}^{(0)}\right)^s \tilde{A}^{(1)} \left(-\tilde{A}^{(0)}\right)^{\tau-s} \psi(t-s) \right) \left(\tilde{U}_{t-\tau-1} + \tilde{\beta} X_{t-\tau-1}\right).$$

Proof. Similarly to (9.45), let us substitute the appropriate expressions for $\{\tilde{A}(t)\}$ into (9.53). From (9.53), we have

$$Y_t^{\varepsilon} = \sum_{\tau=0}^{\infty} \prod_{i=1}^{\tau} \left(-\tilde{A}^{(0)} - \varepsilon \tilde{A}^{(1)} \psi(t-i+1)\right) \tilde{U}_{t-\tau} -$$
$$- \sum_{\tau=0}^{\infty} \prod_{i=1}^{\tau} \left(-\tilde{A}^{(0)} - \varepsilon \tilde{A}^{(1)} \psi(t-i+1)\right) \tilde{\beta} X_{t-\tau} =$$
$$= \sum_{\tau=0}^{\infty} \left(-\tilde{A}^{(0)}\right)^{\tau} \tilde{U}_{t-\tau} - \sum_{\tau=0}^{\infty} \left(-\tilde{A}^{(0)}\right)^{\tau} \tilde{\beta} X_{t-\tau} + \varepsilon V_t^{\varepsilon} + O\left(\varepsilon^2\right) \mathbf{1}_{Np\times 1} =$$
$$= Y_t + \varepsilon V_t^{\varepsilon} + O(\varepsilon^2)\mathbf{1}_{Np\times 1},$$

where the remainder term $O(\varepsilon^2)$ is understood in the sense of almost sure convergence. □

Lemma 9.5. *Under the conditions of Lemma 9.3, assuming $X_t = 0$ for $t \leq 0$, the deviation* (9.43) *satisfies the following asymptotic expansion:*

$$\Delta_{\gamma} = \varepsilon Q^{-1}\left(q_{(1)} Q^{-1} s - s_{(1)}\right) + O(\varepsilon^2)\mathbf{1}_{(N+K)p\times Np}, \tag{9.55}$$

$$s_{(1)} = \begin{pmatrix} \frac{1}{T}\sum_{t=1}^{T}\left(Y_{t-1}V_t'^{\varepsilon} + V_{t-1}^{\varepsilon}Y_t'\right) \\ \hdashline \frac{1}{T}\sum_{t=1}^{T} X_t V_t'^{\varepsilon} \end{pmatrix};$$

$$q_{(1)} = \left(\begin{array}{c:c} \frac{1}{T}\sum_{t=1}^{T}\left(Y_{t-1}V_{t-1}'^{\varepsilon} + V_{t-1}^{\varepsilon}Y_{t-1}'\right) & \frac{1}{T}\sum_{t=1}^{T} V_{t-1}^{\varepsilon} X_t' \\ \hdashline \frac{1}{T}\sum_{t=1}^{T} X_t V_{t-1}'^{\varepsilon} & 0 \end{array}\right),$$

$$V_t^{\varepsilon} = \sum_{\tau=0}^{\infty}\left(\sum_{s=0}^{\tau}\left(-\tilde{A}^{(0)}\right)^{s} \tilde{A}^{(1)} \left(-\tilde{A}^{(0)}\right)^{\tau-s} \psi(t-s)\right)\left(\tilde{U}_{t-\tau-1} + \tilde{\beta} X_{t-\tau-1}\right).$$

The remainder term in (9.55) *is understood in the sense of almost sure convergence.*

Proof. This lemma can be proved similarly to Theorem 9.3. □

Now let us consider distortion in the form of a (fixed) change point τ_0, as defined by (9.26):

$$\psi(t) = \mathbf{1}(t - \tau_0),$$
$$\tilde{A}(t) = \begin{cases} \tilde{A}^{(0)}, & t \leq \tau_0, \\ \tilde{A}^{(0)} + \varepsilon \tilde{A}^{(1)}, & t > \tau_0. \end{cases}$$

Theorem 9.5. *Under the conditions of Lemma 9.5, if* $\psi(t) = \mathbf{1}(t - \tau_0)$*, then in the asymptotics* $T \to \infty$, $\tau_0 \to \infty$, $\lim\limits_{T\to\infty} \frac{\tau_0}{T} = \lambda_0$, $0 < \lambda_0 < 1$*, the deviation* Δ_γ *satisfies the following asymptotic relation:*

$$\Delta_\gamma \xrightarrow{\mathbf{P}} a \neq \mathbf{0}_{(N+K)p\times Np},$$

and the following asymptotic expansion holds for the limit value:

$$a = \varepsilon a_0 + O(\varepsilon^2)\mathbf{1}_{(N+K)p\times Np}, \quad a_0 = \tilde{Q}^{-1}\left(\tilde{q}_{(1)}\tilde{Q}^{-1}\tilde{s} - \tilde{s}_{(1)}\right).$$

The following limit relations hold for $\tilde{Q}$, $\tilde{q}_{(1)}$, $\tilde{s}$, $\tilde{s}_{(1)}$*:*

$$\tilde{Q} = p\lim_{T\to\infty} Q = \left(\begin{array}{c|c} F+H & L_0 \\ \hline L_0' & M_0 \end{array}\right), \quad \tilde{s} = p\lim_{T\to\infty} s = \left(\begin{array}{c} -(F+H)\tilde{A}' - L_0\tilde{\beta}' \\ \hline -L_0'\tilde{A}' - M_0\tilde{\beta}' \end{array}\right),$$

$$\tilde{s}_{(1)} = p\lim_{T\to\infty} s_{(1)} = (1-\lambda_0)\times$$

$$\times\left(\begin{array}{c} \sum\limits_{\tau=0}^{\infty}\sum\limits_{s=0}^{\tau}\left(\left(-\tilde{A}^{(0)}\right)^s \tilde{A}^{(1)}\left(-\tilde{A}^{(0)}\right)^{\tau-s}\left(\tilde{\beta}L_\tau'\left(-\tilde{A}'\right) - M_{-\tau-2}'\tilde{\beta}'\right)\right)' \\ \hline \sum\limits_{\tau=0}^{\infty}\sum\limits_{s=0}^{\tau} M_{-\tau-1}\tilde{\beta}'\left(\left(-\tilde{A}^{(0)}\right)'\right)^{\tau-s}\left(\tilde{A}^{(1)}\right)'\left(\left(-\tilde{A}^{(0)}\right)'\right)^s \end{array}\right) +$$

$$+(1-\lambda_0)\left(\begin{array}{c} \sum\limits_{\tau=0}^{\infty}\sum\limits_{s=0}^{\tau} L_{\tau-1}\tilde{\beta}'\left(\left(-\tilde{A}^{(0)}\right)'\right)^{\tau-s}\left(\tilde{A}^{(1)}\right)'\left(\left(-\tilde{A}^{(0)}\right)'\right)^s \\ \hline \sum\limits_{\tau=0}^{\infty}\sum\limits_{s=0}^{\tau} M_{-\tau-1}\tilde{\beta}'\left(\left(-\tilde{A}^{(0)}\right)'\right)^{\tau-s}\left(\tilde{A}^{(1)}\right)'\left(\left(-\tilde{A}^{(0)}\right)'\right)^s \end{array}\right),$$

$$\tilde{q}_{(1)} = p\lim_{T\to\infty} q_{(1)} = (1-\lambda_0)\times$$

$$\times\left(\begin{array}{c|c} \sum\limits_{\tau=0}^{\infty}\sum\limits_{s=0}^{\tau} L_\tau\tilde{\beta}'\left(\left(-\tilde{A}^{(0)}\right)'\right)^{\tau-s}\left(\tilde{A}^{(1)}\right)'\left(\left(-\tilde{A}^{(0)}\right)'\right)^s & (\cdot)_{21}' \\ \hline \sum\limits_{\tau=0}^{\infty}\sum\limits_{s=0}^{\tau}\left(\left(-\tilde{A}^{(0)}\right)^s \tilde{A}^{(1)}\left(-\tilde{A}^{(0)}\right)^{\tau-s}\tilde{\beta}M_{-\tau-2}'\right) & 0 \end{array}\right) +$$

$$+(1-\lambda_0)\left(\begin{array}{c|c} \sum\limits_{\tau=0}^{\infty}\sum\limits_{s=0}^{\tau}\left(-\tilde{A}^{(0)}\right)^s \tilde{A}^{(1)}\left(-\tilde{A}^{(0)}\right)^{\tau-s}\tilde{\beta}L_\tau' & 0 \\ \hline 0 & 0 \end{array}\right).$$

Proof. Apply Lemmas 9.4, 9.5 as in the proof of Theorem 9.4. □

Theorems 9.4 and 9.5 show that parameter drift generally leads to loss of consistency of the least squares estimator $\hat{\gamma}^{\varepsilon}$. These theorems also allow us to find the limit difference a between the least squares estimator and the true value γ:

$$\hat{\gamma}^{\varepsilon} \xrightarrow{\mathbf{P}} \gamma + a.$$

9.5 Sensitivity of Forecast Risk to Parameter Drift

Let us investigate how parameter drift affects forecasting algorithms based on least squares estimators, with specific attention paid to forecast risk. We are going to start by evaluating how the deviation Δ_{γ} of the estimator $\hat{\gamma}^{\varepsilon}$ from $\hat{\gamma}$ affects the accuracy of a one-step-ahead forecast $y_{T+1} \in \mathbb{R}^N$.

Let us introduce the following matrix notation:

$$C = \begin{pmatrix} 1 & 0 & \dots 0 & 0 & \dots 0 \\ 0 & 1 & \dots 0 & 0 & \dots 0 \\ \vdots & \vdots & \dots \vdots & \vdots & \dots \vdots \\ 0 & 0 & \dots 1 & 0 & \dots 0 \end{pmatrix} \in \mathbb{R}^{N \times Np}, \qquad G = a'_{0A} C' C a_{0A} \in \mathbb{R}^{Np \times Np},$$

$$R = a'_{0\beta} C' C a_{0\beta} \in \mathbb{R}^{Kp \times Kp}, \qquad J = a'_{0A} C' C a_{0\beta} \in \mathbb{R}^{Np \times Kp},$$

$$F_1 = \lim_{T \to \infty} w_T w'_T \in \mathbb{R}^{Np \times Np}, \qquad F_2 = \lim_{T \to \infty} w_T X'_{T+1} \in \mathbb{R}^{Np \times Kp}, \tag{9.56}$$

where a_{0A} and $a_{0\beta}$ are, respectively, $(Np \times Np)$- and $(Np \times Kp)$-matrices defined from the limit expression established in Sect. 9.4:

$$\Delta_{\gamma} = \hat{\gamma}^{\varepsilon} - \hat{\gamma} = \begin{pmatrix} \Delta_A \\ --- \\ \Delta_{\beta} \end{pmatrix} \xrightarrow{\mathbf{P}} a = \varepsilon \begin{pmatrix} a_{0A} \\ --- \\ a_{0\beta} \end{pmatrix} + O(\varepsilon^2) \mathbf{1}_{(N+K)p \times N_p}.$$

The plug-in forecast (9.14) will be chosen for one-step-ahead prediction of the vector y_{T+1} of dependent variables:

$$\hat{y}_{T+1} = \left(\hat{Y}_{T+1}\right)_{(1)} = \left(\hat{\hat{A}} Y_T - \hat{\hat{\beta}} X_{T+1}\right)_{(1)}, \tag{9.57}$$

where $(\cdot)_{(1)}$ is the subvector (block) consisting of the first N components of a vector, $\hat{\hat{A}}$ and $\hat{\hat{\beta}}$ are estimators of the respective matrices, X_{T+1} is the known column vector of nonrandom exogenous variables at a future time $T + 1$. Note that $\hat{y}_{T+1} = C\hat{Y}_{T+1}$.

It has been proved that if the coefficients of the equation (9.31) are known exactly, then the forecast (9.57) is mean square optimal in the family of linear forecasts.

Theorem 9.6. *Under the conditions of Theorem 9.4, let $T \to \infty$, $X_T \to X^0 \in \mathbb{R}^K$. Then the mean square forecast risk*

$$r_\varepsilon(T) = \mathbb{E}\{\|\hat{y}_{T+1} - y_{T+1}\|^2\} = \mathbb{E}\{(\hat{y}_{T+1} - y_{T+1})'(\hat{y}_{T+1} - y_{T+1})\}$$

satisfies the following convergence as $T \to \infty$:

$$\begin{aligned} r_\varepsilon(T) \to r_\varepsilon &= \mathrm{tr}\left(A_0^{-1}\Sigma\left(A_0^{-1}\right)'\right) + \varepsilon^2 b; \\ b &= (X^0)'RX^0 + \mathrm{tr}\left((F + F_1)\,G\right) + 2\mathrm{tr}\left(F_2 J\right) \geq 0. \end{aligned} \tag{9.58}$$

Proof. From (9.57), we can write

$$\begin{aligned} \hat{y}_{T+1} &= \left(-\left(\tilde{A}Y_T + \tilde{\beta}X_{T+1}\right) - \left(\Delta_A Y_T + \Delta_\beta X_{T+1}\right)\right)_{(1)} = \\ &= y_{T+1} - A_0^{-1}u_{T+1} - (\Delta_A Y_T)_{(1)}\left(\Delta_\beta X_{T+1}\right)_{(1)}. \end{aligned}$$

Therefore, we have

$$\hat{y}_{T+1} - y_{T+1} = -A_0^{-1}u_{T+1} - (\Delta_A Y_T)_{(1)} - (\Delta_\beta X_{T+1})_{(1)}.$$

Now let us rewrite the expression for the risk $r_\varepsilon(T)$:

$$\begin{aligned} r_\varepsilon(T) &= \mathbb{E}\left\{(\hat{y}_{T+1} - y_{T+1})'\,(\hat{y}_{T+1} - y_{T+1})\right\} = \\ &= \mathbb{E}\left\{u'_{T+1}A_0'^{-1}A_0^{-1}u_{T+1}\right\} + \mathbb{E}\left\{(\Delta_A Y_T)'_{(1)}(\Delta_A Y_T)_{(1)}\right\} = \\ &= \mathbb{E}\left\{(\Delta_\beta X_{T+1})'_{(1)}(\Delta_\beta X_{T+1})_{(1)}\right\} + 2\mathbb{E}\left\{(\Delta_A Y_T)'_{(1)}(\Delta_\beta X_{T+1})_{(1)}\right\}. \end{aligned}$$

This implies the following convergence [in the notation of (9.56)]:

$$\begin{aligned} r_\varepsilon(T) \to \mathrm{tr}\left(A_0^{-1}\Sigma\left(A_0^{-1}\right)'\right) &+ \varepsilon^2\mathrm{tr}\left(\left(\lim_{T\to\infty}\mathbb{E}\{Y_T Y_T'\}\right)G\right) + \\ &+ \varepsilon^2\mathrm{tr}\left(X^0\left(X^0\right)'R\right) + 2\varepsilon^2\mathrm{tr}\left(\left(\lim_{T\to\infty}\mathbb{E}\{Y_T X'_{T+1}\}\right)J\right). \end{aligned}$$

Applying (9.32) and (9.45) yields the simplified expressions:

$$\mathbb{E}\{Y_T Y_T'\} = F + w_T w_T', \quad \mathbb{E}\{Y_T X'_{T+1}\} = w_T X'_{T+1}.$$

Finally, we can write

$$r_\varepsilon(T) \to \mathrm{tr}\left(A_0^{-1}\Sigma\left(A_0^{-1}\right)'\right) + \varepsilon^2 \mathrm{tr}\left(\left(F + \lim_{T\to\infty} w_T w_T'\}\right) G\right) + \varepsilon^2 (X^0)' R X^0 +$$

$$+ 2\varepsilon^2 \mathrm{tr}\left(\left(\lim_{T\to\infty} w_T X_{T+1}'\right) J\right) = \mathrm{tr}\left(A_0^{-1}\Sigma A_0^{-1}\right) + \varepsilon^2 b. \qquad \square$$

Corollary 9.4. *Under the conditions of Theorem 9.6, the following asymptotic expansion holds for the risk averaged over time:*

$$r_\varepsilon^* = \lim_{T\to\infty} \frac{1}{T} \sum_{t=1}^{T} r_\varepsilon(t) = \mathrm{tr}\left(A_0^{-1}\Sigma A_0^{-1}\right) + \varepsilon^2 b^* + O(\varepsilon^2),$$

where

$$b^* = \mathrm{tr}\,((F + H)G) + \mathrm{tr}\,(M_0 R) + 2\,\mathrm{tr}\,(L_1 J).$$

Proof. This result follows from Theorem 9.6 and the expressions (9.35)–(9.40). $\square$

Corollary 9.5. *Under the conditions of Theorem 9.6, the risk instability coefficient tends to the following limit as $T \to +\infty$:*

$$\kappa_\varepsilon(T) = \frac{r_\varepsilon(T) - r_0}{r_0} \to \varepsilon^2 \frac{b}{\mathrm{tr}\left(A_0^{-1}\Sigma(A_0^{-1})'\right)},$$

and the δ-admissible distortion level $\varepsilon_+(\delta)$ defined by the inequality

$$\lim_{T\to\infty} \kappa_{\varepsilon_+}(T) \le \delta, \quad \delta > 0,$$

satisfies the following approximation:

$$\varepsilon_+(\delta) \approx \sqrt{\frac{\delta\,\mathrm{tr}\,(A_0^{-1}\Sigma(A_0^{-1})')}{b}}.$$

Theorem 9.6, together with its corollaries, allows us to evaluate the robustness of the least squares forecasting statistic (9.57) under parameter drift.

In conclusion, let us mention that certain methods to improve the robustness of the forecasting statistic (9.57) have been developed in [9].

9.6 Numerical Results for the Ludeke Econometric Model

The well-known Ludeke model [3] is the simplest model of macroeconomic dynamics. It is represented as an SSE(4,3):

$$\begin{cases} C_t = a_0 + a_1 Y_t + a_2 C_{t-1} + u_{t1}, \\ I_t = b_0 + b_1 Y_t + b_2 U_{t-1} + u_{t2}, \\ Im_t = c_0 + c_1 Y_t + c_2 Im_{t-1} + u_{t3}, \\ Y_t = C_t + I_t - Im_t + G_t. \end{cases}$$

The model has $N = 4$ dependent (endogenous) variables: consumption C_t, investment I_t, import Im_t, national income Y_t, and $K = 3$ independent (exogenous) variables: dividends U_t, government expenditure G_t, and a dummy variable.

Using the notation (9.22) in the absence of drift, this model can be represented in the following form ($p = 1$):

$$y_t = A_0(t)y_t + A_1(t)y_{t-1} + B(t)x_t + u_t, \tag{9.59}$$

where

$$y_t = (C_t\, I_t\, Im_t\, Y_t)' \in \mathbb{R}^4, \quad x_t = (U_{t-1}\, G_t\, 1)', \quad u_t = (u_{t1}\, u_{t2}\, u_{t3})' \in \mathbb{R}^3,$$

$$A_0(t) = A_0 = \begin{pmatrix} 0 & 0 & 0 & a_1 \\ 0 & 0 & 0 & b_1 \\ 0 & 0 & 0 & c_1 \\ 1 & 1 & -1 & 0 \end{pmatrix}, \quad A_1(t) = A_1 = \begin{pmatrix} a_2 & 0 & 0 & 0 \\ 0 & 0 & 0 & 0 \\ 0 & 0 & c_2 & 0 \\ 0 & 0 & 0 & 0 \end{pmatrix} \in \mathbb{R}^{4\times 4},$$

$$B(t) = B = \begin{pmatrix} 0 & 0 & a_0 \\ b_2 & 0 & b_0 \\ 0 & 0 & c_0 \\ 0 & 1 & 0 \end{pmatrix} \in \mathbb{R}^{4\times 3}.$$

The vector of true model parameters has nine dimensions:

$$\gamma = (a_1, b_1, c_1, a_2, b_2, c_2, a_0, b_0, c_0)'.$$

In our simulation, the value of this vector was the estimate obtained from observing eighteen econometric indicators of German economy [3]:

$$\gamma = (0.353, 0, 0.169, 0.434, 0.793, 0.628, 28.679, 10.717, -29.724)'. \tag{9.60}$$

Future states of the system were simulated based on this "true" model. The experiments used different observation lengths: $T = 20, 50, 100, 150, 200, 250$. Least squares estimates of the parameter vector $\hat{\gamma}$ were found from (9.34). Then a

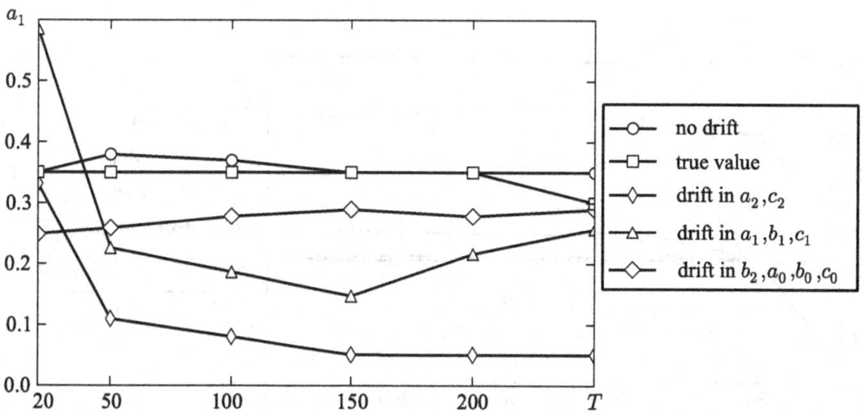

Fig. 9.1 Estimates for the coefficient a_1

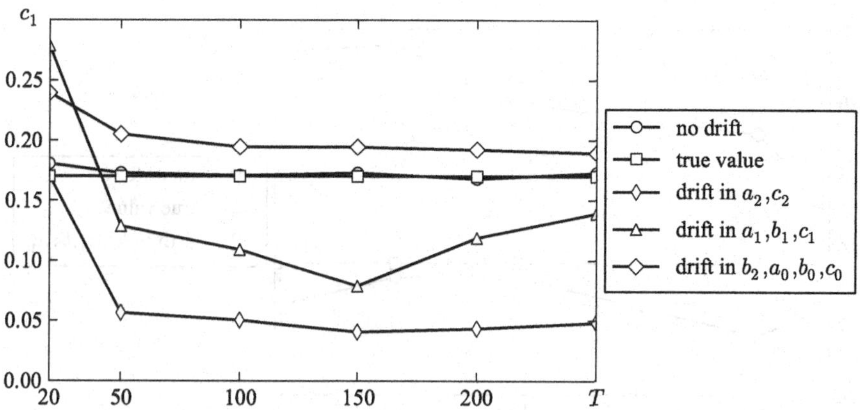

Fig. 9.2 Estimates for the coefficient c_1

change point τ_0 was artificially introduced into the system: the simulated time series used different values of certain parameters for $t \geq \tau_0$. In the first series of experiments, the ratio τ_0/T was chosen as $\lambda = 0.75$, and in the second series it was $\lambda = 0.5$. Least squares estimates of the parameter vector $\hat{\gamma}^\varepsilon$ were calculated from this distorted data by applying (9.34).

Two series of simulations were performed for different distortion levels ε. Presented in Figs. 9.1, 9.2, 9.3 are the results of estimating, respectively, the model coefficients a_1, c_1, c_2 for different observation lengths T, where past the change point $\tau_0 = 0.75T$ the coefficients of the model (9.59) were increased by 50 % compared to the initial values (9.60).

Figures 9.4–9.6 present similar results in a setting where model coefficients were increased by 20 % past the change point $\tau_0 = 0.5T$.

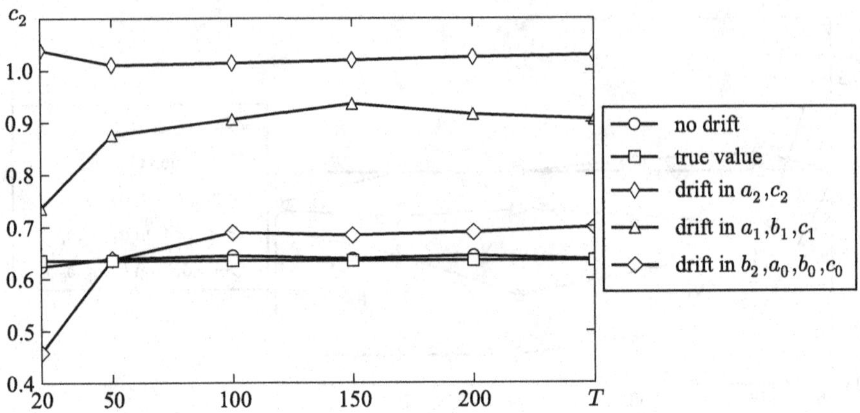

Fig. 9.3 Estimates for the coefficient c_2

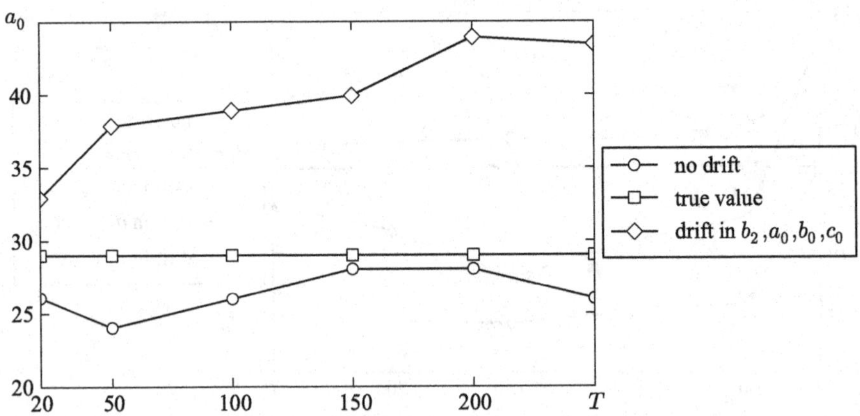

Fig. 9.4 Estimates for the coefficient a_0

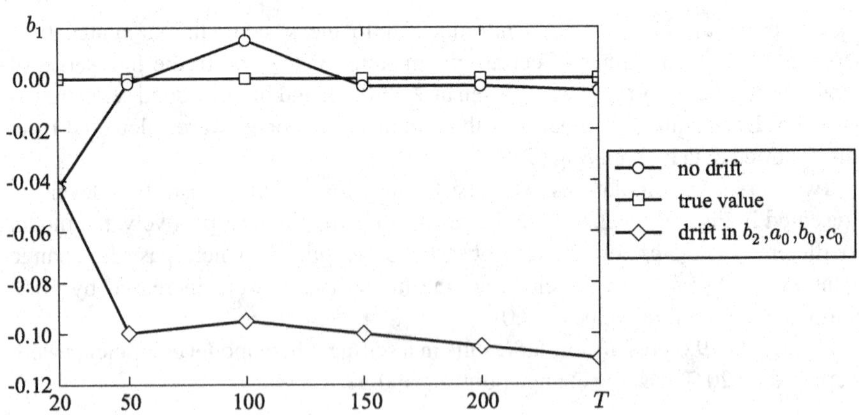

Fig. 9.5 Estimates for the coefficient b_1

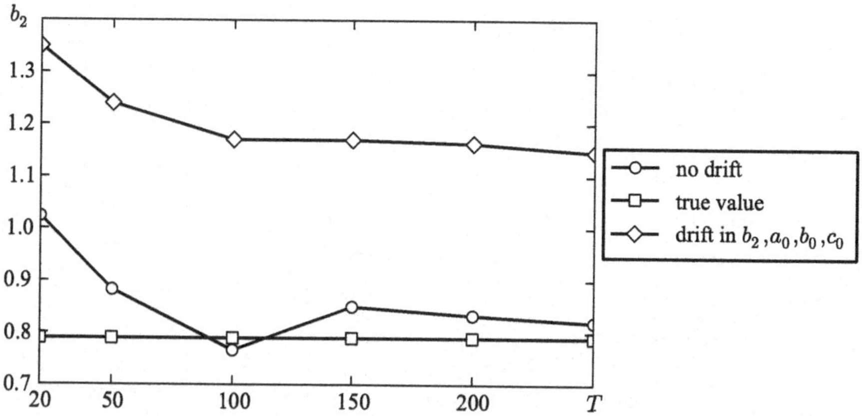

Fig. 9.6 Estimates for the coefficient b_2

The simulation results shown in Figs. 9.1–9.6 illustrate the inconsistency of least squares estimators under parameter (coefficient) drift.

References

1. Anderson, T.: The Statistical Analysis of Time Series. Wiley, New York (1994)
2. Braverman, E., Muchnik, I.: Approximation approach to estimation of systems of simultaneous equations. Autom. Remote Control **39**(11), 120–128 (1978)
3. Greene, W.: Econometric Analysis. Macmillan, New York (2000)
4. Harvey, A.: The Econometric Analysis of Time Series. Oxford Allan, Oxford (1981)
5. Horn, R., Johnson, C.: Matrix Analysis. Cambridge University Press, Cambridge (2010)
6. Kharin, Yu., Malugin, V.: Development of econometric software for macroeconomic modeling and forecasting under transition. Lith. Stat. J. **40**(9), 74–77 (2001)
7. Kharin, Yu., Rogatch, P.: On statistical estimation of systems of simultaneous equations under time-varying coefficients. Res. Memo. Univ. Leic. **98**(7), 1–23 (1998)
8. Kharin, Yu., Staleuskaya, S.: Robustness in statistical analysis of regression and simultaneous-equations models. In: Proceedings of the International Conference on "Prague Stochastics'98", pp. 289–293. UCMP, Prague (1998)
9. Kharin, Yu., Staleuskaya, S.: Robustification of "approximating approach" in simultaneous equation models. In: New Trends in Probability and Statistics, vol. 5, pp. 143–150. TEV, Vilnius (2000)
10. Kharin, Yu., Staleuskaya, S.: Robust modeling and forecasting of stochastic systems dynamics. Concept. Anal. Gest. Syst. Ind. **2**, 999–1005 (2001)
11. Kharin, Yu., Charemza, W., Malugin, V.: On modeling of Russian and Belarusian economies by the LAM-3 econometric model (in Russian). Appl. Econom. **1**(2), 124–139 (2006)
12. Lutkepohl, H.: Introduction to Multiple Time Series Analysis. Springer, Berlin (1993)
13. Quinn, B., Nicholls, D.: The estimation of random coefficient autoregressive models. J. Time Ser. Anal. **2**(3), 185–203 (1981)
14. Raj, B., Ullah, A.: Econometrics: A Varying Coefficients Approach. St. Martin's Press, New York (1981)
15. Stulajter, F.: Predictions in Time Series Using Regression Models. Springer, New York (2002)

Chapter 10
Forecasting of Discrete Time Series

Abstract This chapter is devoted to forecasting in the non-classical setting where the state space of the time series is finite, necessitating the use of discrete-valued time series models. The field of discrete statistics has remained relatively underdeveloped until the recent years, when rapid introduction of digital equipment stimulated the researchers to develop numerous discrete models and techniques. In this chapter, we discuss optimal forecasting statistics and forecast risks for Markov chain models, including high-order Markov chains, and the beta-binomial model.

10.1 Forecasting by Discriminant Analysis of Markov Chains

10.1.1 The Time Series Model

As in Sect. 2.4.1, let us consider a setting where the predicted random variable ν is discrete:

$$\nu \in \{1, 2, \dots, L\},$$

where $2 \leq L < +\infty$ is the number of possible values of ν (and hence, the number of possible forecasts). The statistical data X used to make the forecast $\hat{\nu}$ will be modeled by realizations of finite homogeneous Markov chains (HMCs). This model is used in genetic engineering [36], medical diagnostics [22,30], analytic finance [10], and other applications [15, 16, 20]. Forecasting will be based on the discriminant analysis approach discussed in Sect. 2.4.

Let us reformulate the problem in terms of discriminant analysis. Assume that a sequence of discrete random variables

$$\{X_t\}, \quad X_t \in \mathcal{A} = \{1, 2, \dots, N\}, \quad t = 1, 2, \dots,$$

Y. Kharin, *Robustness in Statistical Forecasting*, DOI 10.1007/978-3-319-00840-0_10,

is observed, and it belongs to one of the $L \geq 2$ classes $\Omega_1, \Omega_2, \ldots, \Omega_L$ with given prior probabilities $q_1, q_2, \ldots, q_L$ ($q_l > 0$, $l = 1, \ldots, L$; $q_1 + \cdots + q_L = 1$). Each class Ω_l is composed of HMCs defined by initial probabilities vectors $\pi^{(l)}$ and matrices of one-step transition probabilities $P^{(l)}$:

$$\begin{aligned} \pi^{(l)} &= \left(\pi_i^{(l)}\right): \quad \pi_i^{(l)} = \mathbb{P}\{X_1 = i \mid \Omega_l\}, \\ P^{(l)} &= \left(p_{ij}^{(l)}\right): \quad p_{ij}^{(l)} = \mathbb{P}\{X_t = j \mid X_{t-1} = i, \Omega_l\}, \end{aligned} \qquad i, j \in \mathcal{A}, \quad l = 1, \ldots, L. \tag{10.1}$$

We are going to assume that Markov chains in all classes are stationary and ergodic: $\pi_i^{(l)} > 0$, $i \in \mathcal{A}$. Without loss of generality, we will also assume that transitions between any two states are possible:

$$p_{ij}^{(l)} > 0, \quad i, j \in \mathcal{A}, \quad l \in \{1, \ldots, L\}, \tag{10.2}$$

otherwise only the possible transitions should be considered. Finally, assume that the classes $\{\Omega_l\}$ differ in the one-step transition probability matrices $\{P^{(l)}\}$.

Let a random realization of the HMC X, which lies in one of the L classes (10.1), be recorded over n units of time:

$$X = (x_1, x_2, \ldots, x_n), \quad x_t \in \mathcal{A}, \quad t \in 1, \ldots, n. \tag{10.3}$$

Let us construct a forecasting statistic or, in other words, the decision rule (DR) to predict the value ν based on the realization (10.3): $\hat{\nu} = d(X)$, $\hat{\nu} \in \{1, 2, \ldots, L\}$, and estimate its performance at different levels of prior information. The performance of the decision rule will be evaluated using the misclassification probability (incorrect forecast probability):

$$r = \mathbb{P}\{d(X) \neq \nu\}, \tag{10.4}$$

where $\nu \in \{1, 2, \ldots, L\}$ is a discrete random variable with the probability distribution $\mathbb{P}\{\nu = l\} = q_l$, which represents the true unobserved number of the class containing the HMC realized as (10.3).

10.1.2 The Bayesian Decision Rule and Its Properties

As mentioned in Sect. 2.4, the decision rule that yields the minimum misclassification error is the Bayesian decision rule (BDR).

Let us define the following frequency estimators for the probabilities of pairs of states based on the observed realization (10.3):

$$\hat{\Pi} = \left(\hat{\Pi}_{ij}\right): \quad \hat{\Pi}_{ij} = \frac{n_{ij}}{n}, \quad n_{ij} = \sum_{t=1}^{n-1} \mathbf{I}\{x_t = i, x_{t+1} = j\}, \quad i, j \in \mathcal{A},$$

where $\mathbf{I}\{\cdot\}$ is the indicator function.

Due to the normalization condition on the probabilities $\{\Pi_{ij} : i, j \in \mathcal{A}\}$, it is sufficient to estimate the parameters $\{\Pi_{ij} : (i, j) \in \mathcal{A}_\Pi\}$, where $\mathcal{A}_\Pi = \mathcal{A}^2 \setminus \{N, N\}$; then we can write

$$\Pi_{NN} = 1 - \sum_{(i,j)\in\mathcal{A}_\Pi} \Pi_{ij}.$$

Theorem 10.1. *In the model* (10.1)–(10.3)*, the minimum incorrect forecast probability* (10.4) *is attained for the BDR written as*

$$\hat{\nu} = d_{BDR}(X) = \arg\max_{1\le l\le L} \left(\frac{1}{n}\log q_l + \frac{1}{n}\log \pi^{(l)}_{x_1} + \sum_{i,j\in\mathcal{A}} \hat{\Pi}_{ij} \log P^{(l)}_{ij} \right), \quad X \in \mathcal{A}^n. \tag{10.5}$$

Proof. It suffices to use the definition of a BDR given in Sect. 2.4 and to construct a log likelihood function for the observed realization X. □

Corollary 10.1. *In a setting where the set $\mathcal{A}$ is limited to only two possible solutions ($L = 2$), the BDR can be written as*

$$\hat{\nu} = d_{BDR}(X) = \mathbf{1}(\Lambda(X)) + 1, \quad X \in \mathcal{A}^n, \tag{10.6}$$

$$\Lambda(X) = \Lambda^*(X) + \frac{1}{n}\log\frac{q_2}{q_1} + \frac{1}{n}\log\frac{\pi^{(2)}_{x_1}}{\pi^{(1)}_{x_1}}, \quad \Lambda^*(X) = \sum_{i,j\in\mathcal{A}} \hat{\Pi}_{ij}\log\frac{p^{(2)}_{ij}}{p^{(1)}_{ij}}, \tag{10.7}$$

where $\Lambda(X)$ is the discriminant function for the classes Ω_1, Ω_2, and $\mathbf{1}(\cdot)$ is the Heaviside step function.

We are going to evaluate the probability of a forecast error for $L = 2$ when the BDR (10.6), (10.7) is used to make a forecast in the hardest case of contigual classes:

$$p^{(2)}_{ij} = p^{(1)}_{ij}(1 + b_{ij}\varepsilon), \quad \varepsilon \to 0, \quad p^{(2)}_{ij} \to p^{(1)}_{ij}, \quad i, j \in \mathcal{A}, \tag{10.8}$$

where $\{b_{ij}\}$ are some constant weights:

$$\sum_{j\in\mathcal{A}} p^{(1)}_{ij} b_{ij} = 0, \quad i \in \mathcal{A},$$

and ε is the contiguity parameter [26, 27].

Let us introduce the following auxiliary variables:

$$a_l = (-1)^l \sum_{i \in \mathcal{A}} \pi_i^{(l)} \sum_{j \in \mathcal{A}} p_{ij}^{(l)} \log \frac{p_{ij}^{(2)}}{p_{ij}^{(1)}} > 0, \tag{10.9}$$

$$s_{ijuv}^{(l)} = \pi_i^{(l)} p_{ij}^{(l)} \left(\delta_{iu}\delta_{iv} - \pi_u^{(l)} p_{uv}^{(l)}\right) + p_{ij}^{(l)} p_{uv}^{(l)} \left(\pi_i^{(l)} c_{ju}^{(l)} + \pi_u^{(l)} c_{vi}^{(l)}\right), \tag{10.10}$$

$$c_{ju}^{(l)} = \sum_{k=0}^{\infty} \left(p_{ju}^{(l)}(k) - \pi_u^{(l)}\right) < \infty,$$

$$\sigma_{ijuv}^{(l)} = \frac{\delta_{iu}}{\pi_i^{(l)}} \left(\delta_{jv} p_{ij}^{(l)} - p_{ij}^{(l)} p_{uv}^{(l)}\right), \quad i, j, u, v \in \mathcal{A}, \quad l \in \{1, 2\}, \tag{10.11}$$

where $a_l > 0$ is a weighted sum of the Kullback–Leibler information values computed for the transition probability matrices $P^{(1)}$ and $P^{(2)}$; $\{s_{ijuv}^{(l)}\}$ and $\{\sigma_{ijuv}^{(l)}\}$ are defined in the form of covariances; $p_{ju}^{(l)}(k) = ((P^{(l)})^k)_{ju}$ is the probability of a k-step transition from state j to state u for a Markov chain belonging to the class Ω_l. Note that the series $c_{ju}^{(l)}$ converges exponentially, and its sum can be easily computed.

The next lemma, which was proved in [22], describes the asymptotic properties of the auxiliary variables introduced in (10.9)–(10.11) and the stationary HMC probability distributions in the asymptotics (10.8).

Lemma 10.1. *In the asymptotics of contigual classes* (10.8) *of stationary HMCs* (10.1), (10.2), *the following expansions are satisfied:*

$$\pi_j^{(2)} = \pi_j^{(1)}(1 + \varepsilon h_j + O(\varepsilon^2)); \quad \sigma_{ijuv}^{(2)} = \sigma_{ijuv}^{(1)} + O(\varepsilon); \quad s_{ijuv}^{(2)} \to s_{ijuv}^{(1)};$$

$$a_l = \frac{\varepsilon^2}{2} \sum_{i,j \in \mathcal{A}} b_{ij}^2 \pi_i^{(l)} p_{ij}^{(1)} + O(\varepsilon^3), \quad l \in \{1, 2\}, \quad \frac{a_2}{a_1} \to 1,$$

where $|h_j| < +\infty$, $i, j, u, v \in \mathcal{A}$.

Let us now estimate the forecast risk (10.4) in the asymptotics (10.8) if the contiguity parameter satisfies the asymptotic condition

$$\varepsilon = O(n^{-1/2}).$$

Denote

$$\mu = \sum_{i,j \in \mathcal{A}} b_{ij}^2 \pi_i^{(1)} p_{ij}^{(1)}, \quad V = \sum_{(i,j),(u,v) \in \mathcal{A}_{\Pi}} (b_{ij} - b_{NN}) s_{ijuv}^{(1)} (b_{uv} - b_{NN}) > 0, \tag{10.12}$$

$$\Delta_l = \Delta + (-1)^l \frac{2 \log(q_2/q_1)}{c\sqrt{V}}, \quad \Delta = \frac{c\mu}{\sqrt{V}} > 0, \quad l \in \{1, 2\}, \tag{10.13}$$

where $0 < c < +\infty$, and the covariances $\{s_{ijuv}^{(1)}\}$ are defined by (10.10).

Theorem 10.2. *Assuming an increasing number n of observations and two contigual classes* (10.8), *for*

$$n \to \infty, \quad \varepsilon = \frac{c}{\sqrt{n}} \to 0, \quad 0 < c < +\infty,$$

the forecast risk (10.4) *of the BDR* (10.6) *converges to a limit value:*

$$r_0 \to \tilde{r}_0 = q_1 \Phi\left(-\Delta_1/2\right) + q_2 \Phi\left(-\Delta_2/2\right),$$

where $\Phi(\cdot)$ is the standard normal distribution function, and the values Δ_1, Δ_2 are defined by (10.13).

Proof. From (10.6), the conditional probabilities of an incorrect decision can be written as

$$r_1 = \mathbb{P}\{d(X) \neq \nu \mid \nu = 1\} = 1 - \mathbb{P}\{\Lambda(X) < 0 \mid \nu = 1\},$$
$$r_2 = \mathbb{P}\{d(X) \neq \nu \mid \nu = 2\} = \mathbb{P}\{\Lambda(X) < 0 \mid \nu = 2\}.$$

Let us find the probability distribution of the statistic $\Lambda(X)$ defined by (10.7). The summand $\Lambda^*(X)$ is a linear combination of the random variables $\{\tilde{\Pi}_{ij}\}$. It is known that if the observed realization X belongs to the class Ω_l, then, as $n \to \infty$, the random variables $\xi_{ij}^{(l)} = \sqrt{n}(\hat{\Pi}_{ij} - \Pi_{ij}^{(l)})$ are asymptotically normally distributed [2] with zero expectations and the covariances $\mathbf{Cov}\{\xi_{ij}^{(l)}, \xi_{uv}^{(l)}\} = s_{ijuv}^{(1)}$ defined by (10.10), where $\Pi_{ij}^{(l)} = \pi_i^{(l)} p_{ij}^{(l)}$. Thus, the conditional probability distribution of $\Lambda^*(X)$ is also asymptotically normal under the condition $\nu = l$. The asymptotic mean of $\Lambda^*(X)$ is a linear combination of the means of $\{\hat{\Pi}_{ij}\}$; it is equal to $(-1)^l a_l$. The asymptotic variance of $\Lambda^*(X)$ is a quadratic form in $\{s_{ijuv}^{(l)}\}$; due to the normalization condition on $\{\Pi_{ij}\}$, the asymptotic variance equals

$$\sigma_l^2 = \sum_{(i,j),(u,v) \in \mathcal{A}_\Pi} \log \frac{p_{ij}^{(2)}}{p_{ij}^{(1)}} \frac{p_{NN}^{(1)}}{p_{NN}^{(2)}} \log \frac{p_{uv}^{(2)}}{p_{uv}^{(1)}} \frac{p_{NN}^{(1)}}{p_{NN}^{(2)}} s_{ijuv}^{(l)} > 0.$$

Observe that $\sigma_l^2 > 0$ since for $l = 1, 2$ the covariance matrices $\left(s_{ijuv}^{(l)}\right)_{(i,j),(u,v) \in \mathcal{A}_\Pi}$ are nonsingular [2], and $P^{(1)} \neq P^{(2)}$. In the asymptotics (10.8), we can write σ_l^2 as

$$\sigma_l^2 = \varepsilon^2 \sum_{(i,j),(u,v) \in \mathcal{A}_\Pi} (b_{ij} - b_{NN}) s_{ijuv}^{(l)} (b_{uv} - b_{NN}) + O(\varepsilon^3). \tag{10.14}$$

Consider the last summand in (10.7): $\zeta = n^{-1} \log(\pi_{x_1}^{(2)} / \pi_{x_1}^{(1)})$. We have

$$\mathbb{P}\{\Lambda(X) < 0 \mid \nu = l\} = \mathbb{P}\left\{\Lambda^*(X) + \frac{1}{n}\log\frac{q_2}{q_1} + \zeta < 0 \mid \nu = l\right\} =$$

$$= \mathbb{P}\left\{\sqrt{n}\frac{\Lambda^*(X) - (-1)^l a_l}{\sigma_l} + \frac{\sqrt{n}\zeta}{\sigma_l} < -\sqrt{n}\frac{(-1)^l a_l}{\sigma_l} - \frac{1}{\sqrt{n}\sigma_l}\log\frac{q_2}{q_1} \mid \nu = l\right\}.$$

From Lemma 10.1 and the expansion (10.14), it follows that $\zeta = O_p(\varepsilon/n)$,

$$\sqrt{n}\frac{a_l}{\sigma_l} = \sqrt{n}\frac{\frac{1}{2}\varepsilon^2 \sum_{i,j\in\mathcal{A}} b_{ij}^2 \pi_i^{(l)} p_{ij}^{(1)} + O(\varepsilon^3)}{\sqrt{\varepsilon^2 \sum_{(i,j),(u,v)\in\mathcal{A}_\Pi} (b_{ij} - b_{NN}) s_{ijuv}^{(l)} (b_{uv} - b_{NN}) + O(\varepsilon^3)}},$$

$$\frac{\log(q_2/q_1)}{\sqrt{n}\sigma_l} = \frac{\log(q_2/q_1)}{\sqrt{n}\sqrt{\varepsilon^2 \sum_{(i,j),(u,v)\in\mathcal{A}_\Pi} (b_{ij} - b_{NN}) s_{ijuv}^{(l)} (b_{uv} - b_{NN}) + O(\varepsilon^3)}},$$

$$-\sqrt{n}\frac{(-1)^l a_l}{\sigma_l} - \frac{\log(q_2/q_1)}{\sqrt{n}\sigma_l} \quad\longrightarrow\quad -(-1)^l\frac{\Delta}{2} - \frac{\log(q_2/q_1)}{c\sqrt{V}},$$

and $\sqrt{n}\zeta/\sigma_l = O_p(n^{-1/2}) \to 0$ in probability. By applying the well-known result on convergence in distribution of a sum of two random variables (Theorem 15 in [32]) to $\sqrt{n}\zeta/\sigma_l = O_P(n^{-1/2}) \xrightarrow{\mathbf{P}} 0$ and the asymptotically normally distributed $\Lambda^*(X)$, we obtain

$$r_1 = 1 - \mathbb{P}\{\Lambda(X) < 0 \mid \nu = 1\} \longrightarrow 1 - \Phi\left(\frac{\Delta}{2} - \frac{\log(q_2/q_1)}{c\sqrt{V}}\right) = \Phi\left(-\Delta_1/2\right),$$

$$r_2 = \mathbb{P}\{\Lambda(X) < 0 \mid \nu = 2\} \longrightarrow \Phi\left(-\frac{\Delta}{2} - \frac{\log(q_2/q_1)}{c\sqrt{V}}\right) = \Phi\left(-\Delta_2/2\right),$$

and $r_0 = q_1 r_1 + q_2 r_2 \to \tilde{r}_0$. □

Corollary 10.2. *Assuming the equiprobability of the classes ($q_1 = q_2 = 0.5$), the limit value of the risk equals*

$$\tilde{r}_0 = \Phi(-\Delta/2).$$

It follows from the proof of Theorem 10.1 that in the asymptotics (10.8), the BDR (10.5) is equivalent to the decision rule

$$\hat{\nu} = d(X) = \arg\max_{1\le l\le L}\left(\frac{1}{n}\log q_l + \sum_{i,j\in\mathcal{A}} \hat{\Pi}_{ij}\log p_{ij}^{(l)}\right), \quad X \in \mathcal{A}^n.$$

10.1.3 The Plug-In Decision Rule and Its Risk

Now let us consider a situation where the class parameters (10.1) are a priori unknown, and a classified training sample is observed:

$$\begin{aligned}&\mathbb{X} = \{X^{(1)}, X^{(2)}, \ldots, X^{(L)}\},\\ &X^{(L)} = \left(x_1^{(l)}, x_2^{(l)}, \ldots, x_{n_l}^{(l)}\right), \quad x_t^{(l)} \in \mathcal{A}, \quad t \in \{1, \ldots, n_l\},\end{aligned} \tag{10.15}$$

where $X^{(l)}$ is a realization of length n_l of an HMC from the class $\Omega_l, l \in \{1, \ldots, L\}$. It is assumed that $X,\ X^{(1)}, \ldots,\ X^{(L)}$ are jointly independent.

Let us construct maximum likelihood estimators (MLEs) for the unknown matrices of one-step transition probabilities $P^{(l)}$ based on the training sample (10.15):

$$\hat{P}^{(l)} = \left(\hat{p}_{ij}^{(l)}\right), \quad \hat{p}_{ij}^{(l)} = \frac{n_{ij}^{(l)}}{n_{i\cdot}^{(l)}}, \quad i, j \in \mathcal{A}, \quad l \in \{1, \ldots, L\},$$

$$n_{ij}^{(l)} = \sum_{t=1}^{n_l-1} \mathbf{I}\left\{x_t^{(l)} = i,\ x_{t+1}^{(l)} = j\right\}, \quad n_{i\cdot}^{(l)} = \sum_{j\in\mathcal{A}} n_{ij}^{(l)}.$$

As mentioned above, due to the normalization condition, the reduced set of transition probabilities is used as the unknown parameters: $\{p_{ij}^{(l)} : (i, j) \in \mathcal{A}_P\}$, where

$$\mathcal{A}_P = \{(i, j) : i \in \mathcal{A}, \quad j \in \mathcal{A} \setminus \{N\}\}; \qquad p_{iN}^{(l)} = 1 - \sum_{j=1}^{N-1} p_{ij}^{(l)}.$$

Substituting the MLE $\{\hat{P}^{(l)}\}$ into (10.5) in place of the unknown values $\{P^{(l)}\}$, let us construct the plug-in Bayesian decision rule (PBDR):

$$d_{\mathrm{PBDR}}(X, \mathbb{X}) = \arg\max_{1\le l\le L} \left(\frac{1}{n}\log q_l + \frac{1}{n}\log \hat{\pi}_{x_1}^{(l)} + \sum_{i,j\in\mathcal{A}} \hat{\Pi}_{ij} \log \hat{p}_{ij}^{(l)}\right),$$

where $\hat{\pi}_i^{(l)} = n_{i\cdot}^{(l)}/n.$

For $L = 2$, the PBDR assumes the form

$$d_{\mathrm{PBDR}}(X, \mathbb{X}) = \mathbf{1}\left(\hat{\Lambda}(X, \mathbb{X})\right) + 1,$$

$$\hat{\Lambda}(X, \mathbb{X}) = \hat{\Lambda}^*(X, \mathbb{X}) + \frac{1}{n} \log \frac{q_2}{q_1} + \frac{1}{n} \log \frac{\hat{\pi}^{(2)}_{x_1}}{\hat{\pi}^{(1)}_{x_1}}, \qquad (10.16)$$

$$\hat{\Lambda}^*(X, \mathbb{X}) = \sum_{i,j \in \mathcal{A}} \hat{\Pi}_{ij} \log \frac{\hat{p}^{(2)}_{ij}}{\hat{p}^{(1)}_{ij}}.$$

Theorem 10.3. *For increasing numbers of observations n, n_1, n_2 and two contigual classes* (10.8)*:*

$$n, n_l \to \infty, \quad n_l/n \to \tilde{\lambda}_l > 0, \quad l = 1, 2; \quad \varepsilon = cn^{-1/2} \to 0, \quad 0 < c < \infty, \qquad (10.17)$$

the error probability (10.4) *of the PBDR* (10.16) *tends to the limit*

$$r \to \tilde{r} = q_1 \Phi\left(-\frac{\tilde{\Delta}_1}{2}\right) + q_2 \Phi\left(-\frac{\tilde{\Delta}_2}{2}\right),$$

$$\tilde{\Delta}_l = \frac{c\mu}{\sqrt{V + \tilde{V}_l}} + (-1)^l \frac{2\log(q_2/q_1)}{c\sqrt{V + \tilde{V}_l}}, \qquad (10.18)$$

$$\tilde{V}_l = \frac{1}{\tilde{\lambda}_{3-l}} \sum_{(i,j),\,(u,v) \in A_P} \pi^{(1)}_i \pi^{(1)}_u (b_{ij} - b_{iN})(b_{uv} - b_{uN}) \sigma^{(1)}_{ijuv} > 0,$$

where μ, V are defined by (10.12), *and the covariances $\{\sigma^{(1)}_{ijuv}\}$—by* (10.11).

Proof. Define a random event

$$Z = \left\{ n^{(l)}_{ij} > 0, \quad l \in \{1, 2\}, \ i, j \in \mathcal{A} \right\}.$$

If the opposite event $\overline{Z}$ occurs, then expressions (10.16) are not defined; let us agree, in that case, to make the decision $d_{\mathrm{PBDR}}(X, \mathbb{X}) = 0$.

Consider the following conditional error probabilities:

$$r_l = \mathbb{P}\{d_{\mathrm{PBDR}}(X, \mathbb{X}) \neq \nu \mid \nu = l\} = \mathbb{P}\{\{d_{\mathrm{PBDR}}(X, \mathbb{X}) \neq \nu\} \cap Z \mid \nu = l\} +$$
$$+ \mathbb{P}\left\{\{d_{\mathrm{PBDR}}(X, \mathbb{X}) \neq \nu\} \cap \overline{Z} \mid \nu = l\right\}, \quad l \in \{1, 2\}.$$

By (10.2) and the HMC stationarity condition, in the asymptotics required by the theorem we can write

$$\mathbb{P}\{\overline{Z}\} \to 0, \quad \mathbb{P}\left\{\{d_{\mathrm{PBDR}}(X, \mathbb{X}) \neq \nu\} \cap \overline{Z} \mid \nu = l\right\} \to 0,$$

and thus it is sufficient to investigate the first summand in r_l and the probability distribution of the statistic $\hat{\Lambda}(X, \mathbb{X})$.

Consider the main term $\hat{\Lambda}^*(X, \mathbb{X})$ in (10.16) assuming that the realization X belongs to Ω_l. From (10.16), we can see that

$$\hat{\Lambda}^*(X, \mathbb{X}) = f(\hat{P}^{(1)}, \hat{P}^{(2)}, \hat{\Pi})$$

is a function of one-step transition probability matrix estimators.

The random variables

$$\theta_{ij}^{(l)} = \sqrt{n_l}\left(\hat{p}_{ij}^{(l)} - p_{ij}^{(l)}\right) = \sqrt{\tilde{\lambda}_l n}\left(\hat{p}_{ij}^{(l)} - p_{ij}^{(l)}\right)$$

are asymptotically normally distributed [2] with mean values equal to zero and the covariances $\mathbf{Cov}\{\theta_{ij}^{(l)}, \theta_{uv}^{(l)}\} = \sigma_{ijuv}^{(l)}$ defined by (10.11). The random variables

$$\xi_{ij}^{(l)} = \sqrt{n}(\hat{\Pi}_{ij} - \Pi_{ij}^{(l)})$$

also have asymptotically normal distributions, zero mean values, and the covariances $\mathbf{Cov}\{\xi_{ij}^{(l)}, \xi_{uv}^{(l)}\} = s_{ijuv}^{(l)}$ defined by (10.10); $\Pi_{ij}^{(l)} = \pi_i^{(l)} p_{ij}^{(l)}$ for $i, j, u, v \in \mathcal{A}$. The independence of X, $X^{(1)}$, $X^{(2)}$ implies the independence of the random variables $\{\xi_{ij}^{(l)}\}$, $\{\theta_{ij}^{(1)}\}$, $\{\theta_{ij}^{(2)}\}$. By Anderson's theorem [1] on functional transformations of asymptotically normal random variables, the random variable $\hat{\Lambda}^*(X, \mathbb{X})$ has an asymptotically normal probability distribution:

$$\mathcal{L}\left\{\sqrt{n}\,\frac{\hat{\Lambda}^*(X, \mathbb{X}) - (-1)^l a_l}{\tilde{\sigma}_l} \mid \nu = l\right\} \to \mathcal{N}(0, 1)$$

with an asymptotic mean equal to

$$f\left(\mathbb{E}\{\hat{P}^{(1)}\}, \mathbb{E}\{\hat{P}^{(2)}\}, \mathbb{E}\{\hat{\Pi}\}\right) = (-1)^l a_l,$$

where a_l is defined by (10.9). The asymptotic variance $\tilde{\sigma}_l^2$ can be written as a quadratic form w.r.t. the partial derivatives of $f(\cdot)$ [1]:

$$\tilde{\sigma}_l^2 = \sum_{(i,j),(u,v)\in\mathcal{A}_P} \frac{\partial f}{\partial \hat{p}_{ij}^{(1)}} \frac{\partial f}{\partial \hat{p}_{uv}^{(1)}} \frac{\sigma_{ijuv}^{(1)}}{\tilde{\lambda}_1} +$$

$$+ \sum_{(i,j),(u,v)\in\mathcal{A}_P} \frac{\partial f}{\partial \hat{p}_{ij}^{(2)}} \frac{\partial f}{\partial \hat{p}_{uv}^{(2)}} \frac{\sigma_{ijuv}^{(2)}}{\tilde{\lambda}_2} + \sum_{(i,j),(u,v)\in\mathcal{A}_\Pi} \frac{\partial f}{\partial \hat{\Pi}_{ij}} \frac{\partial f}{\partial \hat{\Pi}_{uv}} s_{ijuv}^{(l)} =$$

$$= \sum_{(i,j),(u,v)\in \mathcal{A}_P} \left(\frac{\pi_i^{(l)} p_{ij}^{(l)}}{p_{ij}^{(1)}} - \frac{\pi_i^{(l)} p_{iN}^{(l)}}{p_{iN}^{(1)}} \right) \left(\frac{\pi_u^{(l)} p_{uv}^{(l)}}{p_{uv}^{(1)}} - \frac{\pi_u^{(l)} p_{uN}^{(l)}}{p_{uN}^{(1)}} \right) \frac{\sigma_{ijuv}^{(1)}}{\tilde{\lambda}_1} +$$

$$+ \sum_{(i,j),(u,v)\in \mathcal{A}_P} \left(\frac{\pi_i^{(l)} p_{ij}^{(l)}}{p_{ij}^{(2)}} - \frac{\pi_i^{(l)} p_{iN}^{(l)}}{p_{iN}^{(2)}} \right) \left(\frac{\pi_u^{(l)} p_{uv}^{(l)}}{p_{uv}^{(2)}} - \frac{\pi_u^{(l)} p_{uN}^{(l)}}{p_{uN}^{(2)}} \right) \frac{\sigma_{ijuv}^{(2)}}{\tilde{\lambda}_2} +$$

$$+ \sum_{(i,j),(u,v)\in \mathcal{A}_\Pi} \log \frac{p_{ij}^{(2)}}{p_{ij}^{(1)}} \frac{p_{NN}^{(1)}}{p_{NN}^{(2)}} \log \frac{p_{uv}^{(2)}}{p_{uv}^{(1)}} \frac{p_{NN}^{(1)}}{p_{NN}^{(2)}} s_{ijuv}^{(l)}.$$

This asymptotic variance is positive, $\tilde{\sigma}_l^2 > 0$, since $\nabla f(P^{(1)}, P^{(2)}, \Pi) \neq 0$, and the covariance matrices $\left\{s_{ijuv}^{(l)} : (i,j),(u,v) \in \mathcal{A}_\Pi\right\}$, $\left\{\sigma_{ijuv}^{(l)} : (i,j),(u,v) \in \mathcal{A}_P\right\}$ are nonsingular ($l = 1, 2$). In the asymptotics (10.8), the variance $\tilde{\sigma}_l^2$ can be represented as follows:

$$\begin{aligned} \tilde{\sigma}_l^2 = \frac{\varepsilon^2}{\tilde{\lambda}_{3-l}} &\sum_{(i,j),(u,v)\in \mathcal{A}_P} \pi_i^{(l)} \pi_u^{(l)} (b_{ij} - b_{iN})(b_{uv} - b_{uN}) \sigma_{ijuv}^{(l)} + \\ + \varepsilon^2 &\sum_{(i,j),(u,v)\in \mathcal{A}_\Pi} (b_{ij} - b_{NN})(b_{uv} - b_{NN}) s_{ijuv}^{(l)} + O(\varepsilon^3). \end{aligned} \tag{10.19}$$

By applying Lemma 10.1, (10.17), and (10.19), we obtain

$$\sqrt{n}\frac{a_l}{\tilde{\sigma}_l} \longrightarrow \frac{c\mu}{2\sqrt{V + \tilde{V}_l}}, \qquad \frac{\log q_2/q_1}{\sqrt{n}\tilde{\sigma}_l} \longrightarrow \frac{\log(q_2/q_1)}{c\sqrt{V + \tilde{V}_l}}.$$

Finally, let us consider the summand $\hat{\zeta} = n^{-1} \log\left(\hat{\pi}_{x_1}^{(2)} / \hat{\pi}_{x_1}^{(1)}\right)$ in (10.16). The estimators $\{\hat{\pi}_i^{(l)}\}$ for stationary distributions are consistent:

$$\hat{\pi}_i^{(l)} \xrightarrow{P} \pi_i^{(l)} > 0, \quad i \in \mathcal{A}, \quad l = 1, 2.$$

Thus, by the same argument that was used to prove Theorem 10.2, we can evaluate the asymptotic behavior of $\hat{\zeta}$ and conclude the proof. □

Note that if n_1, n_2 are growing faster than n, so that $\tilde{\lambda}_1, \tilde{\lambda}_2 \to +\infty$, then $\tilde{V}_1, \tilde{V}_2 \to 0$ and the PBDR risk tends to the risk of the BDR: $\tilde{r} - \tilde{r}_0 \to 0$.

10.1.4 An Asymptotic Expansion of the PBDR Risk

Let us investigate the convergence of PBDR and BDR risks as the sizes of the training samples tend to infinity, $n_1, n_2 \to +\infty$, and the length n of the realization

X used to construct the forecast $\hat{\nu}$ remains fixed in a setting of two ($L = 2$) equiprobable classes ($q_1 = q_2 = 0.5$).

Let us introduce the following notation: $\tilde{\Lambda}(X) = q_2 L_2(X) - q_1 L_1(X)$ is the discriminant function based on the conditional likelihood functions of the classes Ω_1, Ω_2 for the realization X:

$$L_l(X) = \pi_{x_1}^{(l)} \prod_{i,j \in \mathcal{A}} \left(p_{ij}^{(l)}\right)^{n_{ij}(X)}, \quad l = 1, 2,$$

where $n_{ij}(X)$ is the frequency of the bigram (i, j) computed from the realization X.

By the definition of conditional error probabilities,

$$r_1 = \mathbb{P}\{d_{\text{BDR}}(X) = 2 \mid \nu = l\} = \sum_{X \in \mathcal{A}^n} L_1(X) \mathbf{1}(\Lambda(X)),$$

$$r_2 = 1 - \mathbb{P}\{d_{\text{BDR}}(X) = 2\nu \mid \nu = 2\} = 1 - \sum_{X \in \mathcal{A}^n} L_2(X) \mathbf{1}(\Lambda(X)),$$

we have the following expression for the BDR risk:

$$r_0 = q_1 r_1 + q_2 r_2 = q_2 - \sum_{X \in \mathcal{A}^n} \tilde{\Lambda}(X) \mathbf{1}(\Lambda(X)) = q_2 - \sum_{X \in \mathcal{A}^n, \Lambda(X) \geq 0} \tilde{\Lambda}(X). \tag{10.20}$$

Writing an expression similar to (10.20) and averaging the risk $r_{\text{PBDR}}(\mathbb{X})$ of the PBDR over the possible training samples (10.15) yields the following expression for the unconditional PBDR risk:

$$r = q_2 - \sum_{X \in A^n} \tilde{\Lambda}(X) \mathbb{E}\{\mathbf{1}(\hat{\Lambda}(X, \mathbb{X}))\}. \tag{10.21}$$

Both discriminant functions, $\hat{\Lambda}(X, \mathbb{X})$ and $\Lambda(X)$, depend on the one-step transition probabilities and the stationary distributions. Let us express these discriminant functions in terms of bigram probabilities:

$$\hat{\Lambda}(X, \mathbb{X}) = \sum_{i,j \in \mathcal{A}} \frac{n_{ij}(X)}{n} \log \frac{\hat{\Pi}_{ij}^{(2)}}{\hat{\Pi}_{ij}^{(1)}} - \sum_{i \in \mathcal{A}} \frac{n_{i\cdot}(X) - \delta_{ix_1}}{n} \log \frac{\hat{\Pi}_{i\cdot}^{(2)}}{\hat{\Pi}_{i\cdot}^{(1)}},$$

$$\Lambda(X) = \sum_{i,j \in \mathcal{A}} \frac{n_{ij}(X)}{n} \log \frac{\Pi_{ij}^{(2)}}{\Pi_{ij}^{(1)}} - \sum_{i \in \mathcal{A}} \frac{n_{i\cdot}(X) - \delta_{ix_1}}{n} \log \frac{\Pi_{i\cdot}^{(2)}}{\Pi_{i\cdot}^{(1)}},$$

$$\hat{\Pi}^{(l)} = \left(\hat{\Pi}_{ij}^{(l)}\right), \quad \hat{\Pi}_{ij}^{(l)} = \frac{n_{ij}^{(l)}}{n_l}, \quad \Pi^{(l)} = \left(\Pi_{ij}^{(l)}\right), \quad \Pi_{ij}^{(l)} = \pi_i^{(l)} p_{ij}^{(l)}, \quad i, j \in \mathcal{A},$$

where $\hat{\Pi}_{i\cdot}^{(l)} = \sum_{j \in \mathcal{A}} \hat{\Pi}_{ij}^{(l)}, \Pi_{i\cdot}^{(l)} = \sum_{j \in \mathcal{A}} \Pi_{ij}^{(l)}, l \in \{1, 2\}$.

Let us introduce the following notation:

$$B_n(X) = \sum_{i,j \in \mathcal{A}} \frac{n_{ij}(X)}{n} b_{ij} + \sum_{i \in \mathcal{A}} \frac{\delta_{ix_1}}{n} h_i,$$

$$D_n(X) = \frac{\lambda_1 + \lambda_2}{\lambda_1 \lambda_2} \sum_{(i,j),(u,v) \in \mathcal{A}_\Pi} g_{ij}^{(1)}(X) g_{uv}^{(1)}(X) s_{ijuv}^{(1)},$$

$$g_{ij}^{(l)}(X) = \frac{1}{n}\left(\frac{n_{ij}(X)}{\Pi_{ij}^{(l)}} - \frac{n_{NN}(X)}{\Pi_{NN}^{(l)}}\right) - \frac{1}{n}\left(\frac{n_{i\cdot}(X)}{\Pi_{i\cdot}^{(l)}} - \frac{n_{N\cdot}(X)}{\Pi_{N\cdot}^{(l)}}\right),$$

$$\mathcal{A}^{(l)} = \left\{X \in \mathcal{A}^n : g^{(l)}(X) = 0\right\}, \quad g^{(l)}(X) = \left(g_{ij}^{(l)}(X)\right), \quad (i,j) \in \mathcal{A}_\Pi,$$

where $\lambda_1, \lambda_2 > 0$; $\{h_i\}$ are defined by Lemma 10.1; the functions $g_{ij}^{(l)}(X)$ are partial derivatives of $\hat{\Lambda}(X, \mathbb{X})$ in $\hat{\Pi}_{ij}^{(l)}$ for $\hat{\Pi}^{(1)} = \Pi^{(1)}$, $\hat{\Pi}^{(2)} = \Pi^{(2)}$, $l \in \{1,2\}$.

Theorem 10.4. *Assume that the classes Ω_1, Ω_2 are equiprobable ($q_1 = q_2 = 0.5$), $n_* = \min\{n_1, n_2\}$, and n is fixed. In the asymptotics* (10.8) *of contigual classes and increasing training sample lengths defined as*

$$n_* \to \infty, \quad n_l/n_* \to \lambda_l > 0, \quad l \in \{1,2\}, \quad \varepsilon = c n_*^{-1/2} \to 0, \quad 0 < c < \infty, \tag{10.22}$$

the following expansion holds for the PBDR risk:

$$r = r_0 + \frac{\tilde{c}}{\sqrt{n_*}} + o\left(\frac{1}{\sqrt{n_*}}\right),$$

$$\tilde{c} = \frac{cn}{2} \sum_{X \in \mathcal{A}^n \setminus \mathcal{A}^{(1)}} L_1(X) B_n(X) \Phi(-|\Delta(X)|) > 0, \quad \Delta(X) = \frac{c B_n(X)}{\sqrt{D_n(X)}}. \tag{10.23}$$

Proof. The discriminant function $\hat{\Lambda}(X, \mathbb{X}) = g\big(\hat{\Pi}^{(1)}, \hat{\Pi}^{(2)}\big)$ is a function of bigram statistics; the random variables $\{\sqrt{n}(\hat{\Pi}_{ij}^{(l)} - \Pi_{ij}^{(l)})\}$ are asymptotically normally distributed, as established in the proof of Theorem 10.3. By Anderson's theorem on functional transformations of asymptotically normal random variables [1], $\hat{\Lambda}(X, \mathbb{X})$ has an asymptotically normal distribution with the asymptotic mean value

$$\Lambda(X) = g\big(\mathbb{E}\{\hat{\Pi}^{(1)}\}, \mathbb{E}\{\hat{\Pi}^{(2)}\}\big)$$

and the asymptotic variance $\sigma^2(X)$:

$$\sigma^2(X) = \sum_{(i,j),(u,v) \in \mathcal{A}_\Pi} g_{ij}^{(1)}(X) g_{uv}^{(1)}(X) \frac{s_{ijuv}^{(1)}}{\lambda_1} + \sum_{(i,j),(u,v) \in \mathcal{A}_\Pi} g_{ij}^{(2)}(X) g_{uv}^{(2)}(X) \frac{s_{ijuv}^{(2)}}{\lambda_2};$$

or, in other words,

$$\mathcal{L}\left\{\sqrt{n_*}\,\frac{\hat{\Lambda}(X,\mathbb{X})-\Lambda(X)}{\sigma(X)}\right\}\to\mathcal{N}(0,1).$$

In the asymptotics (10.8), we have $\Pi_{ij}^{(2)}=\Pi_{ij}^{(1)}\left(1+\varepsilon(b_{ij}+h_{ij})\right)+O(\varepsilon^2)$. Let us write a Taylor expansion for $\Lambda(X)$ and $\sigma^2(X)$.

Let us start with the case $X\in\mathcal{A}^n\backslash(\mathcal{A}^{(1)}\cup\mathcal{A}^{(2)})$, i.e., $n_{ij}(X)\neq n_{i\cdot}(X)p_{ij}^{(l)}$, where $l\in\{1,2\}$. Then Taylor expansions of $\Lambda(X)$ and $\sigma^2(X)$ for an arbitrary fixed X yield

$$\Lambda(X)=\varepsilon B_n(X)+O(\varepsilon^2),\quad \sigma^2(X)=D_n(X)+O(\varepsilon),$$

where $B_n(X)=O(1)$, $D_n(X)=O(1)$. We have $D_n(X)>0$ since $g^{(1)}(X)\neq 0$, and the covariance matrix $\{s_{ijuv}^{(1)},\ (i,j),(u,v)\in\mathcal{A}_\Pi\}$ is nonsingular.

Now consider the case $X\in\mathcal{A}^{(1)}$, i.e., $n_{ij}(X)=n_{i\cdot}(X)p_{ij}^{(1)}$. Applying the normalization condition $\sum_{j\in\mathcal{A}}p_{ij}^{(1)}b_{ij}=0$, let us write the following Taylor expansions:

$$\Lambda(X)=\varepsilon\sum_{i\in\mathcal{A}}\frac{\delta_{ix_1}}{n}h_i+O(\varepsilon^2),$$

$$\sigma^2(X)=\varepsilon^2\sum_{(i,j),(u,v)\in\mathcal{A}_\Pi}\left(\frac{n_{i\cdot}(X)}{n\pi_i^{(1)}}b_{ij}+\frac{n_{N\cdot}(X)}{n\pi_N^{(1)}}b_{NN}\right)\times$$
$$\times\left(\frac{n_{u\cdot}(X)}{n\pi_u^{(1)}}b_{uv}+\frac{n_{N\cdot}(X)}{n\pi_N^{(1)}}b_{NN}\right)\frac{s_{ijuv}^{(2)}}{\lambda_2}+O(\varepsilon^3),$$

where $\sigma^2(X)>0$ since the covariance matrix $\{s_{ijuv}^{(1)},\ (i,j),(u,v)\in\mathcal{A}_\Pi\}$ is nonsingular.

A similar argument can be constructed for the case $X\in\mathcal{A}^{(2)}$.

Thus, in the asymptotics (10.8), (10.22), we obtain:

$$\sqrt{n_*}\frac{\Lambda(X)}{\sigma(X)}\to\Delta^*(X)=\begin{cases}\Delta(X), & \text{if } X\in\mathcal{A}^n\backslash\mathcal{A}^{(1)}\cup\mathcal{A}^{(2)},\\ \operatorname{sign}(\Lambda(X))\infty, & \text{if } X\in\mathcal{A}^{(1)}\cup\mathcal{A}^{(2)}.\end{cases}$$

Therefore, under the conditions of the theorem, the expectation in (10.21) can be written asymptotically as

$$\mathbb{E}\left\{\mathbf{1}\left(\hat{\Lambda}(X,\mathbb{X})\right)\right\} = 1 - \mathbb{P}\left\{\hat{\Lambda}(X,\mathbb{X}) < 0\right\} =$$

$$= 1 - \mathbb{P}\left\{\sqrt{n_*}\frac{\hat{\Lambda}(X,\mathbb{X}) - \Lambda(X)}{\sigma(X)} < -\sqrt{n_*}\frac{\Lambda(X)}{\sigma(X)}\right\} \to 1 - \Phi(-\Delta^*(X)).$$

From the above conclusion applied to (10.21), the finiteness of n, and the relation $\Phi(\Delta^*(X)) = 1 - \Phi(-\Delta^*(X))$, we can write

$$r = \frac{1}{2} - \sum_{X\in\mathcal{A}^n} \tilde{\Lambda}(X)\Phi(-\Delta^*(X)) + o(1) =$$

$$= \frac{1}{2} - \sum_{\substack{X\in\mathcal{A}^n\\ \Lambda(X)\geq 0}} \tilde{\Lambda}(X) + \sum_{\substack{X\in\mathcal{A}^n\\ \Lambda(X)<0}} \tilde{\Lambda}(X) - \sum_{X\in\mathcal{A}^n} \tilde{\Lambda}(X)\Phi(\Delta^*(X)) + o(1) =$$

$$= r_0 + \sum_{\substack{X\in\mathcal{A}^n\\ \Lambda(X)\geq 0}} \tilde{\Lambda}(X)(1 - \Phi(\Delta^*(X))) - \sum_{\substack{X\in\mathcal{A}^n\\ \Lambda(X)<0}} \tilde{\Lambda}(X)\Phi(\Delta^*(X)) + o(1).$$

From the relation $\operatorname{sign}\left(\tilde{\Lambda}(X)\right) = \operatorname{sign}(\Lambda(X)) = \operatorname{sign}(\Delta^*(X))$, $X \in \mathcal{A}^n$, we have

$$\sum_{\substack{X\in\mathcal{A}^n\\ \Lambda(X)\geq 0}} \tilde{\Lambda}(X)(1 - \Phi(\Delta^*(X))) =$$

$$= \sum_{\substack{X\notin\mathcal{A}^{(1)}\bigcup\mathcal{A}^{(2)}\\ \Lambda(X)\geq 0}} |\tilde{\Lambda}(X)|\Phi(-|\Delta(X)|) + \sum_{\substack{X\in\mathcal{A}^{(1)}\bigcup\mathcal{A}^{(2)}\\ \Lambda(X)\geq 0}} |\tilde{\Lambda}(X)|\Phi(-|\Delta^*(X)|),$$

$$\sum_{\substack{X\in\mathcal{A}^n,\\ \Lambda(X)<0}} \tilde{\Lambda}(X)\Phi(\Delta^*(X)) =$$

$$= \sum_{\substack{X\notin\mathcal{A}^{(1)}\bigcup\mathcal{A}^{(2)},\\ \Lambda(X)<0}} -|\tilde{\Lambda}(X)|\Phi(-|\Delta(X)|) - \sum_{\substack{X\in\mathcal{A}^{(1)}\bigcup\mathcal{A}^{(2)},\\ \Lambda(X)<0}} |\tilde{\Lambda}(X)|\Phi(-|\Delta^*(X)|);$$

$\Phi(-|\Delta^*(X)|) = 0$ for all $X \in \mathcal{A}^{(1)} \cup \mathcal{A}^{(2)}$. Thus, we obtain

$$r = r_0 + \sum_{X\notin\mathcal{A}^{(1)}\cup\mathcal{A}^{(2)}} |\tilde{\Lambda}(X)|\Phi(-|\Delta(X)|) + o(1). \tag{10.24}$$

Now consider $|\tilde{\Lambda}(X)| = \frac{1}{2}|L_2(X) - L_1(X)|$. By writing a Taylor expansion in ε, $\Lambda(X) = \varepsilon B_n(X) + O(\varepsilon^2)$ for $X \notin \mathcal{A}^{(1)} \cup \mathcal{A}^{(2)}$, we obtain

$$|\tilde{\Lambda}(X)| = \frac{L_1(X)}{2}\left|\frac{L_2(X)}{L_1(X)} - 1\right| = \frac{L_1(X)}{2}\left|e^{n\Lambda(X)} - 1\right| = \frac{nL_1(X)}{2}\left|\varepsilon B_n(X) + O(\varepsilon^2)\right|.$$

Substituting this expansion into (10.24), omitting the terms of order $O(\varepsilon^2)$, and making the substitution $\varepsilon = c/\sqrt{n_*}$ proves (10.23). □

10.2 HMC Forecasting Under Missing Values

Let us consider a common setting of applied statistical forecasting, where the realizations $X = (x_1, \ldots, x_n)$ of an HMC contain missing values. As in Chap. 8, let us consider a missing values template

$$M = (m_1, m_2, \ldots, m_n), \quad m_t \in \{0, 1\}, \quad t \in \{1, \ldots, n\}, \tag{10.25}$$

which is assumed to be deterministic, known and fixed. Here $m_t = 0$ means that the observation x_t is missing, and $m_t = 1$ means that the value x_t has been recorded. We assume that the first and the last values have been observed, $m_1 \equiv m_n \equiv 1$.

The binary vector M is, essentially, the design of the experiment.

10.2.1 *Likelihood Functions for HMCs with Missing Values*

Denote by T the number of segments of the realization X which do not contain missing values; in other words, T is equal to the number of series of ones in the binary vector M ($T \geq 2$). Let us present the observation results (X, M) as consecutive segments:

$$X = (x_1, \ldots, x_n) = \left(X_{(1)} \vdots \overline{X}_{(1)} \vdots X_{(2)} \vdots \ldots \vdots \overline{X}_{(T-1)} \vdots X_{(T)}\right),$$
$$X_{(t)} = (x_{(t),1}, x_{(t),2}, \ldots, x_{(t),M_t^*}), \quad t \in \{1, \ldots, T\},$$

where $X_{(t)}$ is the tth segment of length M_t^* in the realization X corresponding to the tth subsequence of ones in M; $\overline{X}_s$ is the sth missing segment of length $\overline{M}_s^*$ corresponding to the sth subsequence of zeros in M.

Theorem 10.5. *The likelihood function of the HMC parameters (π, P) under missing values has the following form:*

$$L(\pi, P; X, M) = \pi_{x_{(1),1}} \left(\prod_{t=1}^{T} L_t(P, X_{(t)})\right) \left(\prod_{t=1}^{T-1} p_{x_{(t),M_t^*}, x_{(t+1),1}}\left(\overline{M}_t^* + 1\right)\right), \tag{10.26}$$

where $p_{ij}(k) = (P^k)_{ij}$ *is the probability of a* k*-step transition from state* i *to state* j*;* $L_s(P, X_{(s)})$ *is the probability that a segment* $X_{(s)}$ *is observed, taken conditionally on the first element* $x_{(s),1}$ *being observed:*

$$L_s = L_s\big(P; X_{(s)}\big) = \prod_{t=1}^{M_s^*-1} p_{x_{(s),t},\, x_{(s),t+1}}. \tag{10.27}$$

Proof. To find the desired likelihood function, let us apply the total probability formula and sum the likelihood function for a "complete" realization of X over all of the missing segments:

$$L(\pi, P; X, M) = \sum_{\overline{X}_{(1)},\dots,\overline{X}_{(T-1)}} \pi_{x_1} \prod_{u=1}^{n-1} p_{x_u x_{u+1}}.$$

Taking the multipliers $L_s(P; X_{(s)})$ outside the parentheses, calculating a sum over $\overline{X}_{(t)} = (\overline{x}_{(t),1}, \dots, \overline{x}_{(t),\overline{M}_t^*})$, and applying the relation

$$\sum_{\overline{X}_{(t)}} p_{x_{(t),M_t^*},\overline{x}_{(t),1}} \left(\prod_{u=1}^{\overline{M}_t^*-1} p_{\overline{x}_{(t),u},\overline{x}_{(t),u+1}} \right) p_{\overline{x}_{(t),\overline{M}_t^*},\, x_{(t+1),1}} = p_{x_{(t),M_t^*},\, x_{(t+1),1}}(\overline{M}_t^* + 1)$$

yields (10.26). □

From (10.26), we can see that the likelihood function is a nonlinear function which cannot be easily maximized. Let us construct an approximation of this function.

Let

$$\overline{M}_-^* = \min_{1 \le t \le T-1} \overline{M}_t^*$$

be the length of the shortest missing value segment(s).

Theorem 10.6. *If a stationary Markov chain with parameters* (π, P) *is observed through a missing value template* (10.25)*, and there exists a positive integer* M_0 *such that*

$$\overline{M}_-^* \ge M_0, \quad \varrho = 1 - \min_{i,j \in A} p_{ij}(M_0) < 1,$$

then the following multiplicative approximation based on the likelihood functions of the observed segments $\{L(\pi, P; X_{(t)})\}$ *holds for the likelihood function* (10.26)*:*

$$L(\pi, P; X, M) = \prod_{t=1}^{T} L(\pi, P; X_{(t)}) + \delta(\pi, P; X, M),$$
$$\left|\frac{\delta(\pi, P; X, M)}{L(\pi, P; X, M)}\right| = O\left(T\varrho^{\overline{M}_{-}^{*} M_0^{-1}}\right), \quad (10.28)$$

where $L(\pi, P; X_{(t)}) = \pi_{x_{(t),1}} L_t(P; X_{(t)})$, *and* $L_t(\cdot\,;\cdot)$ *is defined by* (10.27).

Proof. Consider the case of a single missing segment ($X = (X_{(1)} \,\vdots\, \overline{X}_{(1)} \,\vdots\, X_{(2)})$, $T = 2$, and estimate the error of the approximation $\delta(\pi, P; X, M)$:

$$|\delta(\pi, P; X, M)| = \left|L(\pi, P; X, M) - L(\pi, P; X_{(1)}) L(\pi, P; X_{(2)})\right| =$$

$$= \left|\pi_{x_{(1),1}} L_1 L_2 p_{x_{(1),M_1^*}, x_{(2),1}}(\overline{M}_{-}^{*} + 1) - L(\pi, P; X_{(1)}) L(\pi, P; X_{(2)})\right| =$$

$$= \left|\pi_{x_{(1),1}} L_1 L_2 \left(p_{x_{(1),M_1^*}, x_{(2),1}}(\overline{M}_{-}^{*} + 1) - \pi_{x_{(2),1}} + \pi_{x_{(2),1}}\right) - \pi_{x_{(1),1}} L_1 \pi_{x_{(2),1}} L_2\right| =$$

$$= \pi_{x_{(1),1}} L_1 L_2 |p_{x_{(1),M_1^*}, x_{(2),1}}(\overline{M}_{-}^{*} + 1) - \pi_{x_{(2),1}}|.$$

Using the inequality

$$\left|p_{x_{(1),M_1^*}, x_{(2),1}}(\overline{M}_{-}^{*} + 1) - \pi_{x_{(2),1}}\right| \leq c\varrho^{[(\overline{M}_{-}^{*}+1)/M_0]-1}$$

which was proved in [3] yields

$$\left|\frac{\delta(\pi, P; X, M)}{L(\pi, P; X, M)}\right| \leq \frac{c}{p_{x_{(1),M_1^*}, x_{(2),1}}(\overline{M}_{-}^{*} + 1)} \varrho^{[(\overline{M}_{-}^{*}+1)/M_0]-1}.$$

The case $T > 2$ can be treated similarly. □

Note that if $M_0 = 1$, then $\varrho = 1 - \min_{i,j \in \mathcal{A}} p_{ij}$, $\varrho \in (0, 1)$.

Corollary 10.3. *Under the conditions of Theorem 10.6, in the asymptotics where the number of segments increases, and the lengths of the segments are increasing,*

$$T \to \infty, \quad \overline{M}_{-}^{*} \to \infty, \quad T\varrho^{\overline{M}_{-}^{*}} \to 0, \quad (10.29)$$

the following almost sure convergence holds:

$$\left|\frac{\delta(\pi, P; X, M)}{L(\pi, P; X, M)}\right| \xrightarrow{\text{a.s.}} 0.$$

Thus, under the conditions of Corollary 10.3, the observed "complete" segments $\{X_{(t)}\}$ can be considered as negligibly dependent subrealizations of the same HMC with the parameters (π, P). We are going to assume the case of asymptotics (10.29), where the approximation error in (10.28) is negligible:

$$L(\pi, P; X, M) = \prod_{t=1}^{T} L(\pi, P; X_{(t)}). \tag{10.30}$$

This allows us to apply the results of Sect. 10.1 in the case of HMCs with missing values.

10.2.2 The Decision Rule for Known $\{\pi^{(l)}, P^{(l)}\}$

Let $M^* = \sum_{t=1}^{T} M_t^*$ be the total number of registered values in the realization (X, M).

Theorem 10.7. *In the asymptotics*

$$M^*, T, \overline{M^*_-} \to \infty, \quad T\varrho_l{}^{\overline{M^*_-}} \to 0, \quad l \in \{1, \ldots, L\}, \tag{10.31}$$

the BDR based on the approximated likelihood function (10.30) *for the model* (10.1), (10.3), (10.25) *can be written as*

$$d(X) = \arg\max_{1 \le l \le L} \left(\frac{1}{M^*} \log q_l + \frac{1}{M^*} \sum_{i \in \mathcal{A}} \nu_i \log \pi_i^{(l)} + \sum_{i,j \in \mathcal{A}} \hat{\Pi}_{ij} \log p_{ij}^{(l)} \right),$$

$$\hat{\Pi}_{ij} = \frac{n_{ij}}{M^*}, \quad n_{ij} = \sum_{t=1}^{n-1} m_t m_{t+1} \mathbf{I}\{x_t = i,\ x_{t+1} = j\}, \quad \nu_i = \sum_{t=1}^{T} \mathbf{I}\{x_{(t),1} = i\}, \tag{10.32}$$

where $i, j \in \mathcal{A}$, $\varrho_l = 1 - \min_{i,j \in \mathcal{A}} p_{ij}^{(l)}$.

Proof. Repeat the proof of Theorem 10.1, substituting the approximation (10.30) for the likelihood function. □

Theorem 10.8. *For* $L = 2$, *in the asymptotics* (10.31) *and the asymptotics of contigual classes* (10.8), *if we have*

$$\varepsilon = \frac{c}{\sqrt{M^*}} \to 0, \quad \varepsilon T \to 0, \quad 0 < c < \infty,$$

then the error probability (10.4) *for the BDR* (10.32) *has the following limit:*

$$r_0 \rightarrow \tilde{r}_0 = q_1 \Phi\left(-\Delta_1/2\right) + q_2 \Phi\left(-\Delta_2/2\right),$$

where Δ_1, Δ_2 *are defined by* (10.13).

Proof. Under the conditions of this theorem, we have the approximation (10.30) for the likelihood function. Thus, the statistical estimators $\{\hat{\Pi}_{ij}\}$ based on incomplete data have the same asymptotic properties as in the case of complete observations. The rest of the proof is the same as in Theorem 10.2. □

10.2.3 The Case of Unknown Parameters

Now assume that the parameters $\{\pi^{(l)}, P^{(l)}\}$ are unknown, and a classified training sample, which (like the observations considered earlier) may contain missing values, has been observed:

$$\mathbb{X} = \left\{\left(X^{(1)}, M^{(1)}\right), \left(X^{(2)}, M^{(2)}\right), \ldots, \left(X^{(L)}, M^{(L)}\right)\right\},$$

where for the lth realization $X^{(l)}$ of length n_l from the class Ω_l we have an a priori known missing values template $M^{(l)} = \left(m_1^{(l)}, m_2^{(l)}, \ldots, m_{n_l}^{(l)}\right)$ with $m_t^{(l)} \in \{0, 1\}$, $t \in \{1, \ldots, n_l\}$, $l \in \{1, \ldots, L\}$.

Let us introduce the following notation: $T_l \geq 2$ is the number of segments without missing values in the realization $X^{(l)}$; $X_{(t)}^{(l)}$ is the tth observed segment of the realization $X^{(l)}$ corresponding to the tth series of ones in the missing values template $M^{(l)}$; $M_{(l)t}^{*}$ is the length of the segment $X_{(t)}^{(l)}$, $t \in \{1, \ldots, T_l\}$; $\overline{X}_{(s)}^{(l)}$ is the sth missing segment of the realization $X^{(l)}$ corresponding to the sth sequence of zeros in $M^{(l)}$; $\overline{M}_{(l)s}^{*}$ is the length of $\overline{X}_{(s)}^{(l)}$, $s \in \{1, \ldots, T_l - 1\}$; $M_{(l)}^{*} = \sum_{t=1}^{T_l} M_{(l),t}^{*}$ is the total number of recorded observations in $X^{(l)}$; $\overline{M}_{(l),-}^{*} = \min_{1 \leq s \leq T_l - 1} \overline{M}_{(l),s}^{*}$ is the minimal length of a missing segment in $X^{(l)}$, $l \in \{1, \ldots, L\}$.

The asymptotics

$$T_l \rightarrow \infty, \quad \overline{M}_{(l),-}^{*} \rightarrow \infty, \quad T_l \varrho_l^{\overline{M}_{(l),-}^{*}} \rightarrow 0, \quad l \in \{1, \ldots, L\}, \tag{10.33}$$

allow us to use the approximation (10.30) for all realizations from $\mathbb{X}$.

As in Sect. 10.1, let us construct a PBDR:

$$d(X,\mathbb{X}) = \arg\max_{1\le l\le L}\left(\frac{1}{M^*}\log q_l + \frac{1}{M^*}\sum_{i\in A} \nu_i \log\hat{\pi}_i^{(l)} + \sum_{i,j\in A}\hat{\Pi}_{ij}\log\hat{p}_{ij}^{(l)}\right),$$

$$\hat{\Pi}_{ij} = \frac{n_{ij}}{M^*}, \quad n_{ij} = \sum_{t=1}^{n-1} m_t m_{t+1}\mathbf{I}\{x_t = i,\ x_{t+1} = j\},$$

$$\hat{p}_{ij}^{(l)} = \frac{n_{ij}^{(l)}}{n_{i\cdot}^{(l)}}, \quad \hat{\pi}_i^{(l)} = \frac{n_{i\cdot}^{(l)}}{n_l}, \quad n_{ij}^{(l)} = \sum_{t=1}^{n_l-1} m_t^{(l)} m_{t+1}^{(l)}\mathbf{I}\left\{x_t^{(l)} = i,\ x_{t+1}^{(l)} = j\right\}, \tag{10.34}$$

where the bigram frequencies $\{n_{ij}^{(l)}\}$ are computed from the observed segments of the realization $X^{(l)}, i, j \in A, l \in \{l, \ldots, L\}$.

Theorem 10.9. *For $L = 2$, in the asymptotics* (10.31), (10.33), (10.8), *if we have*

$$M_{(l)}^* \to \infty, \quad \frac{M_{(l)}^*}{M^*} \to \tilde{\lambda}_l > 0, \quad \varepsilon = \frac{c}{\sqrt{M^*}} \to 0, \quad T\varepsilon \to 0, \quad 0 < c < \infty,$$

then the error probability (10.4) *for the PBDR* (10.34) *tends to a limit:*

$$r \to \tilde{r} = q_1\Phi\left(-\tilde{\Delta}_1/2\right) + q_2\Phi\left(-\tilde{\Delta}_2/2\right),$$

where $\tilde{\Delta}_1, \tilde{\Delta}_2$ are defined by (10.18).

Proof. Under the conditions of the theorem, we have the approximation (10.30) for the likelihood function of $(X, M), (X^{(l)}, M^{(l)})$. Thus, in the considered asymptotics, the statistical estimators $\{\hat{\Pi}_{ij}\}, \{\hat{p}_{ij}^{(l)}\}$ based on incomplete data have the same asymptotic properties as in the case of complete observations. The rest of the proof is the same as in Theorem 10.3. □

10.3 Forecasting in the Beta-Binomial Model Under Additive Distortions

10.3.1 The Beta-Binomial Model, Its Properties and Distortions

The beta-binomial distribution (BBD) is a mixture of the binomial and beta probability distributions introduced by Pearson [31] and formalized by Skellam [34]. The BBD is widely used to model collections of observed sequences of random binary outcomes in medicine, economics, marketing, and other disciplines. Parameters of BBDs are traditionally estimated by applying the moment method, the maximum likelihood method, the χ^2 minimum method, or the Bayesian approach [35].

All of these classical methods have been extensively studied for undistorted BBD models. In this section, we are going to evaluate the robustness of the classical methods under stochastic distortion of the binary data and propose new parameter estimation methods improving the robustness of statistical inferences [23].

Assume that a collection of K objects and a random event A have been defined. A series of n trials is performed on each object in the collection. The outcomes of the trials are described by a binary $(K \times n)$-matrix

$$B = (b_{ij}), \quad i = 1, \dots, K, \;\; j = 1, \dots, n.$$

Here b_{ij} is 1 (or 0) if the event A has occurred (or hasn't occurred) for the ith object in the jth trial. Furthermore, let the following assumptions be satisfied:

A1. Stochastic properties of each object do not change between trials.
A2. Stochastic properties of the objects in the collection are not uniform. For the ith object, the probability p_i of the event A is a random variable following a beta probability distribution with the parameters α, β; the random variables

$$p_1, p_2, \dots, p_K$$

are jointly independent.

Assume that the binary matrix B is subject to binary random errors $\{\eta_{ij}\}$, and a distorted binary matrix $\tilde{B} = (\tilde{b}_{ij})$ is observed:

$$\tilde{b}_{ij} = b_{ij} \oplus \eta_{ij}, \quad \mathbb{P}\{\eta_{ij} = 1 \mid b_{ij} = 0\} = \varepsilon_0, \qquad \mathbb{P}\{\eta_{ij} = 1 \mid b_{ij} = 1\} = \varepsilon_1, \tag{10.35}$$

where $\oplus$ is addition Mod 2, $\{\eta_{ij}\}$ are independent Bernoulli random variables, $\varepsilon_0, \varepsilon_1 \in [0, 1]$ are distortion levels. If we have $\varepsilon_0 = \varepsilon_1 = 0$, then we have the undistorted hypothetical observation model.

In this section, we are going to consider the problem of statistical estimation of the parameters α, β based on the distorted data $\tilde{B}$, study the robustness of classical methods of estimating the parameters α, β under the distortion (10.35), and finally construct new estimators which are robust under the distortion (10.35).

Lemma 10.2. *The random variable $x_i = \sum_{j=1}^{n} \tilde{b}_{ij}$ has a distorted beta-binomial distribution (DBBD) with the parameters n, α, β, ε_0, ε_1: for $i = 1, 2, \dots, K$, we have*

$$\mathbb{P}_r(\alpha, \beta, \varepsilon_0, \varepsilon_1) = P\{x_i = r\} = \sum_{i=0}^{n} w_{ri}(\varepsilon_0, \varepsilon_1)\, C_n^i \frac{\mathrm{B}(\alpha + i, \beta + n - i)}{\mathrm{B}(\alpha, \beta)}, \tag{10.36}$$

$$w_{ri}(\varepsilon_0, \varepsilon_1) = \sum_{l=\max(i,r)}^{\min(n,i+r)} C_i^{l-r} C_{n-i}^{l-i} \varepsilon_0^{l-i} (1 - \varepsilon_0)^{n-l} \varepsilon_1^{l-r} (1 - \varepsilon_1)^{i+r-l}$$

where $r = 0, 1, \dots, n$, and $\mathrm{B}(\cdot, \cdot)$ is the beta function.

Proof. The probability distribution (10.36) can be obtained from (10.35) by the same argument as in the case without distortion ($\varepsilon_0 = \varepsilon_1 = 0$) [14]. □

Note that in the absence of distortion, i.e., for $\varepsilon_0 = \varepsilon_1 = 0$, the above distribution is the same as a BBD with the parameters n, α, β [14].

10.3.2 Robustness of MM Estimators

Let us introduce the following notation: α_0, β_0 are the true values of the parameters α, β; the estimators $\tilde{\alpha}_{\mathrm{MM}}(\varepsilon_0, \varepsilon_1), \tilde{\beta}_{\mathrm{MM}}(\varepsilon_0, \varepsilon_1)$ are the classical MM estimators of the parameters α, β computed from the distorted sample X, where the distortion level is equal to $\varepsilon_0, \varepsilon_1$. Also denote

$$y^{[z-]} ::= y(y-1)\dots(y-z+1), \quad y^{[z+]} ::= y(y+1)\dots(y+z-1), \quad y \in \mathbb{R}, z \in N.$$

Theorem 10.10. *Under binary distortion of levels $\varepsilon_0, \varepsilon_1$ defined by* (10.35)*, the following stochastic expansions can be written for the MM estimators of the BBD parameters:*

$$\tilde{\alpha}_{MM}(\varepsilon_0, \varepsilon_1) = \alpha_0 + (\alpha_0 + 2\beta_0 + 1) \times \varepsilon_0 + \\ + \alpha_0(\alpha_0 + 1)\beta_0{}^{-1}\,\varepsilon_1 + o(\varepsilon_0) + o(\varepsilon_1) + O_P(1/\sqrt{K}),$$

$$\tilde{\beta}_{MM}(\varepsilon_0, \varepsilon_1) = \beta_0 + \beta_0(\beta_0 + 1)/\alpha_0{}^{-1}\,\varepsilon_0 + \\ + (2\alpha_0 + \beta_0 + 1)\,\varepsilon_1 + o(\varepsilon_0) + o(\varepsilon_1) + O_P(1/\sqrt{K}).$$

Proof. By [14], the MM estimators for the BBD are equal to

$$\tilde{\alpha}_{MM} = \frac{(n - \bar{x} - s^2/\bar{x})\bar{x}}{(s^2/\bar{x} + \bar{x}/n - 1)n}, \quad \tilde{\beta}_{MM} = \frac{(n - \bar{x} - s^2/\bar{x})(n - \bar{x})}{(s^2/\bar{x} + \bar{x}/n - 1)n}, \tag{10.37}$$

where $\bar{x}$ and s^2 are, respectively, the sample mean and the sample variance of the observations $\{x_1, \dots, x_k\}$. From Lemma 10.2, we have:

$$m_1(\varepsilon_0, \varepsilon_1) = n\alpha/(\alpha + \beta) + n\beta(\alpha + \beta)^{-1}\,\varepsilon_0 - n\alpha(\alpha + \beta)^{-1}\,\varepsilon_1, \tag{10.38}$$

$$m_2(\varepsilon_0, \varepsilon_1) = m_1 + \\ + \frac{n^{[2+]}}{(\alpha + \beta)^{[2+]}}\left(\alpha^{[2+]} + \beta^{[2+]}\varepsilon_0^2 + \alpha^{[2+]}\varepsilon_1^2 - 2\alpha\beta\varepsilon_0 - \alpha^{[2+]}\varepsilon_1 - 2\alpha\beta\varepsilon_0\varepsilon_1\right). \tag{10.39}$$

Since $\bar{x}$ and s^2 are unbiased strictly consistent estimators as $K \to \infty$, and, from [4], $\mathbb{D}\{\bar{x}\} = d(\varepsilon_0, \varepsilon_1)K^{-1}$, $\mathbb{D}\{s^2\} = O(1/K)$, we have

$$\bar{x} = m(\varepsilon_0, \varepsilon_1) + O_P(1/\sqrt{K}), \quad s^2 = d(\varepsilon_0, \varepsilon_1) + O_P(1/\sqrt{K}).$$

Applying these asymptotic equalities to (10.37) and using the probability convergence properties of the remainder term $O_P(\cdot)$ yields

$$\tilde{\alpha}_{MM}(\varepsilon_0, \varepsilon_1) = \frac{(n - m(\varepsilon_0, \varepsilon_1) - d(\varepsilon_0, \varepsilon_1)/m(\varepsilon_0, \varepsilon_1))m(\varepsilon_0, \varepsilon_1)}{(d(\varepsilon_0, \varepsilon_1)/m(\varepsilon_0, \varepsilon_1) + m(\varepsilon_0, \varepsilon_1)/n - 1)n} + O_P(1/\sqrt{K}),$$

$$\tilde{\beta}_{MM}(\varepsilon_0, \varepsilon_1) = \frac{(n - m(\varepsilon_0, \varepsilon_1) - d(\varepsilon_0, \varepsilon_1)/m(\varepsilon_0, \varepsilon_1))(n - m(\varepsilon_0, \varepsilon_1))}{(d(\varepsilon_0, \varepsilon_1)/m(\varepsilon_0, \varepsilon_1) + m(\varepsilon_0, \varepsilon_1)/n - 1)n} + O_P(1/\sqrt{K}).$$

Now applying (10.38), (10.39), and writing Taylor's linear expansions in ε_0, ε_1 with the remainder terms in the Peano form proves the theorem. □

By Theorem 10.10, MM estimators are inconsistent under distortion (10.35).

10.3.3 Robustness of MLEs

Now let $\tilde{\alpha}_{\text{MLE}}(\varepsilon_0, \varepsilon_1)$, $\tilde{\beta}_{\text{MLE}}(\varepsilon_0, \varepsilon_1)$ be the classical MLEs for the parameters α, β, based on the distorted sample $X = \{x_1, \ldots, x_K\}$ at distortion levels ε_0, ε_1. Let us introduce the following notation:

$$p_j(\varepsilon_0, \varepsilon_1) = \mathbb{P}\{x_1 = j\}, \ j = 0, 1, \ldots, n; \qquad P_i(\varepsilon_0, \varepsilon_1) = \sum_{j=0}^{i} p_j(\varepsilon_0, \varepsilon_1), \ i = 0, 1, \ldots, n;$$

$$S_\alpha = \sum_{i=0}^{n-1} \frac{1 - P_i(0,0)}{(\alpha_0 + i)^2}, \qquad S_\beta = \sum_{i=0}^{n-1} \frac{P_i(0,0)}{(\beta_0 + n - i - 1)^2},$$

$$S_{\alpha\beta} = \sum_{i=0}^{n-1} \frac{1}{(\alpha_0 + \beta_0 + i)^2}, \qquad S_{\alpha p} = -\sum_{i=0}^{n-1} \frac{(n-i)p_i(0,0)}{\alpha_0 + i},$$

$$S_{\alpha p}^{+} = \sum_{i=0}^{n-1} \frac{(i+1)p_{i+1}(0,0)}{\alpha_0 + i}, \qquad S_{\beta p} = \sum_{i=0}^{n-1} \frac{(n-i)p_i(0,0)}{\beta_0 + n - i - 1},$$

$$S_{\beta p}^{+} = -\sum_{i=0}^{n-1} \frac{(i+1)p_{i+1}(0,0)}{\beta_0 + n - i - 1}, \qquad G = (G_{ij}), \ H = (H_{ij}) \in \mathbb{R}^{2\times 2},$$

$$H_{11} = S_{\alpha\beta} - S_\alpha, \qquad H_{12} = H_{21} = S_{\alpha\beta},$$

$$H_{22} = S_{\alpha\beta} - S_\beta, \qquad G_{11} = S_{\alpha p},$$

$$G_{12} = S_{\alpha p}^{+}, \ G_{21} = S_{\beta p}^{+}, \qquad G_{22} = S_{\beta p}.$$

Theorem 10.11. *In the above setting of independent trials* (10.35) *under distortion of levels* ε_0, ε_1, *the following stochastic expansions hold for the differences of the BBD parameter MLEs and the true values:*

$$\begin{pmatrix} \Delta\tilde{\alpha}_{MLE}(\varepsilon_0,\varepsilon_1) \\ \Delta\tilde{\beta}_{MLE}(\varepsilon_0,\varepsilon_1) \end{pmatrix} =$$

$$= \begin{pmatrix} H_{11} & H_{12} \\ H_{21} & H_{22} \end{pmatrix}^{-1} \begin{pmatrix} G_{11} & G_{12} \\ G_{21} & G_{22} \end{pmatrix} \begin{pmatrix} \varepsilon_0 \\ \varepsilon_1 \end{pmatrix} + \begin{pmatrix} o(\varepsilon_0) + o(\varepsilon_1) + O_P(1/\sqrt{K}) \\ o(\varepsilon_0) + o(\varepsilon_1) + O_P(1/\sqrt{K}) \end{pmatrix},$$

where $\Delta\tilde{\alpha}_{MLE}(\varepsilon_0,\varepsilon_1) = \tilde{\alpha}_{MLE}(\varepsilon_0,\varepsilon_1) - \alpha_0$, $\Delta\tilde{\beta}_{MLE}(\varepsilon_0,\varepsilon_1) = \tilde{\beta}_{MLE}(\varepsilon_0,\varepsilon_1) - \beta_0$.

Proof. The system of equations for finding the MLEs can be written as follows [14]:

$$\sum_{i=0}^{n-1} \frac{K - F_i}{\alpha + i} - \sum_{i=0}^{n-1} \frac{K}{\alpha+\beta+i} = 0, \quad \sum_{i=0}^{n-1} \frac{F_i}{\beta + n - i - 1} - \sum_{i=0}^{n-1} \frac{K}{\alpha+\beta+i} = 0, \tag{10.40}$$

where

$$F_i = f_0 + f_1 + \cdots + f_i, \quad f_j = \sum_{t=1}^{K} \delta_{x_t,j}, \quad j = 0,1,\ldots,n.$$

This system has a unique solution, which corresponds to the maximum value of the likelihood function [14]. By construction, the frequencies f_i are binomially distributed:

$$\mathcal{L}\{f_i\} = \mathrm{Bi}\left(K, p_i(\varepsilon_0,\varepsilon_1)\right), \quad i = 0,1,\ldots,n.$$

It is a well-known fact that in the case of discrete distributions, the relative frequencies $\tilde{f}_i = f_i/K$ are unbiased strictly consistent estimators of the corresponding theoretical probabilities [2], and

$$\mathbb{D}\{\tilde{f}_i\} = p_i(\varepsilon_0,\varepsilon_1)\big(1 - p_i(\varepsilon_0,\varepsilon_1)\big)/K,$$

therefore

$$\tilde{f}_i = f_i/K = p_i(\varepsilon_0,\varepsilon_1) + O_P(1/\sqrt{K}), \quad i = 0,1,\ldots,n.$$

This leads to the following representation of the system (10.40):

$$\sum_{i=0}^{n-1} \frac{1 - P_i(\varepsilon_0,\varepsilon_1)}{\alpha + i} - \sum_{i=0}^{n-1} \frac{1}{\alpha+\beta+i} + O_P\left(\frac{1}{\sqrt{K}}\right) = 0,$$

$$\sum_{i=0}^{n-1} \frac{P_i(\varepsilon_0,\varepsilon_1)}{\beta + n - i - 1} - \sum_{i=0}^{n-1} \frac{1}{\alpha+\beta+i} + O_P\left(\frac{1}{\sqrt{K}}\right) = 0.$$

Linearization of this system of equations in ε_0, ε_1 in a neighborhood of the point $(\alpha_0, \beta_0, 0, 0)$ yields the expressions

$$A^0_\alpha \Delta\tilde{\alpha}_{MLE}(\varepsilon_0, \varepsilon_1) + A^0_\beta \Delta\tilde{\beta}_{MLE}(\varepsilon_0, \varepsilon_1) + A^0_{\varepsilon_0}\varepsilon_0 + A^0_{\varepsilon_1}\varepsilon_1 +$$
$$+ o(\varepsilon_0) + o(\varepsilon_1) + O_P(1/\sqrt{K}) = 0,$$
$$B^0_\alpha \Delta\tilde{\alpha}_{MLE}(\varepsilon_0, \varepsilon_1) + B^0_\beta \Delta\tilde{\beta}_{MLE}(\varepsilon_0, \varepsilon_1) + B^0_{\varepsilon_0}\varepsilon_0 + B^0_{\varepsilon_1}\varepsilon_1 +$$
$$+ o(\varepsilon_0) + o(\varepsilon_1) + O_P(1/\sqrt{K}) = 0,$$

where the coefficients of the system are the respective easily computed derivatives. Expressing $\Delta\tilde{\alpha}_{\text{MLE}}(\varepsilon_0, \varepsilon_1)$, $\Delta\tilde{\beta}_{\text{MLE}}(\varepsilon_0, \varepsilon_1)$ in terms of ε_0, ε_1 proves the theorem. □

This theorem shows that under BBD models, introducing distortion into statistical samples leads to inconsistency of MLEs. Let us construct some estimators which remain consistent under distortion.

10.3.4 MM Estimators Under Distortion of A Priori Known Levels

Let us consider the case where the distortion levels ε_0, ε_1 are a priori known. Let us introduce the notation

$$m_1^* = \bar{x} = K^{-1}\sum_{i=1}^K x_i, \quad m_2^* = K^{-1}\sum_{i=1}^K x_i^2.$$

Theorem 10.12. *In the above setting of independent trials* (10.35), *assuming a priori known distortion levels* ε_0, ε_1, *the MM estimators for the parameters* α, β *adapted for the distortion model* (10.35) *can be defined as*

$$\hat{\alpha}_{MM}(\varepsilon_0, \varepsilon_1) = \frac{(m_1^* - \varepsilon_0 n)\left(m_1^* n - m_2^* + \varepsilon_0(n-1)(m_1^* - (1-\varepsilon_1)n) - m_1^*\varepsilon_1(n-1)\right)}{(1-\varepsilon_0-\varepsilon_1)(m_2^* n - m_1^* n - m_1^{*2}(n-1))}, \tag{10.41}$$

$$\hat{\beta}_{MM}(\varepsilon_0, \varepsilon_1) =$$
$$= \frac{(m_1^* - (1-\varepsilon_1)n)\left(m_2^* + m_1^*\varepsilon_1(n-1) - m_1^* n - \varepsilon_0(n-1)(m_1^* - n(1-\varepsilon_1))\right)}{(1-\varepsilon_0-\varepsilon_1)(m_2^* n - m_1^* n - m_1^{*2}(n-1))}. \tag{10.42}$$

Proof. By applying Lemma 10.1 and (10.38), (10.39), it can be shown that the MM parameter estimators for α, β can be found by solving a system of equations:

$$m_1^* = n\alpha/(\alpha+\beta) + n\beta/(\alpha+\beta)\,\varepsilon_0 - n\alpha/(\alpha+\beta)\,\varepsilon_1, \tag{10.43}$$

$$m_2^* = m_1^* + \frac{n^{[2+]}}{(\alpha+\beta)^{[2+]}} \left(\alpha^{[2+]} + \beta^{[2+]}\varepsilon_0^2 + \alpha^{[2+]}\varepsilon_1^2 - \right.$$
$$\left. -2\alpha\beta\varepsilon_0 - \alpha^{[2+]}\varepsilon_1 - 2\alpha\beta\varepsilon_0\varepsilon_1\right). \quad (10.44)$$

By substituting $u = \alpha/(\alpha+\beta)$, $v = (\alpha+1)/(\alpha+\beta+1)$, $\alpha = u(1-v)/(v-u)$, $\beta = (1-v)(1-u)/(v-u)$, this system can be rewritten as

$$m_1^* = n\big(u + (1-u)\varepsilon_0 - u\varepsilon_1\big),$$
$$m_2^* = m_1^* + n(n-1)\big(vu(1-\varepsilon_0-\varepsilon_1) + \varepsilon_0^2 + 2u\varepsilon_0(1-\varepsilon_0-\varepsilon_1)\big).$$

Solving it in u, v and then calculating α, β proves the theorem. □

Corollary 10.4. *The estimators* (10.41), (10.42) *defined in Theorem 10.12 are consistent.*

Proof. Let $h(\cdot)$ be a vector function constructed from the system (10.43), (10.44): $h(\alpha,\beta) = (m_1(\alpha,\beta), m_2(\alpha,\beta))^T$, where $m_1(\alpha,\beta)$, $m_2(\alpha,\beta)$ are, respectively, the first and the second moments of the distorted beta-binomial random variable defined by the parameters n, α, β, ε_0, ε_1 found from Lemma 10.1 and (10.38), (10.39). To prove the consistency of MM estimators in this setting, it suffices to show that the function $h(\cdot)$ is continuous and invertible [2]. The continuity of the function $h(\cdot)$ in its domain of definition $\alpha > 0$, $\beta > 0$ is obvious. Invertibility follows from Theorem 10.12. □

Parameter estimators for α, β defined by (10.41), (10.42) will be referred to as modified MM estimators (MMM estimators).

10.3.5 Joint Estimation of Probability Distribution Parameters and Distortion Levels

Now consider the case where both the parameters α, β and the distortion levels ε_0, ε_1 are unknown. Assume that the first five sample moments m_1^*, m_2^*, m_3^*, m_4^*, m_5^* have been computed for the sample X. Theorem 10.12 implies that the problem of MM estimation is then reduced to solving a system of two nonlinear equations written as

$$\begin{aligned} m_3^* &= m_3\big(\alpha(\varepsilon_0,\varepsilon_1), \beta(\varepsilon_0,\varepsilon_1), \varepsilon_0, \varepsilon_1\big), \\ m_4^* &= m_4\big(\alpha(\varepsilon_0,\varepsilon_1), \beta(\varepsilon_0,\varepsilon_1), \varepsilon_0, \varepsilon_1\big), \end{aligned} \quad (10.45)$$

where $\alpha(\varepsilon_0,\varepsilon_1)$, $\beta(\varepsilon_0,\varepsilon_1)$ are the MMM estimators constructed for given values of ε_0, ε_1 from m_1^*, m_2^* by the formulas (10.41), (10.42). Let us solve (10.45) by applying the modified Newton's method. The Jacobian matrix of the system (10.45) under the condition that the first two moments are fixed will be denoted as J_0^{cond}. The system (10.45) can be solved by the following iterative procedure [7]:

$$\begin{pmatrix} \varepsilon_0^{k+1} \\ \varepsilon_1^{k+1} \end{pmatrix} = \begin{pmatrix} \varepsilon_0^{k} \\ \varepsilon_1^{k} \end{pmatrix} + \lambda(J_0^{\mathrm{cond}}) \begin{pmatrix} m_3^* - m_3\big(\alpha(\varepsilon_0^k, \varepsilon_1^k), \beta(\varepsilon_0^k, \varepsilon_1^k), \varepsilon_0^k, \varepsilon_1^k\big) \\ m_4^* - m_4\big(\alpha(\varepsilon_0^k, \varepsilon_1^k), \beta(\varepsilon_0^k, \varepsilon_1^k), \varepsilon_0^k, \varepsilon_1^k\big) \end{pmatrix}, \tag{10.46}$$

where the parameter $\lambda \in (0, 1]$ ensures convergence for larger values of $\varepsilon_0, \varepsilon_1$.

Let $\left(\frac{\partial f}{\partial x}\right)_0^c$ denote the conditional partial derivative of a function $f(\cdot)$ in x under the conditions $m_1^* = \mathrm{const}$, $m_2^* = \mathrm{const}$, $\varepsilon_0 = 0$, $\varepsilon_1 = 0$. The following lemmas, which follow from direct computations and the properties of ordinary DBBD moments, will be used to compute the Jacobian in the iterative procedure (10.46).

Lemma 10.3. *If* $m_1(\alpha, \beta, \varepsilon_0, \varepsilon_1) = \mathrm{const}$, $m_2(\alpha, \beta, \varepsilon_0, \varepsilon_1) = \mathrm{const}$, *then the following formulas are satisfied for the conditional partial derivatives of the functions* $m_3(\alpha(\varepsilon_0, \varepsilon_1), \beta(\varepsilon_0, \varepsilon_1), \varepsilon_0, \varepsilon_1)$ *and* $m_4(\alpha(\varepsilon_0, \varepsilon_1), \beta(\varepsilon_0, \varepsilon_1), \varepsilon_0, \varepsilon_1)$*:*

$$\left(\frac{\partial m_3}{\partial \alpha}\right)_0^c = n^{[3-]} \frac{\alpha^{[3+]}}{(\alpha+\beta)^{[3+]}} \sum_{i=0}^{2} \left(\frac{1}{\alpha+i} - \frac{1}{\alpha+\beta+i} \right),$$

$$\left(\frac{\partial m_3}{\partial \beta}\right)_0^c = n^{[3-]} \frac{\alpha^{[3+]}}{(\alpha+\beta)^{[3+]}} \sum_{i=0}^{2} \left(-\frac{1}{\alpha+\beta+i} \right),$$

$$\left(\frac{\partial m_4}{\partial \alpha}\right)_0^c = n^{[4-]} \frac{\alpha^{[4+]}}{(\alpha+\beta)^{[4+]}} \sum_{i=0}^{3} \left(\frac{1}{\alpha+i} - \frac{1}{\alpha+\beta+i} \right) + 6\left(\frac{\partial m_3}{\partial \alpha}\right)_0^c,$$

$$\left(\frac{\partial m_4}{\partial \beta}\right)_0^c = n^{[4-]} \frac{\alpha^{[4+]}}{(\alpha+\beta)^{[4+]}} \sum_{i=0}^{3} \left(-\frac{1}{\alpha+\beta+i} \right) + 6\left(\frac{\partial m_3}{\partial \beta}\right)_0^c,$$

$$\left(\frac{\partial m_3}{\partial \varepsilon_0}\right)_0^c = 3n^{[3-]} \frac{\alpha^{[2+]}\beta}{(\alpha+\beta)^{[3+]}}, \qquad \left(\frac{\partial m_3}{\partial \varepsilon_1}\right)_0^c = 3n^{[3-]} \frac{\alpha^{[3+]}}{(\alpha+\beta)^{[3+]}},$$

$$\left(\frac{\partial m_4}{\partial \varepsilon_0}\right)_0^c = 14n^{[4-]} \frac{\alpha^{[3+]}\beta}{(\alpha+\beta)^{[4+]}} + 6\left(\frac{\partial m_3}{\partial \varepsilon_0}\right)_0^c,$$

$$\left(\frac{\partial m_4}{\partial \varepsilon_1}\right)_0^c = 4n^{[4-]} \frac{\alpha^{[4+]}\beta}{(\alpha+\beta)^{[4+]}} + 6\left(\frac{\partial m_3}{\partial \varepsilon_1}\right)_0^c.$$

Lemma 10.4. *Assuming fixed distortion levels* ε_0, ε_1, *consider MM parameter estimators* $\hat{\alpha}(\varepsilon_0, \varepsilon_1)$, $\hat{\beta}(\varepsilon_0, \varepsilon_1)$ *for the DBBD defined in Theorem 10.12. We have:*

$$\left(\frac{\partial \hat{\alpha}}{\partial \varepsilon_0}\right)_0^c = -(\alpha_0 + 2\beta_0 + 1), \qquad \left(\frac{\partial \hat{\alpha}}{\partial \varepsilon_1}\right)_0^c = -\alpha_0(\alpha_0+1)/\beta_0,$$

$$\left(\frac{\partial \hat{\beta}}{\partial \varepsilon_0}\right)_0^c = -\beta_0(\beta_0+1)\alpha_0, \qquad \left(\frac{\partial \hat{\beta}}{\partial \varepsilon_1}\right)_0^c = -(2\alpha_0 + \beta_0 + 1),$$

where α_0, β_0 are true values of the respective parameters.

The estimators $\hat{\alpha}$, $\hat{\beta}$, $\hat{\varepsilon}_0$, $\hat{\varepsilon}_1$ can be computed by the following algorithm.

Computing moment estimators $\breve{\alpha}_{MM}$, $\breve{\beta}_{MM}$, $\breve{\varepsilon}_0$, $\breve{\varepsilon}_1$.

Input: Sample moments $m_1^*, m_2^*, m_3^*, m_4^*, m_5^*$.
Output: Estimates of the parameters $\breve{\alpha}_{\mathrm{MM}}, \breve{\beta}_{\mathrm{MM}}$ and the distortion levels $\breve{\varepsilon}_0, \breve{\varepsilon}_1$.
Step 1. Initialization: set $\breve{\varepsilon}_0 = 0$, $\breve{\varepsilon}_1 = 0$.
Step 2. For the current values of $\breve{\varepsilon}_0$, $\breve{\varepsilon}_1$, use formulas (10.41), (10.42) to compute the estimates $\breve{\alpha}_{\mathrm{MM}}, \breve{\beta}_{\mathrm{MM}}$.
Step 3. Compute J_0^{cond} as the Jacobian matrix of (10.45).
Step 4. Apply the uniform search algorithm to find the best approximation for the estimates $\breve{\varepsilon}_0$, $\breve{\varepsilon}_1$ w.r.t. the minimum norm of the residual vector for the third and the fourth moments m_3^*, m_4^*.
Step 5. If in the previous step, the estimates $\breve{\varepsilon}_0$, $\breve{\varepsilon}_1$ are multiply defined, then choose the values corresponding to the best approximation of the fifth moment m_5^*.
Step 6. Applying the iterative procedure (10.46), find new estimates $\breve{\varepsilon}_0$, $\breve{\varepsilon}_1$.
Step 7. Check the *stop*-criterion (for the residual vector norm), if it is not satisfied, go to Step 3, otherwise return the current values of the estimates and exit.

For brevity, the estimators for parameters α, β obtained by joint estimation of distribution parameters and distortion levels using the above algorithm will be called JMM estimators.

Construction of joint MLEs (abbreviated as JMLEs) can be reduced to solving the following log likelihood function maximization problem under additional constraints:

$$l(\alpha, \beta, \varepsilon_0, \varepsilon_1) = \sum_{j=0}^{n} f_j \ln P_j(\alpha, \beta, \varepsilon_0, \varepsilon_1) \to \max_{\alpha,\beta,\varepsilon_0,\varepsilon_1} \quad \text{for } \alpha, \beta > 0, \ \varepsilon_0, \varepsilon_1 \in [0, 1],$$

where f_j is the frequency of finding the value j in a sample $\{x_1, \ldots, x_K\}$, and $\{P_j(\cdot)\}$ are defined by (10.36). This maximization problem is solved numerically [23].

10.3.6 Robustness of the Classical Bayesian Predictor

In the beta-binomial model, forecasting is equivalent to estimating the probability p_i of the random event A occurring in a future $(n+1)$th experiment for the ith object ($i = 1, \ldots, K$) based on the collected data $\tilde{B}$. Observe that if we are not using the prior information on the probability model p_i (given in the form of a beta distribution with parameters α_i^0, β_i^0), then the nonparametric forecasting statistic is the relative frequency of the event A (cf. [6]):

$$\check{p}_i = \frac{1}{n} x_i^0, \quad x_i^0 = \sum_{j=1}^{n} b_{ij}. \tag{10.47}$$

If the probability model p_i is used, and the parameters α_i^0, β_i^0 are a priori known, then the mean square error (or risk) is minimized for the Bayesian predictor [23]:

$$\tilde{p}_i(x_i^0) = (\alpha_i^0 + x_i^0)/(\alpha_i^0 + \beta_i^0 + n), \quad i = 1, \ldots, K. \tag{10.48}$$

Let us evaluate the change of the forecast risk under distortion (10.35), where instead of x_i^0, the value $x_i = \sum_{j=1}^{n} \tilde{b}_{ij}$ is observed.

Theorem 10.13. *In the beta-binomial model with additive distortion* (10.35), *if the parameters α_i^0, β_i^0 are a priori known, then the mean square risk of forecasting for the Bayesian predictor is equal to*

$$\tilde{r}_i = r_{i0} + \frac{n\left(\beta_i^0\varepsilon_0 + \alpha_i^0\varepsilon_1\right) + n^{[2-]}\left((\beta_i^0)^{[2+]}\varepsilon_0^2 - 2\alpha_i^0\beta_i^0\varepsilon_0\varepsilon_1 + (\alpha_i^0)^{[2+]}\varepsilon_1^2\right)}{\left(\alpha_i^0 + \beta_i^0\right)^{[2+]}\left(\alpha_i^0 + \beta_i^0 + n\right)^2}, \tag{10.49}$$

where r_{i0} is the risk in the absence of distortion, $\varepsilon_0 = \varepsilon_1 = 0$:

$$r_{i0} = \frac{\alpha_i^0\beta_i^0}{\left(\alpha_i^0 + \beta_i^0\right)^{[2+]}\left(\alpha_i^0 + \beta_i^0 + n\right)}.$$

Proof. By definition, we have

$$\begin{aligned} \tilde{r}_i &= \mathbb{E}\left\{\left(p_i - (\alpha_i^0 + x)/(\alpha_i^0 + \beta_i^0 + n)\right)^2\right\} = \\ &= \mathbb{E}\{p_i^2\} - \frac{\alpha_i^0\mathbb{E}\{p_i\} + \mathbb{E}\{xp_i\}}{\alpha_i^0 + \beta_i^0 + n} + \frac{\alpha_i^{02} + 2\alpha_i^0\mathbb{E}\{x\} + \mathbb{E}\{x^2\}}{(\alpha_i^0 + \beta_i^0 + n)^2}. \end{aligned} \tag{10.50}$$

Due to results of [23], we can write

$$\mathbb{E}\{p_i\} = \frac{\alpha_i^0}{\alpha_i^0 + \beta_i^0}, \quad \mathbb{E}\{p_i^2\} = \frac{\alpha_i^0(\alpha_i^0 + 1)}{(\alpha_i^0 + \beta_i^0)(\alpha_i^0 + \beta_i^0 + 1)},$$

$$\mathbb{E}\{x\} = n\varepsilon_0 + n(1 - \varepsilon_0 - \varepsilon_1)\mathbb{E}\{p_i\}, \quad \mathbb{E}\{xp_i\} = n\varepsilon_0\mathbb{E}\{p_i\} + n(1 - \varepsilon_0 - \varepsilon_1)\mathbb{E}\{p_i^2\},$$

$$\mathbb{E}\{x^2\} = \mathbb{E}\{x\} + n(n-1)\left(\varepsilon_0^2 + 2\varepsilon_0(1 - \varepsilon_0 - \varepsilon_1)\mathbb{E}\{p_i\} + (1 - \varepsilon_0 - \varepsilon_1)\mathbb{E}\{p_i^2\}\right).$$

Substituting these expressions into (10.50) yields (10.49). □

From Theorem 10.13 it follows that $\tilde{r}_i - \tilde{r}_{i0} > 0$, and thus the Bayesian predictor (10.48) is no longer optimal under distortion (10.35). The relation (10.49) allows us to estimate the risk instability coefficient.

In the case where the model parameters are a priori unknown, and the forecasts are based on their estimators $\{\hat{\alpha}_i, \hat{\beta}_i\}$, the forecast risk was evaluated in [23].

10.3.7 Robust Forecasting in the Beta-Binomial Model

Theorem 10.14. *In the beta-binomial model with prior knowledge of the parameters α_i^0, β_i^0, the optimal Bayesian predictor under distortion* (10.35) *has the form*

$$\hat{p}_i(x_i) = \mathbb{E}_{\varepsilon_0,\varepsilon_1}\{p_i \mid x_i\} = \sum_{j=0}^{n} v^i_{x_i j} \frac{\alpha_i^0 + j}{\alpha_i^0 + \beta_i^0 + n}, \quad x_i = \sum_{j=1}^{n} \tilde{b}_{ij}, \qquad (10.51)$$

where weight coefficients are defined as

$$v^i_{x_i j} = C_n^j w_{x_i j} B\left(\alpha_i^0 + j, \beta_i^0 + n - j\right) \left(\sum_{l=0}^{n} C_K^l w_{x_i l} B\left(\alpha_i^0 + l, \beta_i^0 + n - l\right)\right)^{-1},$$

and $\{w_{kj}\}$ are defined by Lemma 10.2.

Proof. By applying the Bayes formula and Lemma 10.2, let us find the conditional probability density of p_i under the condition $x_i = s$:

$$f_{p_i}(x \mid s) = \frac{\sum\limits_{j=0}^{n} w_{sj} C_K^j x^j (1-x)^{n-j} \left(B(\alpha_i^0, \beta_i^0)\right)^{-1} x^{\alpha_i^0 - 1} (1-x)^{\beta_i^0 - 1}}{\int\limits_0^1 \sum\limits_{j=0}^{n} w_{sj} C_K^j y^j (1-y)^{n-j} \left(B(\alpha_i^0, \beta_i^0)\right)^{-1} y^{\alpha_i^0 - 1} (1-y)^{\beta_i^0 - 1} dy}.$$

Simplifying the above formula and using the beta distribution properties [14] yields

$$f_{p_i}(x \mid s) = \sum_{j=0}^{n} v^i_{x_i j} B\left(\alpha_i^0 + j, \beta_i^0 + n - j\right) x^{\alpha_i^0 + j - 1} (1-x)^{\beta_i^0 + n - j - 1}.$$

Computing the expectation for this probability distribution proves (10.51). □

Let us introduce the following notation:

$$
\begin{aligned}
&\alpha_{i,-}^0 = \alpha_i^0 - 1, \quad \beta_{i,-}^0 = \beta_i^0 - 1, \quad \alpha_{i,+}^0 = \alpha_i^0 + 1, \\
&\gamma_0 = \frac{(\alpha_i^0 + \beta_i^0)x - \alpha_i^0 n}{\alpha_i^0 + x}, \qquad \gamma_1 = \frac{(\alpha_i^0 + \beta_i^0)x - \alpha_{i,+}^0 n}{\beta_{i,-}^0 + n - x}, \\
&\xi_0 = \frac{(\alpha_{i,-}^0 + \beta_i^0)x - \alpha_{i,-}^0 n}{\alpha_{i,-}^0 + x}, \qquad \xi_1 = \frac{(\alpha_i^0 + \beta_{i,-}^0)x - \alpha_i^0 n}{\beta_{i,-}^0 + n - x}, \\
&G(x, \alpha_i^0, \beta_i^0, \varepsilon_0, \varepsilon_1) = \frac{1 + \gamma_0 \varepsilon_0 - \gamma_1 \varepsilon_1}{1 + \xi_0 \varepsilon_0 - \xi_1 \varepsilon_1}.
\end{aligned}
\tag{10.52}
$$

Corollary 10.5. *Under the conditions of Theorem 10.14, the optimal predictor* (10.51) *can be written asymptotically as*

$$
\hat{p}_i(x_i) = \tilde{p}_i(x_i) G\big(x_i, \alpha_i^0, \beta_i^0, \varepsilon_0, \varepsilon_1\big) + o(\varepsilon_0, \varepsilon_1). \tag{10.53}
$$

Proof. It is sufficient to write asymptotic expansions for the coefficients $\{w_{kj}\}$ in (10.51) and to use the notation (10.52). □

The asymptotic formula (10.53) shows that the optimal predictor under distortion can be approximated by the Bayesian predictor $\tilde{p}_i(x_i)$ for the undistorted model multiplied by a correction coefficient $G(x_i, \alpha_i^0, \beta_i^0, \varepsilon_0, \varepsilon_1)$, and the approximation error has the order $o(\varepsilon_0, \varepsilon_1)$. For $\varepsilon_0, \varepsilon_1 \to 0$ this correction coefficient tends to 1: $G(x_i, \alpha_i^0, \beta_i^0, \varepsilon_0, \varepsilon_1) \to 1$. Thus, the statistic $\hat{p}_i(x_i)$ can be viewed as a certain "robustification" of the Bayesian predictor $\tilde{p}_i(x_i)$ for the a priori known $\alpha_i^0, \beta_i^0, \varepsilon_0, \varepsilon_1$.

A plug-in forecasting statistic can be constructed by applying the estimators described in Sects. 10.3.4, 10.3.5.

Several other robust estimators of $\{\alpha_i^0, \beta_i^0\}$ for a priori unknown parameters ε_0, ε_1 are presented in [23].

10.3.8 Experimental Results

A series of computer simulations was performed to illustrate the theoretical results presented in this section. The true values of the BBD were taken to be $\alpha = \alpha_0 = 0.5$, $\beta = \beta_0 = 9.5$, $n = 10$, and the distortion levels were chosen as $\varepsilon_0 = \varepsilon_1 = 0.01$. The bias, the standard deviation, and histograms were compared for the following BBD parameter estimators:

- Classical estimators $\tilde{\alpha}_{\text{MM}}, \tilde{\beta}_{\text{MM}}$ and $\tilde{\alpha}_{\text{MLE}}, \tilde{\beta}_{\text{MLE}}$ were compared with MMM-estimators $\hat{\alpha}_{\text{MM}}, \hat{\beta}_{\text{MM}}$;
- Classical estimators $\tilde{\alpha}_{\text{MM}}, \tilde{\beta}_{\text{MM}}$ and $\tilde{\alpha}_{\text{MLE}}, \tilde{\beta}_{\text{MLE}}$ were compared with JMM-estimators $\breve{\alpha}_{\text{MM}}, \breve{\beta}_{\text{MM}}$ and JMLE-estimators $\hat{\alpha}_{\text{JMLE}}, \hat{\beta}_{\text{JMLE}}$.

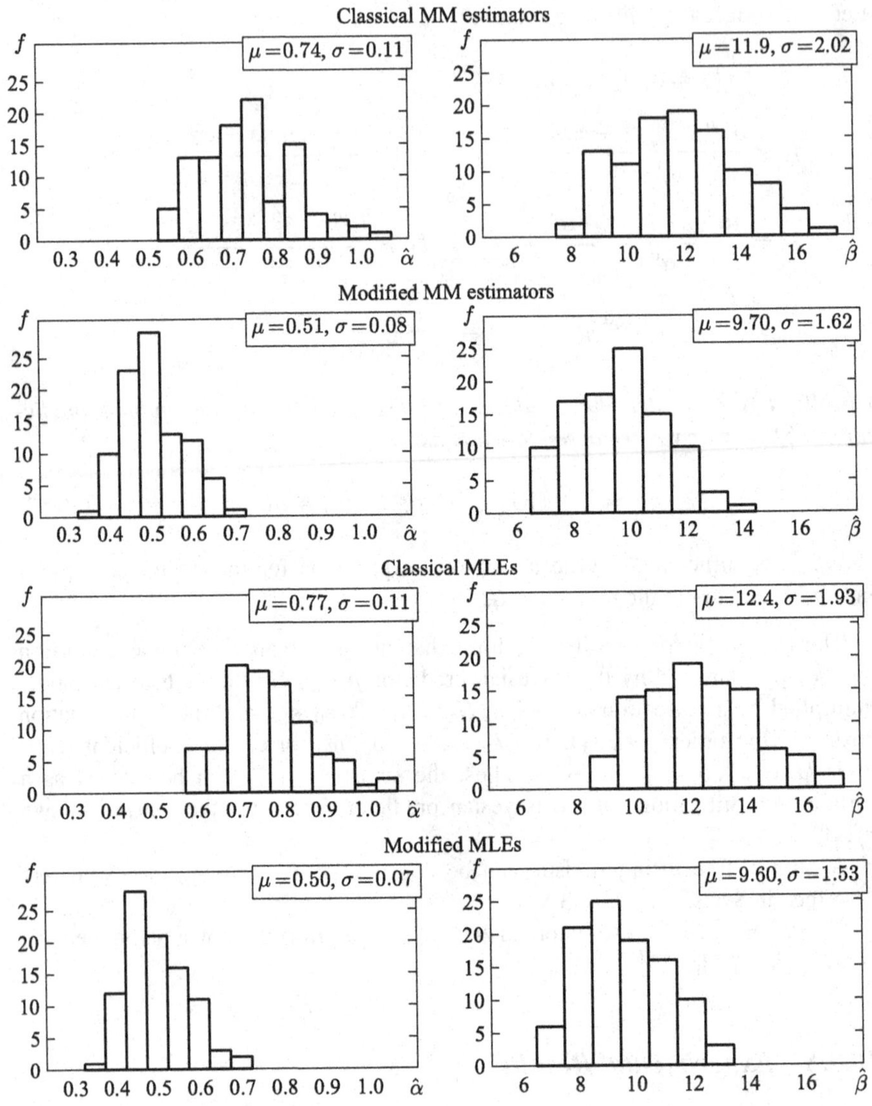

Fig. 10.1 Histograms of the distributions of classical and modified estimators. Estimates of α (true value $\alpha_0 = 0.5$) are shown on the *left*, and estimates of β (true value $\beta_0 = 9.5$)—on the *right*

In the simulations, $L = 100$ independent random samples X of size $K = 1{,}000$ from the DBBD with the parameters n, α, β, ε_0, ε_1 were generated, and the above estimators were computed for each sample. Then a histogram was plotted for each of the estimators, and the respective means and standard deviations were computed.

Figure 10.1 presents the histograms of the classical (based on the prior knowledge of ε_0, ε_1) and modified (for a priori unknown ε_0, ε_1) MM estimators and MLE

Table 10.1 Performance of joint parameter estimation compared to the classical methods

Parameters	α (true value 0.5)			β (true value 9.5)		
Method	MM	MLE	JMM	MM	MLE	JMM
Mean	0.73	0.76	0.49	11.81	12.28	9.20
Std. dev.	0.12	0.11	0.21	2.01	1.97	2.63

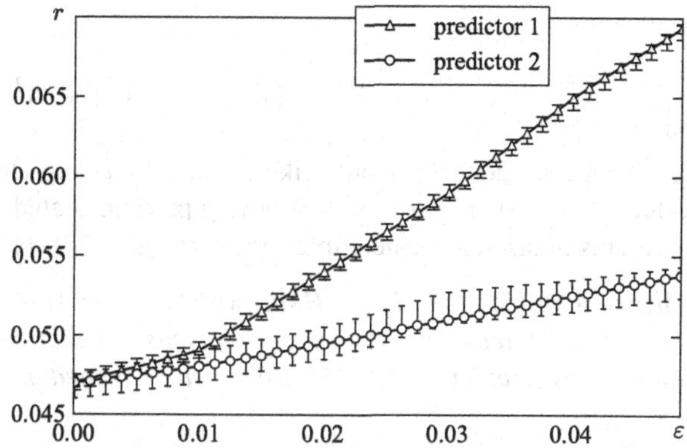

Fig. 10.2 A comparison of classical (predictor 1) and "robustified" (predictor 2) methods. *Solid lines* represent the theoretical risk, and *error bars* indicate 95 % confidence intervals

estimators. Results of comparing MM, MLE, and JMM estimators are presented in Table 10.1. It is easy to see that the proposed modifications result in a significant decrease of the bias and the standard deviation of the estimators for α, β.

The same simulated distribution was used to compare two forecasting statistics:

1. The classical predictor (10.38), where the values $\{\alpha_i^0, \beta_i^0\}$ are replaced by the classical MLEs, and the model is assumed to be free of distortion;
2. The predictor (10.50) from Theorem 10.14, where $\{\alpha_i^0, \beta_i^0, \varepsilon_0, \varepsilon_1\}$ were replaced by JMLEs.

In Fig. 10.2, the risks of these predictors are compared for different levels of simulated distortion $\varepsilon = \varepsilon_0 = \varepsilon_1$. It can be seen that the "robustified" predictor (10.50) based on the JMLE approach is much less sensitive to distortion.

10.4 Forecasting of HMCs

Let us consider a setting mentioned in Sect. 3.6.1, where the time series x_t, $t \in \mathbb{N}$, is an HMC with a finite state space $\mathcal{A} = \{0, 1, \ldots, N-1\}$ of size N, $2 \leq N < +\infty$, the initial probability distribution $\pi = (\pi_0, \pi_1, \ldots, \pi_{N-1})'$, and the one-step transition probability matrix $P = (p_{ij})$:

$$\mathbb{P}\{x_1 = i\} = \pi_i, \quad \mathbb{P}\{x_{t+1} = j \mid x_t = i\} = p_{ij}, \quad i, j \in \mathcal{A}. \qquad (10.54)$$

The following normalization conditions are satisfied:

$$\sum_{i \in \mathcal{A}} \pi_i \equiv 1; \quad \sum_{j \in \mathcal{A}} p_{ij} \equiv 1, \quad i \in \mathcal{A}. \tag{10.55}$$

The time series has been observed at $T \geq 1$ consecutive time points, and a realization

$$X_1^T = (x_1, x_2, \ldots, x_T)' \in \mathcal{A}^T$$

has been recorded. We would like to make a τ-step-ahead forecast of the future value $x_{T+\tau} \in \mathcal{A}$, $\tau \geq 1$. The forecasting problem stated above is one of the topical problems in discrete-valued time series analysis [5, 9–11, 17, 18, 28].

Theorem 10.15. *If x_t is an HMC with a priori known parameters π, P, then the optimal forecasting statistic in the sense of minimum error probability (i.e., minimum risk under the* $(0 - 1)$ *loss function defined in Sect. 2.4) has the form*

$$\hat{x}_{T+\tau} = \arg\max_{j \in \mathcal{A}} \left(P^\tau\right)_{x_T, j}, \tag{10.56}$$

where $(P^\tau)_{ij}$ is the (i, j)th element of the matrix P^τ. The attained minimum risk equals

$$r_0(\tau) = 1 - \sum_{i \in \mathcal{A}} \left(P^T \pi\right)_i \max_{j \in \mathcal{A}} \left(P^\tau\right)_{ij}. \tag{10.57}$$

Proof. Let

$$P_{X_1^T, x_{T+\tau}}(K, j) = \mathbb{P}\{X_1^T = K, \ x_{T+\tau} = j\}, \quad K \in \mathcal{A}^T, \ j \in \mathcal{A},$$

be the joint probability distribution of the observed realization X_1^T and the future value $x_{T+\tau}$. By the discrete analogue of Theorem 2.1, the optimal forecast is found by maximizing the posterior probability:

$$\hat{x}_{T+\tau} = \arg\max_{j \in \mathcal{A}} \mathbb{P}\{x_{T+\tau} = j \mid X_1^T\} = \arg\max_{j \in \mathcal{A}} \frac{P_{X_1^T, x_{T+\tau}}\left(X_1^T, j\right)}{P_{X_1^T}\left(X_1^T\right)}. \tag{10.58}$$

By the Markov property and (10.54), we have

$$P_{X_1^T, x_{T+\tau}}(K, j) = P_{X_1^T}(K)\,(P^\tau)_{x_T, j}, \quad K \in \mathcal{A}^T, \ j \in \mathcal{A}.$$

Substituting this relation into (10.58) yields (10.56).

From the total probability formula, the risk of the forecasting statistic (10.56) can be written as

$$r_0(\tau) = \mathbb{P}\{\hat{x}_{T+\tau} \neq x_{T+\tau}\} = 1 - \sum_{i \in \mathcal{A}} \mathbb{P}\{x_T = i\}\, r_i(\tau), \qquad (10.59)$$

where

$$r_i(\tau) = \mathbb{P}\{\hat{x}_{T+\tau} = x_{T+\tau} \mid x_T = i\}$$

is the conditional probability of making the correct decision if $x_T = i$. By (10.56) and the Markov property, we have

$$r_i(\tau) = \mathbb{P}\left\{\arg\max_{j \in \mathcal{A}} (P^\tau)_{ij} = x_{T+\tau} \mid x_T = i\right\} =$$

$$= \sum_{k \in \mathcal{A}} (P^\tau)_{ik} \mathbf{I}\left\{\arg\max_{j \in \mathcal{A}} (P^\tau)_{ij} = k\right\} = \max_{j \in \mathcal{A}} (P^\tau)_{ij}. \qquad (10.60)$$

Applying the fact that $\mathbb{P}\{x_T = i\} = (P^T \pi)_i$ and substituting (10.60) into (10.59) proves (10.57). □

Note that if a row x_T of the matrix P^τ contains several maximum elements, then we obtain several equivalent forecasts with the same risk (10.57).

Corollary 10.6. *If x_t is a stationary Markov chain, then the formula for the minimum risk can be simplified:*

$$r_0(\tau) = 1 - \sum_{i \in \mathcal{A}} \pi_i \max_{j \in \mathcal{A}} (P^\tau)_{ij}. \qquad (10.61)$$

Proof. The probability distributions of elements in a stationary Markov chain are independent of time:

$$\mathbb{P}\{x_T = i\} = (P^T \pi)_i = \pi_i.$$

Thus, (10.61) follows from (10.57). □

Corollary 10.7. *Under the ergodicity condition, as the forecast horizon increases to infinity, the minimum error probability tends to the following limit:*

$$r_0(\tau) \to 1 - \max_{j \in \mathcal{A}} \pi_j, \quad \tau \to +\infty. \qquad (10.62)$$

Proof. By ergodicity [3], we have

$$(P^\tau)_{ij} = \pi_j + O(\varrho^\tau) \to \pi_j, \quad j \in \mathcal{A}, \quad \tau \to +\infty,$$

for some $\varrho \in (0, 1)$. Applying this convergence to (10.61) yields (10.62). □

Note that the right-hand side of (10.62) contains the risk of the forecasting statistic that is only based on prior information:

$$\mathbb{P}\{x_{T+\tau} = j\} = \pi_j, \quad j \in \mathcal{A}.$$

To illustrate the dependence $r_0 = r_0(\tau)$ of the forecast risk on the forecasting horizon τ, let us consider the case of forecasting a binary ($N = 2$) Markov chain with a bistochastic matrix of one-step transitions:

$$\mathcal{A} = \{0, 1\}, \quad P = \frac{1}{2}\begin{pmatrix} 1+\varepsilon & 1-\varepsilon \\ 1-\varepsilon & 1+\varepsilon \end{pmatrix}, \quad -1 \le \varepsilon \le +1.$$

For $\varepsilon > 0$, we have a Markov chain with attraction, and for $\varepsilon < 0$—with repulsion. Since the matrix P is bistochastic, the stationary probability distribution is uniform:

$$\pi_i = \mathbb{P}\{x_t = i\} = 1/2, \quad i \in \mathcal{A}.$$

By direct calculations, we have

$$P^\tau = \frac{1}{2}\begin{pmatrix} 1+\varepsilon^\tau & 1-\varepsilon^\tau \\ 1-\varepsilon^\tau & 1+\varepsilon^\tau \end{pmatrix}, \quad \tau \in \mathbb{N},$$

and thus (10.61) yields an explicit expression for the forecast risk:

$$r_0 = r_0(\tau) = \left(1 - |\varepsilon|^\tau\right)/2, \quad \tau \in \mathbb{N}. \tag{10.63}$$

Figure 10.3 presents a plot of the dependence (10.63) for $\varepsilon = 0.9; 0.5$, showing exponentially fast convergence of the error probability to the error probability of a purely random ("coin tossing") forecast as $\tau \to +\infty$:

$$r_0(\tau) \to 1/2.$$

If the matrix $P = (p_{ij})$ is a priori unknown, then we can use (10.54), (10.55) to construct an MLE $\hat{P} = (\hat{p}_{ij})$ based on the observed realization X_1^T:

$$\hat{p}_{ij} = \begin{cases} \sum\limits_{t=1}^{T-1} \delta_{x_t,i}\delta_{x_{t+1},j} \left(\sum\limits_{t=1}^{T-1} \delta_{x_t,i}\right)^{-1}, & \text{if } \sum\limits_{t=1}^{T-1} \delta_{x_t,i} > 0, \\ N^{-1}, & \text{if } \sum\limits_{t=1}^{T-1} \delta_{x_T,i} = 0. \end{cases} \tag{10.64}$$

Then a plug-in forecasting statistic can be obtained from (10.56):

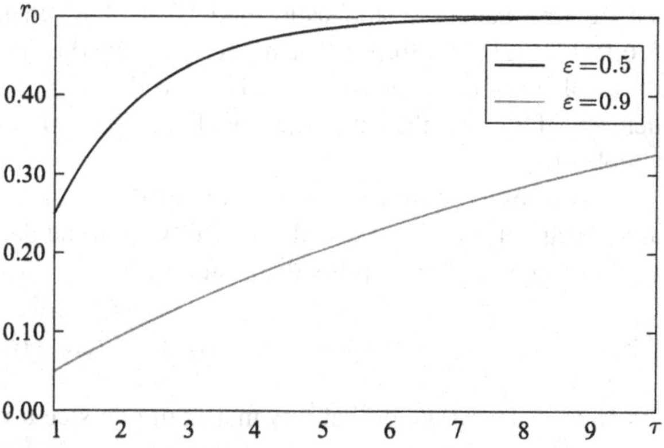

Fig. 10.3 Dependence of the forecast risk $r_0(\tau)$ on forecast depth

$$\hat{x}_{T+\tau} = \arg\max_{j\in\mathcal{A}} \left(\hat{P}^\tau\right)_{x_T,j}. \tag{10.65}$$

Since (10.56) contains the conditional τ-step transition probability from x_T to j,

$$p_{x_T,j}(\tau) = (P^\tau)_{x_T,j}, \tag{10.66}$$

this probability can be estimated directly by applying the frequency method:

$$\tilde{p}_{x_T,j}(\tau) = \begin{cases} \sum_{t=1}^{T-\tau} \delta_{x_t,x_T}\delta_{x_{t+\tau},j} \left(\sum_{t=1}^{T-\tau} \delta_{x_t,x_T}\right)^{-1}, & \text{if } \sum_{t=1}^{T-\tau} \delta_{x_t,x_T} > 0, \\ N^{-1}, & \text{if } \sum_{t=1}^{T-\tau} \delta_{x_t,x_T} = 0. \end{cases} \tag{10.67}$$

This leads to another plug-in forecasting statistic:

$$\tilde{x}_{T+\tau} = \arg\max_{j\in\mathcal{A}} \tilde{p}_{x_T,j}(\tau). \tag{10.68}$$

Since the estimators (10.64), (10.67) are strongly consistent for $T \to +\infty$, the risks of forecasting statistics (10.65), (10.68) converge to $r_0(\tau)$—the forecast risk for stationary HMCs defined by (10.61).

10.5 Higher Order Markov Chain Forecasting

10.5.1 Optimal Forecasting Under HMC(s) Models

Mathematical modeling of complex systems and processes in economics, engineering, medicine, sociology, genetics, and other applications often requires constructing adequate stochastic models of "long memory" discrete time series [5,6,9–11,17,18,28,36,37]. The higher order Markov chain model is a well-studied universal discrete stochastic model; the order $s \in \mathbb{N}$ defines the length of the memory. If $s = 1$, then the Markov chain is called simple [19], and if $s > 1$—complex [8].

Let the observed time series $x_t \in \mathcal{A} = \{0, 1, \ldots, N-1\}, t \in \mathbb{N}, 2 \leq N < \infty$, be an HMC of order s, denoted as HMC(s) (see the definition in Sect. 3.6.2), with the s-variate initial probability distribution

$$\pi_{i_1,\ldots,i_s} = \mathbb{P}\{x_1 = i_1, \ldots, x_s = i_s\}, \quad i_1, \ldots, i_s \in \mathcal{A}, \tag{10.69}$$

and an $(s+1)$-variate probability matrix of one-step transitions

$$P = (p_{i_1,\ldots,i_{s+1}}), \quad p_{i_1,\ldots,i_{s+1}} = \mathbb{P}\{x_{t+1} = i_{s+1} \mid x_t = i_s, \ldots, x_{t-s+1} = i_1\}, \tag{10.70}$$

where $i_1, \ldots, i_{s+1} \in \mathcal{A}, t \geq s$.

Similarly to (10.55), the following normalization conditions are satisfied:

$$\sum_{i_1,\ldots,i_s \in \mathcal{A}} \pi_{i_1,\ldots,i_s} \equiv 1, \quad \sum_{i_{s+1} \in \mathcal{A}} p_{i_1,\ldots,i_s,i_{s+1}} \equiv 1, \quad i_1, \ldots, i_s \in \mathcal{A}. \tag{10.71}$$

Assume that a realization $X_1^T = (x_1, \ldots, x_T)' \in \mathcal{A}^T$ of length $T \geq s$ has been observed.

As in the previous section, let us consider the problem of forecasting the element $x_{T+\tau} \in \mathcal{A}, \tau \geq 1$.

Let us introduce the following notation for $n > m$:

$$J_m^n = (j_m, j_{m+1}, \ldots, j_n)' \in \mathcal{A}^{n-m+1}$$

is a multiindex over the set $\mathcal{A}$,

$$\begin{aligned} p_{J^0_{-s+1}, j_\tau}(\tau) &= \mathbb{P}\left\{x_{T+\tau} = j_\tau \mid X^T_{T-s+1} = J^0_{-s+1}\right\} = \\ &= \sum_{j_1,\ldots,j_{\tau-1} \in \mathcal{A}} p_{J^0_{-s+1}, j_1} p_{J^1_{-s+2}, j_2} \cdots p_{J^{\tau-1}_{-s+\tau}, j_\tau}, \quad J^0_{-s+1} \in \mathcal{A}^\tau, \; j_\tau \in \mathcal{A}, \end{aligned} \tag{10.72}$$

is the conditional probability of a τ-step transition between the states $X^T_{T-s+1} = J^0_{-s+1}$ and $x_{T+\tau} = j_\tau$ in the HMC(s) model. This formula is similar to the expression (10.66) for a simple Markov chain, $s = 1$.

Similarly to Theorem 10.15, we can obtain an expression for the optimal forecasting statistic:

$$\hat{x}_{T+\tau} = \arg\max_{j \in \mathcal{A}} p_{X^T_{T-s+1}, j}(\tau). \tag{10.73}$$

For a stationary HMC(s), we can also prove that the minimum forecast risk equals

$$r_0(\tau) = 1 - \sum_{J \in \mathcal{A}^s} \pi_J \max_{j \in \mathcal{A}} p_{J,j}(\tau). \tag{10.74}$$

A plug-in forecasting statistic is obtained from the optimal statistic (10.73), (10.72) by substituting statistical estimators of $p_{J,j}$ based on the observed realization X_1^T.

However, as noted in Sect. 3.6, the number of parameters D in the model HMC(s) grows exponentially as the order s increases: $D = N^s(N-1)$, and identification (parameter estimation) of the model requires observing a realization $X_1^T = (x_1, \ldots, x_T)'$ of a very large length $T > D$, which is often impossible. This motivates the search for parsimonious (small-parametric) higher order Markov chain models mentioned in Sect. 3.6: the Jacobs–Lewis model, Raftery's MTD model, and the MC(s, r) model proposed by the author.

10.5.2 Identification of the Jacobs–Lewis Model

Let a discrete time series x_t described by the Jacobs–Lewis model (see Sect. 3.6) be defined on the probability space $(\Omega, F, \mathbb{P})$:

$$x_t = \mu_t x_{t-\eta_t} + (1 - \mu_t)\xi_t, \quad t > s, \tag{10.75}$$

where $s \geq 2$, $\{x_1, \ldots, x_s\}$, $\{\xi_t, \eta_t, \mu_t : t > s\}$ are jointly independent random variables with probability distributions

$$\begin{aligned}
&\mathbb{P}\{\xi_t = i\} = \pi_i, \quad i \in \mathcal{A}; \quad \sum_{i \in \mathcal{A}} \pi_i = 1; \\
&\mathbb{P}\{\eta_t = j\} = \lambda_j, \quad j \in \{1, \ldots, s\}; \quad \sum_{j=1}^{s} \lambda_j = 1, \quad \lambda_s \neq 0; \\
&\mathbb{P}\{\mu_t = 1\} = 1 - \mathbb{P}\{\mu_t = 0\} = \varrho; \\
&\mathbb{P}\{x_k = i\} = \pi_i, \quad i \in \mathcal{A}, \ k \in \{1, \ldots, s\}.
\end{aligned} \tag{10.76}$$

The original paper by Jacobs and Lewis [13] which introduced this model presents only the simplest probabilistic properties of (10.75), and doesn't touch upon statistical problems.

Theorem 10.16. *The model* (10.75), (10.76) *defines an HMC* $\{x_t\}$ *of order* s *with the initial probability distribution* $\pi_{i_1,\ldots,i_s} = \pi_{i_1} \cdot \cdots \cdot \pi_{i_s}$ *and an* $(s+1)$*-dimensional probability matrix of one-step transitions*

$$P(\pi, \lambda, \varrho) = (p_{i_1,\ldots,i_{s+1}}),$$

$$p_{i_1,\ldots,i_s,i_{s+1}} = (1-\varrho)\pi_{i_{s+1}} + \varrho \sum_{j=1}^{s} \lambda_j \delta_{i_{s-j+1},i_{s+1}}, \quad i_1,\ldots,i_{s+1} \in \mathcal{A}. \quad (10.77)$$

Corollary 10.8. *Under the model* (10.75), (10.76), *MLEs for the* $d = N + s - 1$ *model parameters* π, λ, ϱ *based on an observed time series* $X_1^T = (x_1, \ldots, x_T)'$ *are solutions of the following optimization problem:*

$$l(\pi, \lambda, \varrho) = \sum_{t=1}^{s} \ln \pi_{x_t} + \sum_{t=s+1}^{T} \ln \left((1-p)\pi_{x_t} + \varrho \sum_{j=1}^{s} \lambda_j \delta_{x_{t-j},x_t} \right) \to \max_{\pi,\lambda,\varrho}. \quad (10.78)$$

Let us introduce the following notation: for a matrix $P = P(\pi, \lambda, \varrho)$, $\|P\|^2$ is the sum of squares of its elements; $F(P)$ is the sum of squares of the elements of the matrix $P = (p_{i_1,\ldots,i_{s+1}})$ that satisfy one of the conditions: $\{i_1 = \cdots = i_s = i_{s+1}\}$ or $\{i_1 \neq i_{s+1}, \ldots, i_s \neq i_{s+1}\}$; $\tilde{P}$ is the sample matrix of transition probabilities computed from X_1^T.

Theorem 10.17. *If* $\varrho \neq 1$, *then the statistics* $\tilde{\pi}$, $\tilde{\lambda}$, $\tilde{\varrho}$ *defined below are consistent estimators for the parameters of the model* (10.75), (10.76) *for* $T \to \infty$*:*

$$\tilde{\pi}_i = \frac{1}{T} \sum_{t=1}^{T} \delta_{x_t,i}, \quad i \in \mathcal{A}; \qquad \tilde{\varrho} = \arg\min_{\varrho} F\left(\tilde{P} - P(\pi, \lambda, \varrho)\right);$$

$$\tilde{\lambda} = \arg\min_{\lambda} \left\| \tilde{P} - P(\pi, \lambda, \varrho) \right\|.$$

Theorems 10.16, 10.17 are proved in [25]. Note that the statistics $\tilde{\lambda}$, $\tilde{\varrho}$ can be written explicitly. The estimators $(\tilde{\pi}, \tilde{\lambda}, \tilde{\varrho})$ are used as the initial approximation in the iterative solution of the optimization problem (10.78).

10.5.3 Identification of Raftery's MTD Model

Let x_t be an HMC of order s defined on the probability space $(\Omega, F, \mathbb{P})$ with an $(s+1)$-dimensional transition probability matrix $P = (p_{i_1,\ldots,i_{s+1}})$ defined by (10.70). Raftery's MTD model [33] defines a special small-parametric (parsimonious) representation of the matrix P:

$$p_{i_1,\ldots,i_{s+1}} = \sum_{j=1}^{s} \lambda_j q_{i_j,i_{s+1}}, \quad i_1,\ldots,i_{s+1} \in \mathcal{A}, \quad (10.79)$$

where $Q = (q_{ik})$ is a stochastic $(N \times N)$-matrix, $i, k \in \mathcal{A}$, and $\lambda = (\lambda_1, \ldots, \lambda_s)'$ is an s-column vector such that

$$\lambda_1 > 0, \quad \lambda_2, \ldots, \lambda_s \geq 0, \quad \lambda_1 + \cdots + \lambda_s = 1.$$

This model has $d = N(N-1)/2 + s - 1$ parameters.

The MTD model (10.79) can be generalized to obtain the MTDg model, where a separate transition probability matrix is defined for each of the s time lag values:

$$p_{i_1,\ldots,i_{s+1}} = \sum_{j=1}^{s} \lambda_j q^{(j)}_{i_j, i_{s+1}}, \quad i_1, \ldots, i_{s+1} \in \mathcal{A}, \tag{10.80}$$

where $Q^{(j)} = (q^{(j)}_{ik})$ is the jth stochastic matrix corresponding to the time lag $s-j$. The number of parameters in the MTDg model is $d = s\,(N(N-1)/2 + 1) - 1$.

Lemma 10.5. *The ergodicity of an order s Markov chain defined by the MTD model* (10.79) *is equivalent to the existence of a positive integer $K \in \mathbb{N}$ such that all elements of the matrix Q^K are nonnegative.*

Proof. The lemma can be proved constructively by applying the Markov property and the Kolmogorov–Chapman equation. □

Let us introduce the following notation:

$$\pi^{(t)} = \left(\pi^{(t)}_0, \ldots, \pi^{(t)}_{N-1}\right)'$$

is the one-dimensional probability distribution of the Markov chain at time $t \in \mathbb{N}$, where $\pi^{(t)}_i = \mathbb{P}\{x_t = i\}$, $i \in \mathcal{A}$;

$$\Pi^{(t)} = \left(\pi^{(t)}_{i_1,\ldots,i_s}\right)$$

is the s-variate probability distribution of the vector $(x_{t-(s-1)}, \ldots, x_t)' \in \mathcal{A}^s$, where

$$\pi^{(t)}_{i_1,\ldots,i_s} = \mathbb{P}\{x_{t-(s-1)} = i_1, \ \ldots, \ x_t = i_s\}, \quad i_1, \ldots, i_{s+1} \in \mathcal{A};$$

the distribution $\Pi^* = (\pi^*_{i_1,\ldots,i_s})$, $i_1, \ldots, i_{s+1} \in \mathcal{A}$, is the s-variate stationary probability distribution of the ergodic Markov chain; $\pi^* = (\pi^*_0, \ldots, \pi^*_{N-1})'$ is the respective univariate stationary probability distribution.

Theorem 10.18. *Under the model* (10.80), *if for some $K \in \mathbb{N}$ every element of the matrix $(Q^{(1)})^K$ is nonnegative, the stationary probability distribution Π^* is*

$$\pi^*_{i_1,\ldots,i_s} = \prod_{l=0}^{s-1} \left(\pi^*_{i_{s-l}} + \sum_{j=l+1}^{s} \lambda_j \left(q^{(j)}_{i_{j-l}, i_{s-l}} - \sum_{r=0}^{N-1} q^{(j)}_{r, i_{s-l}} \pi^*_r \right) \right), \quad i_1, \ldots, i_{s+1} \in \mathcal{A}.$$

Corollary 10.9. *Under the model* (10.79)*, assuming that the ergodicity criterion of Lemma 10.5 is satisfied, the stationary bivariate marginal probability distribution of the random vector* $(x_{t-m}, x_t)'$ *satisfies the relation*

$$\pi^*_{ki}(m) = \pi^*_k \pi^*_i + \pi^*_k \lambda_{s-m+1}(q_{ki} - \pi^*_i), \quad 1 \leq m \leq s, \quad k \in \mathcal{A}. \qquad (10.81)$$

The proofs of Theorem 10.18 and its corollary are similar to the proof of Theorem 10.16.

Let us construct estimators for the parameters of the MTD model by applying the property (10.81).

From the observed realization $X_1^T = (x_1, \ldots, x_T)'$, define the following statistics for $i, k \in \mathcal{A}$, $j = 1, \ldots, s$:

$$\tilde{\pi}_i = \frac{1}{T-2s+1} \sum_{t=s+1}^{T-s+1} \delta_{x_t,i}; \qquad \tilde{\pi}_{ki}(j) = \frac{1}{T-2s+1} \sum_{t=s+j}^{T-s+j} \delta_{x_{t-j},k} \delta_{x_t,i};$$

$$z_{ki}(j) = \tilde{\pi}_{ki}(s-j)/\tilde{\pi}_k - \tilde{\pi}_i; \qquad d_{ki} = \tilde{q}_{ki} - \tilde{\pi}_i;$$

$$\tilde{q}_{ki} = \begin{cases} \sum_{j=1}^{s} \tilde{\pi}_{ki}(j)/\tilde{\pi}_k - (s-1)\tilde{\pi}_i, & \tilde{\pi}_k > 0, \\ 1/N, & \text{otherwise}; \end{cases}$$

$$\tilde{\lambda} = \arg\min \sum_{i,k \in \mathcal{A}} \sum_{j=1}^{s} \left(z_{ki}(j) - \lambda_j d_{ki}\right)^2. \qquad (10.82)$$

Theorem 10.19. *In the MTD model* (10.79)*, assuming that the ergodicity criterion of Lemma 10.5 is satisfied, the statistics* (10.82) *are asymptotically unbiased and consistent estimators for, respectively, Q and λ as $T \to \infty$.*

Proof. It is easy to show that the definitions of consistency and unbiasedness are satisfied. □

The estimators $\tilde{Q}$, $\tilde{\lambda}$ defined by (10.82) are a good initial approximation for iterative maximization of the log likelihood function, which yields the MLEs $\hat{Q}, \hat{\lambda}$:

$$l(Q, \lambda) = \sum_{t=s+1}^{T} \ln \sum_{j=1}^{s} \lambda_j q_{x_{t-s+j-1}, x_t} \to \max_{Q,\lambda}.$$

10.5.4 *Identification of the MC(s, r) Model*

To conclude the section, let us define Markov chains of order s with r partial connections—a small-parametric model family proposed by the author in 2003 [21,24,25].

As in Sect. 10.5.3, assume that x_t is an HMC of order s defined on $(\Omega, F, \mathbb{P})$ with an $(s+1)$-dimensional transition probability matrix

$$P = (p_{i_1,\ldots,i_{s+1}}), \quad i_1,\ldots,i_{s+1} \in \mathcal{A};$$

$r \in \{1,\ldots,s\}$ is the parameter called the number of partial connections;

$$M_r^0 = (m_1^0,\ldots,m_r^0) \in M$$

is an arbitrary integer r-vector with ordered components $1 = m_1^0 < m_2^0 < \cdots < m_r^0 \le s$ which is called the connection template; M is a set of cardinality $K = |M| = C_{s-1}^{r-1}$ which is composed of all possible connection templates with r partial connections; and $Q^0 = (q^0_{j_1,\ldots,j_{r+1}})$ is some $(r+1)$-dimensional stochastic matrix, where the indices $j_1,\ldots,j_{s+1}$ lie in $\mathcal{A}$.

A Markov chain of order s with r partial connections [21], denoted as MC(s,r), is defined by specifying the following one-step transition probabilities:

$$p_{i_1,\ldots,i_s,i_{s+1}} = q^0_{i_{m_1^0},\ldots,i_{m_r^0},i_{s+1}}, \quad i_1,\ldots,i_{s+1} \in \mathcal{A}. \tag{10.83}$$

The relation (10.83) implies that the probability of the process entering a state i_{s+1} at time $t > s$ does not depend on every previous state of the process $i_1,\ldots,i_s$, but is affected only by the r chosen states $i_{m_1^0},\ldots,i_{m_r^0}$. Thus, instead of $D = N^s(N-1)$ parameters, the model (10.83) is defined by $d = N^r(N-1)+r-1$ independent parameters that determine the matrices Q^0, M_r^0. The reduction in the number of parameters can be very significant: for instance, if $N = 2$, $s = 32$, $r = 3$, then we have $D \approx 4.1 \times 10^9$, and $d = 10$.

Note that if $s = r$, $M_r^0 = (1,\ldots,s)$, then $P = Q^0$, and MC(s,s) is a Markov chain of order s. As a constructive example of an MC(s,r) model, let us consider a binary ($N = 2$) autoregression of order s with r nonzero coefficients. A special case of such autoregression model is a linear recursive sequence defined in the ring $\mathbb{Z}_2$ and generated by a degree s polynomial with r nonzero coefficients.

Let us introduce the following notation:

$$J_s = (j_1,\ldots,j_s) = (J_{s-1}, j_s) \in \mathcal{A}^s$$

is a multiindex of order s; δ_{J_s,J'_s} is the Kronecker symbol defined for pairs of multiindices J_s, $J'_s \in \mathcal{A}^s$; the function

$$S_t : \mathcal{A}^T \times M \to \mathcal{A}^r, \quad (X_1^T; M_r) \mapsto (x_{t+m_1-1},\ldots,x_{t+m_r-1}) \in \mathcal{A}^r$$

is called a selector of order r with the parameters M_r and t, where $M_r \in M$, $t \in \{1,\ldots,T-s+1\}$; $\Pi_{K_s} = \mathbb{P}\{X_s = K_s\}$ is the initial s-variate probability distribution of the Markov chain MC(s,r);

$$\nu_{r+1}(J_{r+1}; M_r) = \sum_{t=1}^{T-s} \mathbf{I}\{S_t(X_1^T; M_{r+1}) = J_{r+1}\}$$

is the frequency of the $(r+1)$-gram $J_{r+1} \in \mathcal{A}^{r+1}$ corresponding to the connection template $M_{r+1} = (M_r, s+1)$, where J_{r+1} satisfies the normalization condition

$$\sum_{J_{r+1} \in \mathcal{A}^{r+1}} \nu_{r+1}(J_{r+1}; M_r) = T - s.$$

Finally, an index replaced by a dot denotes summation over all possible values of this index:

$$\nu_{r+1}(J_r \cdot; M_r) = \sum_{j_{r+1} \in \mathcal{A}} \nu_{r+1}(J_{r+1}; M_r), \qquad \nu_{r+1}(\cdot j_{r+1}) = \sum_{J_r \in \mathcal{A}^r} \nu_{r+1}(J_{r+1}; M_r)$$

are the "accumulated" statistics.

Theorem 10.20. *The model MC(s, r) defined by* (10.83) *is ergodic if and only if there exists an integer $l \geq 0$ such that*

$$\min_{J_s, J_s' \in \mathcal{A}^s} \sum_{K_l \in \mathcal{A}^l} \prod_{i=1}^{s+l} q^0_{S_i\left((J_s, K_l, J_s'); M^0_{r+1}\right)} > 0.$$

Proof. The proof is based on transforming MC(s, r) into a special s-vector Markov chain of order one. □

Let us apply the plug-in principle to construct the information functional $\hat{I}_{r+1}(M_r)$ from the observed realization X_n. In other words, a sample-based estimator for the Shannon information about the future symbol $x_{t+s} \in \mathcal{A}$ contained in the r-tuple $S_t(X_1^T, M_r)$ will be constructed [24].

Theorem 10.21. *The MLEs $\hat{M}_r$, $\hat{Q} = (\hat{q}_{J_{r+1}})$, where $J_{r+1} \in \mathcal{A}^{r+1}$, for the parameters M_r^0, Q^0 can be defined as*

$$\hat{M}_r = \arg\max_{M_r \in M} \hat{I}_{r+1}(M_r),$$

$$\hat{q}_{J_{r+1}}(M_r) = \begin{cases} \nu_{r+1}(J_{r+1}; \hat{M}_r)/\nu_{r+1}(J_r \cdot; \hat{M}_r), & \textit{if } \nu_{r+1}(J_r \cdot; \hat{M}_r) > 0, \\ 1/N, & \textit{if } \nu_{r+1}(J_r \cdot; \hat{M}_r) = 0. \end{cases} \tag{10.84}$$

Let us introduce the following notation: for $K_s \in \mathcal{A}^s$, $\Pi^*_{K_s}$ is the stationary probability distribution of MC(s, r);

$$\mu_{r+1}(J_{r+1}; M_r, M_r^0) = \sum_{K_{s+1} \in \mathcal{A}^{s+1}} \mathbf{I}\{S_1(K_{s+1}; M_{r+1}) = J_{r+1}\} \Pi^*_{K_s} p_{K_{s+1}}, \quad J_{r+1} \in \mathcal{A}^{r+1}.$$

Theorem 10.22. *If $MC(s, r)$ defined by* (10.83) *is stationary, and the connection template $M_r^0 \in M$ satisfies the identifiability condition [24, 25], then the MLEs $\hat{M}_r$, $\hat{Q}$ defined by* (10.84) *are consistent for $T \to \infty$:*

$$\hat{M}_r \xrightarrow{\mathbf{P}} M_r^0, \quad \hat{Q} \xrightarrow{\text{m.s.}} Q^0,$$

and the following asymptotic expansion holds for the mean square error of $\hat{Q}$:

$$\Delta_T^2 = \mathbb{E}\{\|\hat{Q} - Q^0\|^2\} = \frac{1}{T-s} \sum_{J_{r+1} \in A^{r+1}} \frac{\left(1 - q_{J_{r+1}}^0\right) q_{J_{r+1}}^0}{\mu_{r+1}(J_r \cdot; M_r^0, M_r^0)} + o\left(\frac{1}{T}\right). \tag{10.85}$$

Theorems 10.21, 10.22 have been proved in [21].

The estimators (10.84) can be used to construct a statistical test for a null hypothesis $H_0 : Q^0 = Q_0$ against an alternative of a general form $H_1 = \overline{H}_0$, where $Q_0 = (q_{0J_{r+1}})$ is some given stochastic matrix. The decision rule of a given asymptotic size $\varepsilon \in (0, 1)$ can be written as follows:

$$\text{decide in favor of } \{H_0, \text{ if } \varrho \leq \Delta; \quad H_1, \text{ if } \varrho > \Delta\}, \tag{10.86}$$

where

$$\varrho = \sum_{\substack{J_{r+1}: \\ q_{0J_{r+1}} > 0}} \nu_{r+1}(J_r \cdot; \hat{M}_r) \frac{\left(\hat{q}_{J_{r+1}} - q_{0J_{r+1}}\right)^2}{q_{0J_{r+1}}}$$

and $\Delta = G_L^{-1}(1 - \varepsilon)$ is the $(1 - \varepsilon)$-quantile of the χ^2 distribution with L degrees of freedom [24].

Performance of the statistical estimators (10.84) and the test (10.86) was evaluated by Monte-Carlo simulation experiments with fixed model parameters: $N = 2$, $s = 256$, $r = 6$; the chosen values of Q^0 and M_r^0 are omitted due to space considerations. For each simulated observation length, 10^4 simulation rounds were performed.

Figure 10.4 illustrates the numerical results obtained for this MC(256,6) model: the mean square error Δ_T^2 of the estimator $\hat{Q}$ is plotted against the observation length T; the curve has been computed theoretically from the leading term of the expansion (10.85), and the circles are the experimental values computed in the simulations.

The models $\text{MC}(s, r)$ can be applied to real-world statistical forecasting problems. To illustrate this, let us take a real-world econometric time series presented in Fig. 10.5: the market price of copper collected from 1800 to 1997 [12,29]. Based on first differences, this data was classified into three categories ($N = 3$), yielding a discrete sequence with three possible states:

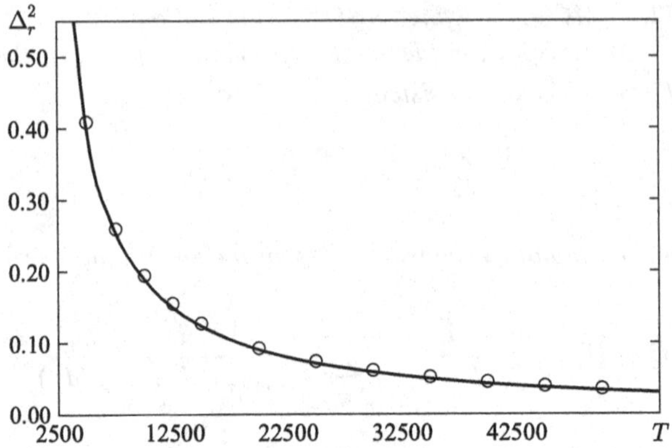

Fig. 10.4 Dependence of Δ_T^2 on T for $N = 2, s = 256, r = 6$

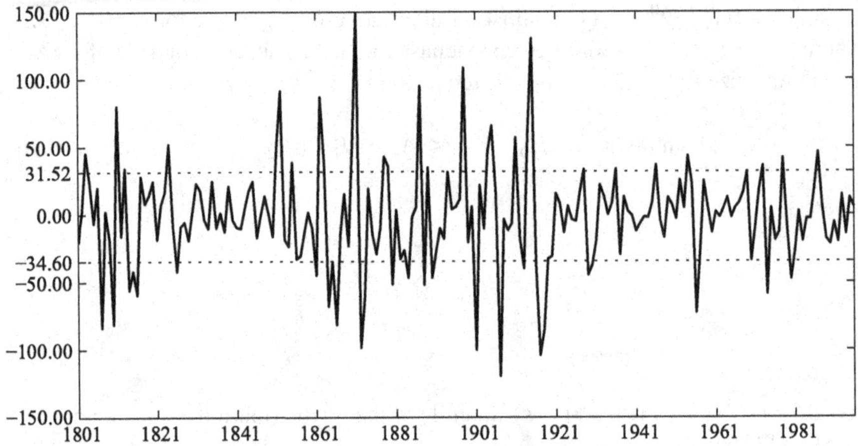

Fig. 10.5 Time series of first differences for the price of copper

(0) The price has decreased by over 34.6 dollars;
(2) The price has increased by over 31.52 dollars;
(1) Other variations.

For the obtained discrete sequence, the best fitted model in the family of Markov chains $\{MC(0, 0);\ MC(s, r),\ 1 \le s \le 3,\ 1 \le r \le s\}$ was of the type MC(2, 1); the obtained estimate for the matrix Q was

$$\hat{Q} = \begin{matrix} & 0 & 1 & 2 \\ 0 & 0.1481 & 0.7778 & 0.0741 \\ 1 & 0.0647 & 0.7626 & 0.1727 \\ 2 & 0.5000 & 0.4286 & 0.0714 \end{matrix} .$$

The constructed model can be used for price forecasting.

References

1. Anderson, T.: An Introduction to Multivariate Statistical Analysis. Wiley, Hoboken (2003)
2. Basawa, I., Rao, B.: Statistical Inference for Stochastic Processes. Academic, New York (1980)
3. Billingsley, P.: Statistical methods in Markov chains. Ann. Math. Stat. **32**, 12–40 (1961)
4. Borovkov, A.: Mathematical Statistics. Gordon & Breach, Amsterdam (1998)
5. Brillinger, D.: An analysis of an ordinal-valued time series. In: Robinson, P., Rosenblatt, M. (eds.) Papers in Time Series Analysis: A Memorial Volume to Edward J. Hannan. Athens Conference, vol. 2. Lecture Notes in Statistics, vol. 115, pp. 73–87. Springer, New York (1996)
6. Collet, D.: Modeling Binary Data. Chapman & Hall, London (2002)
7. Demidovich, B., Maron, I.: Computational Mathematics. Mir Publishers, Moscow (1973)
8. Doob, J.: Stochastic Processes. Wiley, New York (1953)
9. Fokianos, K., Fried, R.: Interventions in ingarch models. J. Time Ser. Anal. **31**, 210–225 (2010)
10. Fokianos, K., Kedem, B.: Prediction and classification of nonstationary categorical time series. J. Multivar. Anal. **67**, 277–296 (1998)
11. Fokianos, K., Kedem, B.: Regression theory for categorical time series. Stat. Sci. **18**, 357–376 (2003)
12. Hyndman, R.: Time series data library. http://www.robhyndman.info/TSDL (2008)
13. Jacobs, P., Lewis, A.: Discrete time series generated by mixtures. J. R. Stat. Soc. B **40**(1), 94–105 (1978)
14. Johnson, N., Kotz, S., Kemp, A.: Univariate Discrete Distributions. Wiley, New York (1996)
15. Ju, W., Vardi, Y.: A hybrid high-order Markov chain model for computer intrusion detection. J. Comput. Graph. Stat. **10**, 277–295 (2001)
16. Kazakos, D.: The Bhattacharyya distance and detection between Markov chains. IEEE Trans. Inf. Theory **IT-24**, 727–745 (1978)
17. Kedem, B.: Time Series Analysis by Higher Order Crossings. IEEE Press, New York (1994)
18. Kedem, B., Fokianos, K.: Regression Models for Time Series Analysis. Wiley, Hoboken (2002)
19. Kemeny, J., Snell, J.: Finite Markov Chains. Springer, New York (1976)
20. Kharin, A.: Minimax robustness of Bayesian forecasting under functional distortions of probability densities. Austrian J. Stat. **31**, 177–188 (2002)
21. Kharin, Yu.: Markov chains with r-partial connections and their statistical estimation (in Russian). Trans. Natl. Acad. Sci. Belarus **48**(1), 40–44 (2004)
22. Kharin, Yu., Kostevich, A.: Discriminant analysis of stationary finite Markov chains. Math. Methods Stat. **13**(1), 235–252 (2004)
23. Kharin, Yu., Pashkevich, M.: Robust estimation and forecasting for beta-mixed hierarchical models of grouped binary data. Stat. Oper. Res. Trans. **28**(2), 125–160 (2004)
24. Kharin, Yu., Piatlitski, A.: A Markov chain of order s with r partial connections and statistical inference on its parameters. Discrete Math. Appl. **17**(3), 295–317 (2007)
25. Kharin, Yu., Piatlitski, A.: Statistical analysis of discrete time series based on the MC(s, r)-model. Austrian J. Stat. **40**(1, 2), 75–84 (2011)
26. Koopmans, L.: Asymptotic rate of discrimination for Markov processes. Ann. Math. Stat. **31**, 982–994 (1960)
27. LeCam, L., Yang, G.: Asymptotics in Statistics. Springer, New York (1990)
28. MacDonald, I., Zucchini, W.: Hidden Markov and other Models for Discrete-valued Time Series. Chapman & Hall, London (1997)
29. Makridakis, S., Hyndman, R., Wheelwright, S.: Forecasting: Methods and Applications. Wiley, New York (1998)
30. Malytov, M., Tsitovich, I.: Second order optimal sequential discrimination between Markov chains. Math. Methods Stat. **10**, 446–464 (2001)
31. Pearson, E.: Bayes theorem in the light of experimental sampling. Biometrika **17**, 338–442 (1925)
32. Petrov, V.: Sums of Independent Variables. Springer, New York (1972)

33. Raftery, A.: A model for high-order Markov chains. J. R. Stat. Soc. Ser. B **47**(3), 528–539 (1985)
34. Skellam, J.: A probability distribution derived from the binomial distribution by regarding the probability of success as a variable between the sets of trials. J. R. Stat. Soc. Ser. B **10**, 257–261 (1948)
35. Tripathi, R., Gupta, R., Gurland, J.: Estimation of parameters in the beta binomial model. Ann. Inst. Stat. Math. **46**, 317–331 (1994)
36. Waterman, M.: Mathematical Methods for DNA Sequences. CRC Press, Boca Raton (1989)
37. Yaffee, R., McGee, M.: Introduction to Time Series Analysis and Forecasting. Academic, New York (2000)

Index

Y. Kharin, *Robustness in Statistical Forecasting*, DOI 10.1007/978-3-319-00840-0,

Printed in the United States
By Bookmasters